Human Well-Being Research and Policy Making

Series Editors

Richard J. Estes, School of Social Policy & Practice, University of Pennsylvania, Philadelphia, PA, USA

M. Joseph Sirgy , Department of Marketing, Virginia Polytechnic Institute & State University, Blacksburg, VA, USA

W0193201

This series includes policy-focused books on the role of the public and private sectors in advancing quality of life and well-being. It creates a dialogue between well-being scholars and public policy makers. Well-being theory, research and practice are essentially interdisciplinary in nature and embrace contributions from all disciplines within the social sciences. With the exception of leading economists, the policy relevant contributions of social scientists are widely scattered and lack the coherence and integration needed to more effectively inform the actions of policy makers. Contributions in the series focus on one more of the following four aspects of well-being and public policy:

- Discussions of the public policy and well-being focused on particular nations and worldwide regions
- Discussions of the public policy and well-being in specialized sectors of policy making such as health, education, work, social welfare, housing, transportation, use of leisure time
- Discussions of public policy and well-being associated with particular population groups such as women, children and youth, the aged, persons with disabilities and vulnerable populations
- Special topics in well-being and public policy such as technology and well-being, terrorism and well-being, infrastructure and well-being.

This series was initiated, in part, through funds provided by the Halloran Philanthropies of West Conshohocken, Pennsylvania, USA. The commitment of the Halloran Philanthropies is to "inspire, innovate and accelerate sustainable social interventions that promote human well-being." The series editors and Springer acknowledge Harry Halloran, Tony Carr and Audrey Selian for their contributions in helping to make the series a reality.

Bazyli Czyżewski • Łukasz Kryszak

Sustainable Agriculture Policies for Human Well-Being

Integrated Efficiency Approach

 Springer

Bazyli Czyżewski
Poznań University of Economics
and Business
Poznań, Poland

Łukasz Kryszak
Poznań University of Economics
and Business
Poznań, Poland

ISSN 2522-5367   ISSN 2522-5375 (electronic)
Human Well-Being Research and Policy Making
ISBN 978-3-031-09798-0   ISBN 978-3-031-09796-6   (eBook)
https://doi.org/10.1007/978-3-031-09796-6

This Springer imprint is published by the registered company Springer Nature Switzerland AG
The registered company address is: Gewerbestrasse 11, 6330 Cham, Switzerland

Introduction

Agriculture is a sector that contributes significantly to climate change, although it is difficult to accurately estimate the magnitude of this contribution and make precise projections. Globally, agriculture is estimated to have been responsible for 9.3 trillion tonnes of carbon dioxide equivalent – 9.3 Gt COeq2., in 2018[1] (FAOSTAT, 2020). Total emissions from the sector have declined from 9.6 Gt in 2000, which has translated into a decline in agriculture's share of global emissions from 24 to 17%.

In the EU this share is lower—the overall share of agriculture in total greenhouse gases (GHG) emissions in Europe is about 10.1% (EEA, 2021). Despite the overall decrease in emissions, agriculture still remains one of the major emitters of GHG. Furthermore, it is important to note that while emissions from activities related to land use have decreased, emissions related to crop and livestock activities have increased by 14% between 2000 and 2018.

Agriculture in particular is the source of methane (CH_4) and nitrous oxide (N_2O) emissions – 49% and 66% of global emissions, respectively. In contrast, only 15% of carbon dioxide (CO_2) comes from anthropogenic management in agriculture. CH_4 emissions from enteric fermentation and N_2O emissions from agricultural soils stand for about 80% of total agricultural GHG emissions, of which CH_4 emissions from manure management have a share of about 10%. CH_4 and N_2O have very high conversion factors to their CO_2 equivalent (21 and 310, respectively). Therefore, emission reductions can be primarily achieved by improving the efficiency of resource management in agriculture—by this means alone it is possible to achieve as much as 20% of the required reduction in agricultural emissions.

This book is an attempt to address the problem, which affects countries all over the world, that particular sustainable agriculture strategies are generally evaluated as ineffective. The question is whether it is really necessary to look for new and better policy measures for sustainable agriculture, or whether the key issue is the

---

[1] This includes direct agricultural emission (mainly CH_4 and N_2O) and emission arising from related land-use change. Without the latter, total emissions are between 5 and 6 $GtCO_2e$ per year.

divergence of the applied effectiveness criteria and the different understanding of the idea of sustainable development in agriculture. The discrepancies occur at different levels—especially in the juxtaposition of global targets with policies built at the national or the local level. The creation of a sustainable agriculture policy has to deal with various fallacies of composition.

We begin by organizing the definition of sustainable agriculture, its systemic origins, and implications for land economics. We then consider the possibilities for operationalizing the trade-offs that occur in a comprehensive approach to sustainability, between economic, social, and environmental dimensions. A promising line of research in this regard is the concept of eco-efficiency, which has been steadily developing, in a better way integrating trade-offs occurring in the definition of sustainable agriculture. We attempt to contribute to this development by testing a new approach called "integrated efficiency".

From "eco-efficiency", through "green efficiency" to "integrated efficiency"—this is how one could describe the evolution of the concept, which aims at a multidimensional optimization of farming practices, taking into account the criteria of sustainable agriculture that seem difficult to reconcile. There is a clash of microeconomic optimization optics—i.e. the maximization of the individual utility function, with the necessity of taking into account the collective utility which is addressed by providing public goods. This necessity is becoming increasingly unavoidable in the context of a worsening climate crisis, biodiversity loss, natural resource degradation, and widespread environmental capacity overruns that may already become irreversible on a global scale within the next 50 years. Integrated efficiency combines individual and collective (social) criteria through an environmentally and socially adjusted classical production function (ESAF), which seeks the most efficient combination of agricultural production resources, agricultural output, socially desirable output, and undesirable output, and then assesses whether the target values meet the condition of not exceeding environmental capacity.

Meanwhile, authors argue that the implementation of environmental and social goals will always be ineffective when it contradicts the Pareto-optimal allocation of resources. In other words, a reduction of an input can be accepted only if it fulfils the Pareto improvement condition, i.e. making a farmer better off in one aspect while the others remain at least unchanged.

Therefore, targeting agricultural policy to improve efficiency under an environmentally and socially adjusted production function can be seen as the necessary conditions of policy effectiveness in terms of implementing sustainable practices in agriculture. Thus, a general postulate should be considered—let us design environmental policy in agriculture in such a way that it reduces the level of inefficiencies in the realization of the production function modified by the consideration of collective utilities. In this context, we should consider whether it is worth focusing on the optimization of environmental subsidies in the context of single ecological indicators, or whether only holistic evaluations using synthetic environmental indicators, or even synthetic measures of sustainability, make sense?

The multidimensional (holistic) perspective can be closer to farmers. Given that the problems of adverse selection and moral hazard occur in the case of individual measures, the effectiveness of environmental policy in agriculture should be assessed from a broader perspective. Hence, a condition for the design of socially and environmentally effective policy measures for environmental services should be a balance above zero of positive and negative external effects in the provision of public goods in agriculture. We believe the assessment of the impact of agricultural policy on inefficiencies under the ESAF should be an important criterion of the environmental policy evaluation.

As aforementioned, despite the adoption of collective utility criteria by policymakers, farmers in practice can still optimize individual production functions and choose Pareto-optimal solutions from the microeconomic point of view. It should be remembered that the optimal provision of public goods depends on the aggregated willingness to pay (WTP), which may significantly differ from the individual WTP. If the proposed policy measures do not qualify as Pareto optimal from the farmers' perspective, then public goods will not be delivered or moral hazard and adverse selection will occur, which will give only the appearance of sustainable practices.

Therefore, to show the validity of the above approach, the book discusses different sustainable agricultural policies focusing on synergies and the trade-offs between them, and the empirical part estimates what is the potential impact of different policy schemes on allocative efficiency under the ESAF. The potential for improving allocative efficiency in the sense of Pareto progress is examined using the well-established concept of "slacks", which reflect room for improvement in terms of the particular input/output assuming that others' resources or agricultural products are not affected.

The review of available sustainable agriculture strategies was divided into sections on direct GHG emissions, carbon sequestration, bioenergy, biodiversity, and other landscape goods. It was observed that the implementation of sustainable practices in agriculture in the world has a dual character, as it focuses on either a crops system or a grassland system accompanied by public goods focused areas. In both directions, specific strategies were identified, e.g. The Good Agricultural and Environmental Conditions (GAEC) supporting CO_2 sequestration in field crops, reducing the use of fertilizers by command and control measures (under crops farming systems), or the development of permanent grassland and public goods focused areas (under grassland farming systems). Then, the potential effectiveness and side effects of the mentioned strategies were evaluated. In this respect, a comprehensive procedure of assessing the holistic cost-effectiveness of agri-environmental policies was proposed and tested on several clusters of countries on a global scale.

To sum up, the book consists of theoretical and review chapters devoted to the concept of sustainable agriculture, effectiveness of public policies, and the methodology of assessing this effectiveness. The conclusions drawn were subjected to empirical verification in the author's concept of "integrated efficiency" and slack-based effectiveness assessment. At the end, the sustainable agriculture policies of the EU, the USA, and China are discussed in detail.

The empirical research is made up of several steps. First, we present descriptive statistics regarding inputs and outputs used in the agricultural sector in different parts of the world. We present data on labour and land productivity as well as average dietary energy supply adequacy since we treat agricultural production and the degree to which food needs are met as desirable outputs of agricultural activity. Data on GHG emission are then presented to illustrate the problem of undesirable outputs that are the side effect of production. Furthermore, we discuss inputs intensity, such as nutrient use, the number of livestock units, or pesticides use. We present all data for the EU, China, the USA, developing states, and other countries. Developing economies are further divided into subgroups following the FAO classification.

In the second step, we perform a cluster analysis based on partial productivity indicators (i.e. gross production value divided by chosen inputs) to obtain groups of countries that share a similar model of agriculture. The result revealed two big clusters (comprised of 29 and 50 countries), while the EU Members are treated as a separate third group.

For these three groups, an integrated efficiency analysis is performed, which is third step of the research. More specifically, we employ a super-efficiency hybrid DEA model with GHG as an undesirable output and dietary adequacy as a desirable social output. The hybrid approach imposes that some variables are treated as radial, while others are assumed to be non-radial meaning that they are changing independently. We obtain the integrated efficiency scores, but special attention is paid to the calculation of slack level and its dynamics. By slack analysis, we search possibilities that countries have to decrease bad output level and input use or increase their good output level. Finally, we employ Malmquist index to evaluate the dynamics of integrated efficiencies in the study period (2005–2018). Following the assumption that technical regress is not possible a modified sequential Malmquist index is used.

The next (fourth) step of the empirical research is devoted to the cost-effectiveness of agri-environmental policy. We construct a panel tobit model in which previously identified slacks are treated as a dependent variable, while monetary outlays from environmental policy are treated as an explanatory variable. The analysis is performed for all three clusters. For the EU a more detailed analysis including more categories of environmental expenditures is additionally carried out. The analysis in step four is complemented by a principal component analysis which attemtps to search for synergies and trade-offs between different slacks. Knowing which slack is (positively or negatively) correlated can be very useful for an adequate policy design. The final 5th step of the empirical strategy is also related to the problem of policy effectiveness. We investigate whether the inclusion of an uncontrollable variable (i.e. environmental policy expenditures) to the hybrid DEA model affects the integrated efficiency scores.

The structure of the book is as follows. The focus of the **first chapter** is the concept of sustainable agriculture. We review different approaches and definitions of what exactly sustainability means. Special attention is paid to the problem of the market treadmill which could be seen as a vicious circle in which farmers are trapped. They would need to increase their effort to overcome the income disparity problem, but it is not possible under intensive agriculture. We describe how sustainable agriculture may contribute to the solving of the market treadmill problem and present main implications of a new farming paradigm for the land rent theory. In the **second chapter**, we show how different aspects of sustainability (economic, social, and environmental) can be integrated in synthetic measures, such as eco-efficiency. Special focus is paid to the problem of trade-offs between different aspects of sustainability and practical way of eco-efficiency and integrated efficiency calculations. We also show how to study the drivers of efficiency.

The first part of the **third chapter** is devoted to calculations of the level and change of basic indicators related to different aspects of farm sustainability. This overview is based on the clustering of countries by the FAO. In the second part of this chapter, we group all studied countries based on the relation between production factors to distinguish main agricultural models. Then we describe our methodological approach for integrated efficiency calculations and present the results of calculations with a special focus on the identification of slacks, i.e. on the aspects where agricultural activity could improve in the Pareto sense.

In the **fourth chapter,** we discuss the problem of the effectiveness of agri-environmental policy. We review different methods used for effectiveness assessment and strategies concerning greenhouse gas emissions, carbon sequestration, bioenergy, biodiversity and landscape public goods. Then, pointing at synergies and trade-offs of policy measures, we explore why analyses of policy effectiveness often fail to produce clear-cut results. We propose a Pareto-inefficiencies criterion as a holistic tool for evaluating the effectiveness of public policies.

The content of the **fifth chapter** is an attempt to assess policy cost-effectiveness using the proposed slack-based approach. We address the call made in the previous part for a holistic way to identify cost-effective policies. We refer to the thesis that such a policy can be effective only if it has a positive influence on the efficiency of production by creating conditions for progress in the Pareto-like sense. This is a necessary, though not sufficient, condition for an effective policy. Therefore, the first section of the chapter elaborates a three-stage strategy for holistically assessing the cost-effectiveness of agri-environmental policy. Next, each stage is empirically illustrated based on the clusters of countries from different parts of the world (using the FAO database).

The **sixth chapter** is entirely devoted to the detailed description of agricultural policy in three major agricultural systems: the EU, China, and the USA. We compare different philosophies of agricultural support and try to establish some general guidelines that could be implemented elsewhere.

The last **chapter** concludes the analysis. We show the potential benefits of implementing the proposed approach on a wider scale. There is a discussion of directional and systemic solutions for agricultural policies, ones that take into account the trade-offs and synergies between different aspects of sustainability.

This book is intended for both academics and those responsible for agricultural policy. It can be used by researchers working in the field of sustainable development, and can also serve as an advanced textbook for students and PhD students working in agricultural or natural resource economics. For policymakers, this book can provide an inspiration for solutions designed in agricultural policy in developed as well as developing countries.

The **main findings** included in this book can be summarized as follows:

- **Economic theory should evolve towards the economics of sustainable development, and the agricultural sector is a very important part of this process. Agriculture must meet the demand for food production while reducing its pressure on the environment, taking into account technical and biological progress, the need for food security, and global economic, social, and environmental rationality.**
- **"Sustainability" is a concept that follows multi-objective logic, which implies efficiency indicators can be useful for its assessment. However, traditional indicators and eco-efficiency measures do not encompass all dimensions of sustainability. We propose integrated efficiency measures that take into account the economic, social, and environmental viability of the sector.**
- **Slack-based integrated efficiency analysis shows that a significant reduction of inputs use and GHG emission is possible to achieve under existing technology, just by removing slacks.**
- **Increasing efficiency by removing slacks in agricultural production could contribute to meeting at least one-third of global GHG emission reduction targets that were designed to limit global warming to 2 °C by 2100.**
- **Removing slacks could also result in significant savings in the use of inputs such as nitrogen or pesticides. Supposing that the new EU goals were treated as general guidelines, it would be possible to reach 46% of the reduction target for nitrogen and 15% for pesticides on a global scale.**
- **On a global scale, current environmental expenditures contribute to the reduction of slacks on GHG, land use, and employment. These expenditures also have an impact on integrated efficiency, especially in more developed countries.**
- **Designing efficient agricultural policy requires that different peculiarities of agriculture are taken into account: high level of risk, provision of public goods, and ensuring food security. What is more, policymakers should take into account that there exist both trade-offs and synergies between different aspects of policy. The objectives of policy should be multi-faceted. The measure of integrated efficiency is an example of such an objective.**

Contents

Chapter 1
Definitions and Origins of the Concept of Sustainable Agriculture

1.1 Towards Pluralism or Synthesis?

Can the concept of sustainable agriculture be unambiguously defined? Many authors doubt this, pointing out that the vast number of diverse conceptions of sustainable agriculture are competitive or even contradictory (Levidow et al., 2012; Hermans et al., 2010; Thompson & Scoones, 2009; Robinson, 2009; Tait & Morris, 2000; Rezaei-Moghaddam & Karami, 2008;), which has made and will make the debate on sustainable agriculture more difficult. However, Velten et al. (2015) recently have concluded that such pluralism of approaches is beneficial: "For complex problems of the modern world such as sustainability challenges in agriculture, ambiguous terms may indeed be more useful than precise and supposedly unambiguous concepts." The quoted authors have even expressed concern about reduction in the variety of frequently discussed strategies.

Definitions of sustainable agriculture can tentatively be divided into **goals-oriented**, **strategy-oriented** (Velten et al., 2015) and **holistic**. Among **the goals-oriented**, the U.S. Farm Bill definition is one of the most quoted: Sustainable agriculture is an "integrated system of plant and animal production practices having a site specific application that will, over the long term: (a) satisfy human food and fiber needs; (b) enhance environmental quality; (c) make efficient use of non-renewable resources and on-farm resources and integrate appropriate natural biological cycles and controls; (d) sustain the economic viability of farm operations; and (e) enhance the quality of life for farmers and society as a whole" (1990 U.S. Farm Bill).

As we can see, more space is devoted in this definition to what sustainable agriculture should be, with little written about the strategy for achieving it. In terms of how to achieve the mentioned goals, there is actually only the general statement "make efficient use of non-renewable resources". The contemporary definition presented by the FAO can also be included in this group: "To be

© The Author(s), under exclusive license to Springer Nature Switzerland AG 2022
B. Czyżewski, Łu. Kryszak, *Sustainable Agriculture Policies for Human Well-Being*,
Human Well-Being Research and Policy Making,
https://doi.org/10.1007/978-3-031-09796-6_1

sustainable, agriculture must meet the needs of present and future generations for its products and services, while ensuring profitability, environmental health and social and economic equity; The global transition to sustainable food and agriculture will require major improvements in the efficiency of resource use, in environmental protection and in systems resilience; Sustainable agriculture requires a system of global governance that promotes food security concerns in trade regimes and trade policies, and revisits agricultural policies to promote local and regional agricultural markets" (FAO, 2021a). In the group of goals-oriented approaches, no major contradictions can be seen. This is understandable, as points of disagreement generally arise during the implementation phase of the postulates contained therein.

Strategies-oriented definitions focus on the efficient management of resources in agricultural production. For example, let have a comparison of some definitions. In the early 1990s it was stated that sustainable agriculture is: "a management system that uses inputs... both those available as natural resources on the farm and those purchased externally... in the most efficient manner possible to obtain productivity and profitability from farming while minimizing adverse effects on the environment" (Agronomy News, 1989). Similarly, Fretz said that: "sustainable agriculture has been defined as an overarching, interconnected framework of technologies, practices and systems developed in response to environmental, social, economic, political, agronomic and horticultural issues. Sustainability achieves maximum efficiency, while enhancing or maintaining environmental quality. Sustainable agricultural systems are resource conserving. Sustainable practices maximize nutrient recycling. Sustainable systems protect groundwater and surface water resources and reduce soil erosion to a minimum. Sustainable systems must view the economics of production not only as short term, but also as long term. 'Sustainable' must be defined as forever" (Fretz, 1992).

Wallace (1994) listed the conditions to be met for agricultural practices to be considered truly sustainable. These are worth quoting in full here: "soil organic matter is not being gradually lost over the years; it is at a steady state or accumulating; Soil erosion is not taking place faster than the soil can be renewed; Substances are not accumulating in the soil that someday will become very severe limiting factors to crop growth if that accumulation continues; Substances are not accumulating in the soil that eventually could be toxic to consumers of the crops grown, even if they have no effect on the crop itself; Undesirable substances also are not accumulating in ground waters; The agriculture is not disruptive or destructive of the local and total environment, including that part of the environment which supports the agriculture, either physically or socially; The inputs required can be available on a permanent basis" (Wallace, 1994).

Although in the group of strategies-oriented definitions at first glance no contradiction can be seen, after a deeper analysis, it is possible to divide scientists into those who want to maintain current yields while reducing pressure on the environment and those who intend to increase yields and at the same time make environmental improvements. The second way of thinking leads to the so-called "holistic definition of sustainable agriculture".

Holistic definitions began to appear in the literature on the subject in the 1980s, e.g.: "Sustainable agriculture involves those farming systems that maintain and enhance the ability of agriculture to meet human and environmental needs now and in the future. It is also a production system that is profitable and competitive within the global economy. It protects the environment by reducing soil erosion from wind and water. It keeps pollutants out of surface and groundwater by employing fertilizer and pest management practices that result in optimum crop response with minimum 'spillage'. It is a systems approach to crop production that optimizes the effectiveness of inputs, including producer management. It is characterized by high yield and low unit cost" (Johnsrud, 1988). Likewise, Dibb (1990) said: "true sustainability must be broadened to include the necessity of increasing production to meet the expanding food, fiber and fuel needs of a continually increasing population. At the same time, it must address efficiency of input use, the economic aspects of production (profitability) and an appropriate concern for the environment". This approach has remained valid over time.

A modern, holistic definition of sustainable agriculture defines farming as a system of producing food and fiber that: firstly, satisfies human food and fiber needs; secondly, is profitable and uses on-farm resources efficiently to minimize adverse effects on the environment and people; thirdly, preserves the natural productivity and quality of soil, air and water; and, fourthly, sustains vibrant rural communities (UCS, 2005; Hendrickson et al., 2008; Reytar et al., 2014). However, it can be seen that the focus of the holistic approach is evolving to address hunger and malnutrition in the world.

The world challenges the need of balancing the food gap, i.e. 6500 trillion kilocalorie per year is missing, taking into account the food available in 2006 and that required in 2050—that means growth of 69%. Equally important is the need of economic and social development in agriculture in less developed countries and the third unquestionable need claims agriculture has to reduce its impact on climate, water and ecosystems, as unsustainable human activities in farming endanger the Earth's carrying capacity. Simultaneously, it shall be recalled that over one-third of produced food is globally lost or wasted each year throughout the supply chain (Reytar et al., 2014).

To address these concerns, FAO has expanded the concept of sustainable agriculture to include a third component—food: "To be sustainable, agriculture must meet the needs of present and future generations, while ensuring profitability, environmental health, and social and economic equity. Therefore, sustainable food and agriculture (SFA) contributes to all four pillars of food security—availability, access, utilization and stability—and the dimensions of sustainability (environmental, social and economic). FAO promotes SFA to help countries worldwide achieve Zero Hunger and the Sustainable Development Goals" (FAO, 2021b).

This definition of SFA has been further developed in 'FAO's Vision', emphasizing both objectives and strategies for achieving them. Productivity and efficiency are at the forefront of the latter: "By modifying current practices, much can be done in terms of improving the productivity of many food and agricultural production systems. Productivity will need to continue to increase in the future to ensure sufficient supply of food and other agricultural products. However, this must be done while limiting the expansion of agricultural land, as well as safeguarding and enhancing the environment (...) Efficiency in productivity has, in the past, been mostly expressed in terms of yield (kg per hectare of production) but future productivity increase should consider more dimensions. Water and energy-smart production systems will become increasingly important as water scarcity increases and as agriculture will need to seek ways to reduce emission of greenhouse gas. This will also have an effect on the use of fertilizers and other agricultural inputs" (FAO 2021b).

Conluding the thread of definitions of sustainable agriculture, it should be noted that: (1) all approaches have common points, although they are expressed in slightly different ways; (2) it is difficult to find postulates which are completely mutually exclusive—technological development of precision agriculture makes it possible to reconcile the low-input approach and the growing output approach, while country-specific objectives and global objectives also have a common ground—e.g. climate crisis and social resilience of rural areas; (3) with regard to the latter, contemporary definitions focus on social capital issues—which have previously been neglected, with particular reference to high levels of hunger and malnutrition; (4) in the light of the above, agricultural production efficiency viewed from an economic perspective is gaining in importance.

These observations are confirmed by the cited above literature review by Velten et al. (2015). Most persistent and relevant topics since the early nineties are as follows: economic viability, provision of products, environment conservation and improvement, natural resource conservation, stability and resilience in terms of objectives, and in the field of implementation strategies there is one: efficiency. Therefore, in further analyses, we will take into account these observations and, when measuring the level of sustainability, we will consider both the aspect of production efficiency, environmental undesirable outputs, as well as socio-economic well-being—including food issues.

1.2 The Background to the Concept of Sustainable Agriculture: Systemic Approach

There is no point in writing again about the historical roots of the concept of "sustainable agriculture", which lay in German forestry practices in the early nineteenth century. Let us focus on the systemic premises that created the need for a new paradigm. We will put forward the thesis that the paradigm of sustainable agriculture was created by the intrinsic economic mechanisms of market economy. Therefore, when writing about the genesis of the concept of sustainable agriculture, we must reach deeper into economic theory.

The developmental challenges of agriculture are integrally linked to the negation of the assumptions of neoclassical economics. The assumption that people, in pursuing personal interests, inevitably foster the achievement of the general well-being is an unrealistic assumption. Continuing to ignore the environment as a provider of public goods in economic calculation may lead humanity to cross the point of no return in terms of depleting the natural capacities of environment and, as a result, the quality of life will drastically decrease. Entities oriented solely towards profit maximizing, according to the logic of neoclassical economics, have a strong tendency to externalize environmental costs, many of which cannot be expressed in monetary terms.

The market mechanism based on private property and supply and demand curves creates demand for money through prices. In the real sphere, this means concentration of production for lowering its unit costs, as well as obligation to increase labor productivity as a condition for gaining competitive advantage. In relation to agriculture, this means an increase in production of agricultural raw materials under conditions of growing pressure on the natural environment. In addition, ensuring food security for consumers requires not only an increase in food supply, but also a decrease in the price level of agricultural raw materials. This, in turn, has had a negative impact on the incomes of farmers who, by producing more but cheaper food, achieve incomes significantly lower than the average for households outside agriculture. However, due to the immobility of the basic production factor, i.e. land, and the assets invested in farms, they were not able to pursue an efficient allocation in the Pareto sense. Therefore, it was left to increase labor productivity in the conditions, however, of decreasing prices of agricultural products. Such effect required progress in production technology, including: machines, equipment and innovations—i.e. continuous farm modernization (Czyżewski & Czyżewski, 2014, 2016).

This peculiar compulsion for efficiency and farm modernization did not take into account the full costs of production processes. Unfortunately, the balancing of such unfavorable production factors as soil depletion, carbon dioxide production, eutrophication of water bodies and soil erosion did not take place. The well-being of the environment with its inherent rarity was not valued and therefore the need to internalize costs was not reported. The technological treadmill described above, to which we will return in the next section, triumphed.

Moreover, another mechanism of economic depreciation of farms interacted with it. It turned out that despite the increase in modernization expenditures, as well as the application of new technologies and technical progress, the share of economic surplus in the product price was decreasing. This happened due to concentration of purchase, processing and trade sectors on the market and, as a result, growing, processing and trade corporations appeared which competed with each other and gained consumers through lowering prices. In this way, the assumption was implemented that the relatively larger commodity turnover would bring such a mass of profit that minimizing unit price would not become an obstacle to development of agriculture.

The paradigm of industrial agriculture based on the mechanism described above was accompanied by market failure, i.e. stimulating the development of oligo- and monopolistic structures in the supplier-consumer relations. As a result, this model of agricultural development did not meet two basic objectives of modern management: first, it was not able to secure agricultural income parity for farmers; second, as the scale of industrialized agricultural production increased, it depreciated environmental conditions by the external costs of agricultural production. The well-being of the environment and its ecological balance was increasingly violated. It should also be stressed that it was happening in conditions that the economic surplus was transferred from agriculture through purchasing intermediaries, processors and sales to consumers, while its retransfer back to agriculture could only be ensured by state interventionism (agricultural policy), which many countries simply could not afford.

It was an illusion that by accelerating industrialization, the so-called "agrarian issue" would be resolved. At the same time, the driving forces of industrial agriculture encountered an environmental barrier that made it impossible to continue ignoring natural conditions. As a result, sustainable agriculture faces the challenge of solving both issues: the market treadmill and the environmental hurdle. These problems are inextricably linked, because if the treadmill is not broken, the paradigm of sustainable agriculture will be not internally consistent—this thread will be developed in the next sub-section (Zegar, 2012).

In the highly developed countries at the end of the 1980s, the need was recognized to acknowledge the social and environmental external costs of agricultural production, which had hitherto been ignored. This resulted in the necessity to move away from the hitherto "industrial" economics of agriculture, towards a new economic paradigm in which the limits to growth are primarily set by the ecosystem; then, through rules and instruments, the desired distribution of resources is determined—the execution of which is entrusted to the market mechanism, but within a specific institutional framework of agricultural policy. This framework enriches the market allocation with an additional criterion that checks whether the marginal utility of economic growth is still greater than the scale of externalities.

It turns out that agriculture must be subject to the supremacy of the ecological system, while the economic system needs to be regulated by adequate social solutions. The fact that the capacity of ecosystems has already been exceeded by 1/3 undermines the hitherto prevailing principle that nature is worth as much as men want to pay for it. Even accelerated scientific and technological progress cannot reverse this assertion. The shift of agricultural development from an industrial to a sustainable model is inevitable in the long run, and putting ethical and social brakes on the mechanism of industrial agricultural development becomes a current necessity.

In the situation of spreading the paradigm of sustainable agriculture and supply limitations, it will be easier to overcome the barrier of demand for food. Of course, its limitations will not disappear, and its elasticity with regard to income will remain low. However, the adaptation mechanism of agriculture will be more oriented towards allocation of production factors in accordance with requirements of the natural environment and its well-being. Internalization of full costs of agricultural production in the conditions of increase in prices of agricultural products (with lower supply) will allow for improvement in the income situation of farms.

In conditions where there is no self-compensating mechanism for income depreciation in the industrial model of agricultural development, the sustainable model is more favorable to farms because of the relationship with the market and the environment. The question arises, however, as to whether it can also effectively struggle with the global food gap.

It can be said that the methodological challenge of the new paradigm of agriculture lies in the fact that this paradigm assigns an intrinsic value to natural capital, going beyond the classical theory of land rent. This is because it is unacceptable to assume the inexhaustibility of natural resources and the limitlessness of the global ecosystem and to treat them as constants term in the production function. Estimating the demand for environmental and social public goods at non-market prices is therefore a necessity that arises from the confrontation of market competitiveness vs. social competitiveness, and microeconomic vs. macroeconomic criteria.

Since the eighteenth century, there has been disagreement in economics about the sources of land rent. Essentially, the debate boils down to whether the substance of the rent is created by the productivity of land or by the subjective perception of the exchange value of this resource related to its scarcity. Nowadays, however, new dilemmas have arisen: if in most highly developed countries on average more than 60% of the economic surplus comes from subsidies, is the land rent not a political rent; does land have intrinsic utility and thus intrinsic productivity due to environmental amenities (i.e. without involving the stock of capital and labor)?

In the literature of mainstream economics, D. Ricardo is recognized as the founder of the theory of land rent. The Ricardian strand assumed the existence of differential rents alone and denied the existence of an absolute rent. K. Marx and K. Rodbertus undertook a polemic with D. Ricardo, proving the existence of an absolute rent. They argued that the rent from marginal land is not the consequence of the increase in grain prices, but, on the contrary, the fact that the worst land must provide rent in order to be cultivated is the cause of the increase in grain prices up to the level where this condition could be fulfilled.

The American economist, H. George, however, defined the land factor much more broadly than the classics, i.e. as a resource that is neither capital nor labor (Backhaus, 1997). In this approach, land rent is somehow detached from the land and differs from other factors in such a way that it cannot be withdrawn from production like labor and capital—it can only be owned or dispossessed by transferring it to another entity. Neoclassical economic theory has resulted in the concepts put forward by A. Marshall that are still considered valid today and are present in economics handbooks. In fact, A. Marshall reactivated the Ricardian rent concept, changing only the mechanism of value formation from labor-based theory to marginal cost theory.

Early- to mid- twentieth century mainstream economics has built upon A. Marshall's notions, focusing on market forces. Hence, according to J. Robinson, the existence of a land rent is determined solely by the elasticity of land supply, which as only factor with completely inelastic supply receive an economic rent (Robinson, 1948); otherwise, the supply of a given factor would increase, and the rent would fall to zero. This reasoning also applies to the land factor—each cultivated acre receives a so-called "transfer price", which is determined by the demand for land with an inelastic land supply function plus a differential rent when the income from the factor exceeds the transfer price.

In summary, the theories of land rent, which originated in the 18th and 19th centuries, are still at work in agricultural economics, even though farming conditions in agriculture have changed dramatically. These theories have increasingly diverged from economic reality because: Ricardian theory ignores the problem of the monopoly of property rights and the current tendency to equalize prices of land from different soil classes; absolute theory rejects the market mechanism of valorization of rent; residual rent theory underestimates the intrinsic productivity of land and reduces all its functions to a locational factor; neoclassical theory limits the sources of rent to the inelasticity of land supply and defines the rent analogously to the remuneration of capital (Czyżewski, 2009). Thus, the concept of sustainable agriculture has to face the lack of an adequate concept of land rent. We will develop this theme in Sect. 1.5.

1.3 Market Treadmill as a Trigger to Move into Sustainable Agriculture

It is rarely pointed out that the concept of sustainable agriculture emerged as a response to the mechanism of market treadmill originally described by W. Cochrane. Good understanding of the market treadmill in agriculture allows for more precise identification of sustainable agriculture goals and for building adequate political solutions supporting implementation of sustainable practices.

The theory of market treadmill in agriculture derives from the work of the American economist W. Cochrane that was published in the 1950s (Cochrane, 1958, 1979). It is about the principle of a machine known from a fitness club, which boils down to the fact that you are running but not moving forward. The treadmill spins faster and faster, so you have to run faster and faster, but you do not move forward. If you slow down, even for a moment—you fall. That is the idea behind the original concept of the treadmill in agriculture.

The assumption that you have to run faster and faster took on a real situation in American agriculture during the Great Depression of the 1930s. It started with the fact that in the 1920s, during a period of exceptional prosperity in agriculture, farmers took out loans *en masse* to buy modern farm machinery or land in order to expand their farms. However, when the speculative bubble burst and the stock market crashed (the famous Black Thursday of 24 October 1929), harder times came and many farmers, especially the smaller ones, were driven off their farms by banks demanding repayment of loans. The reason for this was not so much the economic health of the economy, but nature herself, which took revenge for the over-intensive use of agricultural land. In the pursuit of income, the fields were exhausted and at the beginning of the 1930s, the states of New Mexico, Colorado, Kansas, Texas and especially Oklahoma were hit by massive sandstorms brought about by soil erosion and drought. This caused a tragic exodus of farm families, made famous by John Steinbeck's prose ("The Grapes of Wrath"), migrating west to California in search of home and work. At the same time, the new landowners became even more dependent on mechanization, chemization and the scale of production.

The economic sense of the market treadmill theory comes down to the fact that agricultural incomes do not increase with increasing farm productivity. The direct cause is the drain on the productivity surplus by flexible prices, which means that despite the constant efforts of farmers to increase productivity (e.g. by implementing new technologies), the market wears out the benefits. However, if farmers choose not to implement progress, they become laggards and fall off the treadmill and out of the market.

The concept and underlying rationale for treadmill agriculture is evolving. The problem of agricultural income not keeping up with productivity growth was first called the "technological treadmill" (or "product price treadmill"), and, later, the "land market treadmill" (Levins & Cochrane, 1996). The market treadmill is a secular process occurring in the agricultural sector worldwide. In order to increase agricultural productivity, investments are made in new technologies and the scale of production is increased. However, the increase in productivity does not give the expected increase in income, because the demand for food is inelastic and agricultural commodity prices are highly flexible. As a result, the marginal revenue of agricultural production falls and producers have to invest in increasingly efficient technologies to maintain profitability. Only early adopters benefit in the short term, and the main gains from implementing these new technologies finally go to the consumer, or are captured by non-agricultural agribusiness chains. Many farmers are unable to withstand downward pressure on agricultural prices later on, and fall off the treadmill in the end. The result is a long-term decline in the number of family-owned farms.

Typically, in rich countries, food demand is described by low price elasticity, and these countries have more elastic agricultural supply curves. In developing countries, on the other hand, supply curves are more inelastic, which means that in developed countries with inelastic food demand, farmers benefit less from technological progress, while in developing countries with more elastic food demand, these benefits flow more to farmers (De Gorter & Swinnen, 2002). At the same time, the demand for land, and hence its prices and rental rates, is increasing. These processes also reduce the surplus income from productivity growth. Farmers are thus still under pressure to increase productivity, which translates into rising land prices and falling marginal revenue, and so on indefinitely. This happens assuming that there is no corrective state policy.

Cohrane's starting point was the statement that it is a myth that agriculture automatically returns to equilibrium. According to the author, this fantasy was widespread due to the dynamic growth in other sectors of the economy observed since the nineteenth century, as well as an exaggerated belief in the efficiency of the market mechanism. Additionally, it was observed that particular price fluctuations in the economy accompany periods of war. Hence, the conclusion was drawn that in peacetime, prices would stabilize. Meanwhile, it was in periods of peace that real differences in agricultural and non-agricultural prices were revealed. This is because agricultural prices grew at a faster pace and were subject to stronger oscillations. In turn, in the periods of overproduction, agricultural prices decreased relatively faster and to such a level that more than compensated for earlier increases. Thus, instability resulting from strong price fluctuations is a "normal state" in agriculture.

In the case of increasing supply, there is a disproportionate fall in prices and this ultimately proves to be detrimental to the income of the average farmer. An attempt is made to compensate for the fall in income by increasing production through the use of technological and organizational innovations. This, in turn, causes a further fall in prices. This phenomenon, called the "King effect", is well described in the literature (Heberton, 1967; Barett, 1993; Tweeten & Zulauf, 2008; Chen et al., 2011).

The concept of treadmill develops King's theorem by showing that in the long run, income decreases in conditions of overproduction are not compensated by income increases in conditions of negative supply shocks in agriculture. Research indicates that price and output fluctuations and, as a result, the "surplus drain", is stronger in countries with fragmented agriculture (Majchrzak & Stępień, 2016).

The review of literature carried out by Ward (1993) indicates that one can also speak about the environmental dimension of treadmill in agriculture related to the increasing use of pesticides and other chemicals, as a manifestation of technical progress. Other works indicate the existence of an analogous treadmill mechanism in the systems of management, credit or market competition. In each case, the principle is similar. It is about the necessity of constantly increasing various types of outlays and organizational activities, which do not translate into the expected increase in income.

The concept can also be generalized on the grounds of sociology as a sociological treadmill, which should be understood as the need for farms to constantly adapt to market conditions so that they can survive in a changing environment. This is especially true for small farms, where labor and capital flows between agricultural and non-agricultural uses, and the separation of household and farm enterprise is difficult. Due to this specificity, the reaction of these farms to market impulses is atypical. In the case of an economic downturn in agriculture, its negative effects are cushioned at the level of a farm by, inter alia, decreasing household consumption in favor of maintaining farm-operating expenditures. At the same time, in the case of economic upturn and higher income obtained from agricultural activity, it is possible to partially spend them on household consumption (Czyżewski et al., 2017). Thus, the behavior of small family farms during a downturn reinforces income decreases (in the conditions of flexible prices), and in the situation of improvement of price, it is difficult to compensate for the surplus drain.

The effects of market treadmill are mainly found in farmer-type agriculture. Hayami and Ruttan (1970a, b, 1971) identified the prototypes of different agricultural development paths. They distinguished four development models: the conservation model, the urban-industrial impact model, the diffusion model and the high payoff input model.

For the most part, American agriculture is an example of the rapid diffusion of technological progress in breeding and tillage, which enabled the implementation of modern technologies and agrotechnical practices (diffusion model). This strategy resulted in the development of industrial agriculture, in which we can observe the process of industrialization of animal and plant production based on mechanization, genetic modifications and organizational solutions aimed at achieving the highest possible economies of scale. This model is condemned to the market treadmill, since allocation decisions are made under the primacy of private property and based on the criterion of maximizing profit and asset value. The hunger for land and the lust for possession create a strong market competition that ruthlessly eliminates less efficient actors from production (and from the rural space in general), systematically tightening the links with the market for successive generations of farmers.

In "peasant" agriculture, which has developed in Europe since feudal times, the occurrence and effects of the market treadmill have never been as obvious as in American agriculture, because the motives behind the decision-making of European farmers also appear to be more complex. However, the genesis of the mechanism of treadmill in the sense of fruitless attempts at development can already be seen in the work of the French Physiocrats of the mid-eighteenth century. In their view, only agriculture creates an economic surplus. However, in reading Quesnay's diagram of the input-output table, it is immediately apparent that the farmers transfer this entirely to landowners in the form of rent. The increase in peasants' productivity therefore increases the income of landowners and generates economic growth, but this does not mean that agriculture benefits from the effects of this growth. Therefore, peasants were not guided by the profit motive. This was confirmed by Alexander W. Czajanow's (1888–1937) research on the Russian peasantry before collectivization.

Czajanow argued that the level of activity of the peasant-farmer, measured by the number of hours and days of work and the area of land cultivated, was determined by the number of people in the household whom the peasant had to feed. Under these conditions, we are dealing with an inverted supply curve, i.e. an increase in the price of agricultural commodities paradoxically leads to a decrease in their supply, since fewer products have to be sold to feed the farming family. It follows from this that the agricultural income does not have to fully cover the market valuation of own labor and the opportunity cost of capital.

Theoretically, in this situation, the factors of production should not provide their services, but if the labor remuneration, the cost of land and property are not treated in terms of market costs, this situation can persist for a long time. Czajanow drew attention to this in many publications (1931, 1966, 1991), proving that peasant farms do not maximize profit, but optimize income per family member and leisure. Thus, the motive of their action is not profit maximization from farm activities, which makes them not fall into the market treadmill, because they react differently than the neoclassical model of demand and supply assumes. Therefore, in developing countries, small farms are better at limiting poverty, because they involve a lot of people from their families and equally poor neighbors, and they spend their income on locally produced products and services, thus supporting the local economy and employment. Small farms, hence, create a social safety net enabling the rural poor to exist even when they do not produce for market (Zegar, 2009, p. 260 after Lipton, 2005; Poulton et al., 2005; Hazel et al., 2007).

But is mere "subsistence" enough? If small farms, of which, de facto, are a majority in the world, do not have greater aspirations than "survival", then there is no problem of the treadmill in Cochrane's understanding, and the negative pressures on the environment are limited, but there still remains the agrarian issue in the form of income disparity of the farming population. In the long run, this situation may become a brake on the overall development of the economy. Therefore, a strategy of gradual achievement of a parity situation between agriculture and other sectors of national economy is necessary.

This strategy, in general, should include four stages. The first one consists in creating institutional conditions for investments in production infrastructure of agricultural holdings (the so-called "getting agriculture moving"; Woś, 2003). It then becomes possible to move to the second phase, which consists in industrialization of the agricultural sector. In the third phase, "agriculture is to shrink, but not to weaken". Thus, the resources released from agriculture (mainly labor) are engaged in processing, and agriculture is more strongly integrated with the market environment. In the last, fourth phase, the model of industrialized agriculture is already functioning, and the state policy has only a corrective character as regards the long-term tendency to broaden price gap, and tries to maintain the existing status quo.

Nevertheless, the experience of highly developed countries has shown that it is not possible to stop at the fourth phase, as it causes negative externalities in the environmental and social aspect that are unacceptable in the long run. Therefore, a fifth phase becomes necessary, consisting in the pursuit of balance encompassing simultaneously economic, ecological and social criteria—i.e. implementation of sustainable agriculture. Thus, even if there is no common profit motive in family farming, its ignoring leads to the increase of relative income deprivation of this sector, which causes a number of undesirable effects from the macroeconomic perspective, and damages social well-being. Therefore, sooner or later, family farming has to enter the path of the treadmill.

In the age of globalization, however, it is becoming increasingly difficult to persevere in such conditions. The Canadian economist T. Weiss argues that in their relentless pursuit of expanding markets and increasing profits, large transnational corporations are making farmers more and more dependent on inputs and on standardizing agricultural production. They are making animal husbandry more brutal and poisoning soil and water, externalizing environmental costs, changing dietary habits, breaking local links between production and consumption, and diminishing the value of labor by replacing it with technology (Weis, 2011). Hence, the importance of a proper agricultural policy for sustainable agriculture, the priority of which should be the pursuit of a remedy for the market treadmill in agriculture.

1.4   How Does Sustainable Agriculture Solve the Treadmill Issue?

Let us remember that the market treadmill theory was developed in the middle of the last century on the basis of the situation in American agriculture. If we consider it in modern times, we must bear in mind the changes that have taken place in the agricultural policies of highly developed countries in recent decades. In the twenty-first century, investments are not as difficult for European and American farmers to make as they were in the 1960s. For example, the EU Common Agricultural Policy provides funding to support investment and the implementation of progress. However, the division of farms into "early adopters" and "laggards" is still relevant. The technological direction is mainly expressed by dual development and increasing polarization in the agricultural sector (Boháčková, 2014). In the perspectives of the whole economy, however, the effect is the relative deprivation of this sector, which is secular in nature and limits the participation of farmers in the processes of economic development.

Still, it would be simplistic to limit the causes of the treadmill, understood as the lack of a positive relationship between productivity and income, exclusively to the flexibility of agricultural prices. The reality is more complex. The approaches to this issue of various researchers can be divided into three categories: (1) supply-oriented, (2) demand-oriented and, (3) behavioral.

In the supply-oriented approach, differences in the relative income growth of agriculture (compared to other sectors) are attributed to the distinctive features of agricultural production (Heinrichsmayer & Witzke, 1991; Gardner, 1992), which result in low labor mobility and the occurrence of an "inverted supply curve", which, as mentioned, is derived from the work of Czajanow (1931, 1966, 1991). In addition, the peculiarity of agricultural production stems from the linkage of the agricultural enterprise and the household, which often leads to a non-market hierarchy of objectives. This non-market hierarchy reduces the likelihood of achieving full land rents. What is more, work in agriculture is characterized by lower mobility and limits the choice of education courses. As a result, a process of ageing of farmers is observed, although this problem is relatively less significant in countries with a fragmented agrarian structure (Siudek & Zawojska, 2016).

Another symptom of the production and social specificity of farms is the so-called "high profit trap". Farmers invest in conditions when the expected income from the investment is higher than the costs of its implementation. On the other hand, a situation discouraging investment manifests itself in expected revenue being lower than the resale value of purchased fixed assets. However, if the expected income is lower than the acquisition costs, but higher than the resale value, capital becomes completely immobile, i.e. "locked in" to agriculture (Johnson, 2000).

The behavioral approach, on the other hand, seeks an explanation for falling into the treadmill in the behavior of agricultural producers. Why does maximizing productivity not bring the same results as in other sectors? Older studies point out that the farmer maximizes sales, not the marginal productivity of the factors of production. Other studies question this approach by attaching more importance to farmers' limited adaptability to changing market conditions (Vergopoulos, 1978) and delayed response to these changes (Boháčková, 2014).

Thus, the reasons for the market treadmill in agriculture manifested by the lack of income growth in response to increasing productivity are multidimensional. At the same time, several premises described above overlap:

• peculiarities of production and forms of conducting business activity in farming,
• demand factors on the food market,
• asymmetric market structures,
• behavioral constraints.

An undeniable effect is the outflow of a part of the economic surplus due to the growing real productivity to other sectors and, as a result, insufficient capitalization of land rent in the property owned by farmers.

It would seem that the treadmill loses its significance when farmers' incomes are no longer so strongly dependent on a market competition that determines the scale of production and the prices of raw materials and means of production. If the market process of price and volume formation is distorted by direct subsidies, as well as intervention prices, the treadmill does not work. From this point of view, intervening in the market in the agricultural sector, which is inherently imperfect, would seem to be justified.

Not every model of agricultural interventionism has, however, so far delivered the expected benefits. The unreliability of the state redistributive policy in agriculture is a well-known fact. In the literature on the subject there is a concept of "budget support leakage". Direct payments were supposed to improve the economic situation of "professionally active farmers", meanwhile, to a great extent, they go in the form of increased rent to landowners who are not professionally active farmers, but have leased their land (Góral & Kulawik, 2015). It is estimated that only 20% of all support in agriculture in OECD countries creates a net surplus and the rest flows to other sectors of the economy (OECD, 2000). Still, subsidies affect agricultural productivity and thus the treadmill mechanism, but the results in this regard are not clear.[1]

The impact of support on farm productivity in the EU has already been extensively researched (Hennessy, 1998; Ciaian & Swinnen, 2009; Rizov et al., 2013; Banga, 2014). These studies indicate that subsidies before the introduction of the so-called "decoupling reforms" (the June 2003 Agreement in Luxembourg) had a

---

[1] The problems of budget leakage and many others will be more extensively discussed in Chap. 6.

positive impact on production, but a negative impact on productivity, while, after this reform, the conclusions are ambiguous, but with the indication that the negative impact occurs much less frequently (in the sense of the impact of subsidies on the total productivity index TFP), or does not occur at all (in the sense of the impact on TFP growth dynamics) (Rizov et al., 2013). If this is indeed the case, then the decoupling of subsidies from production should alleviate the problem of market treadmill, because it does not stimulate production and capital intensity, but supports income.

In global context, however, a significant heterogeneity of agriculture is observed. In countries with higher land concentration, area payments may be a much more effective tool for income stabilization than in countries with a fragmented agrarian structure, where they additionally affect the growth of land prices, maintaining its unfavorable distribution among farms. Therefore, it can be said that mechanisms of agricultural policy may potentially both help and hinder overcoming the market treadmill in agriculture.

Nowadays, multifunctional development of rural areas is a chance for overcoming this vicious circle. On the one hand, there are rural areas with advanced, highly efficient production. On the other hand, there are also areas where rapid modernization of agriculture is rather impossible. An opportunity for the development of these areas, however, is the development of a market for high quality traditional food and agri-tourism, which does not require increasing productivity in the traditional sense (Marsden, 1998).

This brings us to the question of whether there is an agricultural development model in which the described market and policy failures are reduced. There is not much choice in this respect. As a matter of fact, the only coherent answer to the problem of contradiction of environmental, social and economic goals are activities that constitute the paradigm of sustainable agriculture. One of its key assumptions is the provision of public goods by agriculture and rural areas (such as natural values, landscape, rural culture, biodiversity, traditional food, food security in a broad sense), on condition of sustainable farming.

Public goods are not subject to market valuation in the strict sense. However, it can be considered that their institutional valuation takes place, and this results in subsidizing certain ways of management. Such a model of valuation is not without flaws, but it gains social acceptance more easily than subsidizing market goods.

There are reasons to believe that financing public goods mitigates market treadmill because:

- a higher share of payments for public goods in subsidies should favor the sustainable development of farms, due to the fact that it stimulates them to move towards multifunctional development and diversification of sources of income; thus, the pressure on productivity growth in the classical sense decreases, since the growth rate of household income can be maintained through non-agricultural (yet agriculture-related) activities;
- a higher share of payments for public goods favors activities with lower product price elasticity, e.g. organic food production, agro-tourism services;

- the provision of public goods more or less relieves market orientation, as it reduces pressure on agricultural prices;
- subsidies for the provision of public goods are less vulnerable to "support leakage", due to the fact that they capitalize to a lesser extent in rents, due to the lack of market valuation of public goods (Czyżewski & Matuszczak, 2017).

A side effect of subsidizing public goods in rural areas may be an extensification of agricultural production. A topic for separate discussion is the question to what extent this is acceptable in the context of global food shortages? Perhaps the financing of public goods should concern only semi-subsistence farms, whose share in commodity production is small. The benefits brought about by subsidies that serve exclusively to finance public goods is debatable. Do, for example, area subsidies contribute to the creation of these goods? In the EU, for instance, cross-compliance rule is a step in this direction, but one could say that it serves more to preserve the suitability of private land and other assets to produce safe food in the long term. The receipt of area payments, on the one hand, is poorly secured by restrictions on the chemization of agriculture and on excessive increase in intensity—both of which have negative impacts on the environment. On the other hand, programmes typically aimed at rural development indisputably contribute to the creation of new or to the care of existing public goods (Czyżewski & Matuszczak, 2016a).

This is the main objective of policies for sustainable agriculture. A common method used by these policies to value public goods is institutional valorization by subsidizing specific activities that create public goods or compensate for abstention from activities that compromise common goods. For example, EU CAP agrienvironmental payments support the following actions: biodiversity protection in rural areas, protection of landscape, promotion of organic farming and protection of genetic resources in agriculture. The implementation of these objectives comes down, inter alia, to measures contributing to improving the structure of the landscape, protecting ponds, enabling afforestation, maintaining rural cultural values, allowing for renaturalization of transformed meadows, enabling soil and water protection and creating open areas and buffer zones, as well as maintaining traditional orchards.

On the other hand, the so-called subsidies to less favored areas are aimed primarily at maintaining the vitality of rural areas, preserving landscape values, promoting environmentally-friendly agriculture and preventing depopulation. What is more, programmes supporting fallow land serve to diversify crops and maintain ecological areas. Other types of rural development subsidies can support reforestation or the preservation of the forest in good conditions, which undoubtedly increases public benefits. In addition to institutional valuation, market mechanisms may indirectly discount public goods in prices of resources and products. We, therefore, deal with such a valorization on the agricultural land market or in the case of organic food.

There are few attempts in the literature to estimate the demand for public goods in rural areas, for which data are generally lacking (Czajkowski et al., 2014; Carson & Czajkowski, 2014). The research of Delbecq et al. (2014) indicates that the value of agricultural land is only partly explained by farm income. The mentioned authors point to non-agricultural attributes of agricultural land that determine its market value. Among these are attributes that are closely related to public goods, such as the availability of water recreation or woodland.

Market and institutional (i.e. by policy) valorization of public goods related to agricultural land has, however, another side to the coin. On the one hand, it has been demonstrated that it mitigates market imperfections, but on the other hand, it translates into higher land prices, leading to a growing gap between its use value (derived from agricultural income) and market value. Agricultural land prices in highly developed countries break away from the discounted stream of rents from agricultural production. This is a kind of land value gap, understood here as the difference between market value and income value, which is magnified by subsidizing the production of public goods in agriculture. This gap will particularly affect land with small production capacity. Thus, if small, unprofitable farms were to be "guardians of the landscape", it is necessary, first of all, to enable the flow of agricultural land to non-agricultural activities, e.g. agrotourism, or residential services, i.e. to create institutional (legislative) conditions for the outflow of land from agriculture. Thus, the larger the gap in the value of agricultural land, the fewer restrictions should be placed on its circulation. Simultaneous subsidizing of public goods supply in agriculture and restriction of agricultural land turnover is contradictory in its logic, because these actions mutually reduce their effectiveness.

The literature generally confirms the thesis of the existence of an agricultural land value gap. The value of agricultural land exceeding the benefits from its use in agricultural production makes it possible to estimate approximately the value of non-market goods and services provided by the land factor. If there is no significant environmental or agglomeration potential in an area, the excess value of land over its productive values can be a measure of a speculative land bubble (Delbecq et al., 2014). This inflated value, regardless of its source, is considered to be the tax base for agricultural property (O'Dea, 2013; Sherrick & Kuethe, 2014).

There is evidence that in many areas across the United States, the market value of farmland exceeds its use value in agricultural production (Barnard, 2000; Flanders et al., 2004). Recent empirical findings suggest that farm profitability in highly developed countries will decline in the coming years in favor of non-farm and non-market determinants of income (Delbecq et al., 2014), which are becoming increasingly important for the financial health of agriculture. Since non-agricultural determinants of economic surplus very often take the form of public goods, agricultural labor and land markets may suffer from free-riding (Kaminski et al., 2012). As we mentioned, agricultural land provides various public goods such as biodiversity, climate balance, rural culture, open space and functions that indirectly affect food quality and human health.

Wasson et al. (2013) argue that the attributes of a plot of land, which include its recreational, perceptual and environmental amenities, are necessary to be able to explain the value of agricultural land. Excluding these attributes from the set of variables that define the value of agricultural land makes it impossible to completely explain the price fluctuations of this resource. According to the authors cited above, the rewarding (by agricultural policy) of the above attributes, as well as the penalties for their degradation, play huge roles, especially in areas abundant in them. For example, in western Wyoming (USA), non-agricultural land utility values accounted for between 5% and 60% of the parcel value (one third on average). However, several European studies contradict these observations. According to Nilsson and Johansson (2013), environmental charges related to agriculture in Sweden had a negative impact on land prices. They argue that municipalities receiving agro-environmental support have highly fragile ecosystems in which it is difficult to farm. A similar conclusion was reached in earlier work by Rutherford et al. (1990).

Other authors argue that land value is driven more by a combination of different macroeconomic factors such as agricultural prices, low interest rates or agglomeration pressures (Weber & Key, 2014) than by heterogeneous quality attributes. These factors have driven large increases in farmland prices in both Europe and the US, where nominal farmland value doubled between 2004 and 2012 (USDA, 2017). Plaxico and Kletke (1979) and Lowenberg-DeBoer and Boehlje (1986) have shown that a long-term increase in agricultural land prices increases a farmer's creditworthiness. Indeed, agricultural real estate accounts for over 80% of the total value of assets in US agriculture, being the main source of collateral for farmers' loans (Nickerson et al., 2012).

Theoretically, under such conditions, an increasing demand for land (feedback demand) could arise, driving up the price of land (MacDonald et al., 2013). Breustedt and Habermann (2011) argue that a speculative bubble in the agricultural land market is possible if rising creditworthiness helps farmers obtain more or cheaper financing for land purchases, thereby strengthening the demand for and price of land, resulting in further increases in landowners' wealth and creditworthiness (Adrian & Shin, 2010; Rajan & Ramchara, 2012).

1.5 Implications of Sustainable Agriculture for Land Rent Theory

Sustainable agriculture raises the need to formulate a new concept of land rent. Since the beginning of man's civilization, the land has created certain utilities to satisfy his needs. These are created without other factors of production, being an undeniable gift of nature. In the encyclical "Caritas in Veritate", His Holiness Benedict XVI described them as the miraculous fruit which man can use responsibly to satisfy his legitimate needs—material and immaterial—with respect for his inner balance. In tribal (natural) economies, when there was no agricultural land in today's sense, examples of the above-mentioned usefulness were forest fruits, hunted animals, access to water or firewood. The creative role of the land factor in their generation was dominant in relation to the labor and capital outlays necessary for their acquisition. Thus, it can be said that most of the utility of land was created spontaneously. With the beginning of the cultivation of land and the domestication of animals, the part attributed to nature diminished slightly in favor of the causal role of man. Nevertheless, the mass growth of plants, trees etc., building material or the widely understood living space was still largely obtained without any input.

In the feudal system, the so-called "servitudes" can be regarded as a specific legitimization of the intrinsic utility of land, treating them as the right to use the natural utility of the land belonging to the lord (in the form of brushwood, fruit, clay or fish). As the commodity-money economy developed, that part of the utility of the land factor that was created without the participation of capital and labor began to be perceived as its "intrinsic productivity" (in monetary terms). This is reflected, for example, in the eighteenth century concept of pure product, presented by the Physiocrats. Accordingly, only in agriculture can there remain a monetary surplus over the incurred inputs (capital and labor)—precisely thanks to the causal role of nature. The notion of "pure product" in F. Quesnay's input-output table is thus the first attempt to valorize the intrinsic productivity of land.

Thus, in the peasant economy, the part of utility attributed to the exclusive action of natural forces was relatively large and was also expressed in a certain part of the monetary productivity of the holding (as it created part of the product without any input). Its significance began to decrease under the conditions of the industrialization of agriculture and the activation of the law of diminishing marginal utility and returns.

In industrial agriculture, the spontaneous contribution of land to the creation of utility declined in favor of capital and hired labor. The intrinsic monetary productivity of land also largely disappeared. Over time, however, the productive functions of agricultural land, which were subordinated to microeconomic optimization, and the necessity of satisfying human existential needs through it, became competitive with each other. This gave rise to the need to search for a new concept of economic development—a paradigm of sustainable development.

One of the reasons for applying this paradigm to agricultural economics as sustainable agriculture is that the environment in highly developed countries has become almost entirely anthropogenic. Under such conditions, the use of natural resources must also change. This is enforced by the necessity to ensure the renewability of natural resources, as well as to maintain the pro-social and pro-environmental criteria of allocation of production factors. This rediscovery of the "utilities" of the land factor marginalized in industrial agriculture, gives them the character of public goods for which the whole society should pay. This, however, cannot be the same intrinsic utility of agricultural land as in the eighteenth century, because, at least in the highly developed countries, the environment has been radically altered by man.

Still, an increasing proportion of the utility of land is again spontaneously generated, but under conditions of far-reaching and irreversible capital accumulation. Thus, it can be said that in sustainable agriculture, many new land factor utilities are created spontaneously, i.e. without additional capital and labor inputs, (but not without their causal role at all), and in some cases without increasing the total sum of capital and labor inputs. As they have the nature of public goods, they are paid for largely through taxation, and this payment goes to the owners of the land resource. Hence, the intrinsic utility of land takes the form of a monetary product and can be called a form of "intrinsic productivity" that raises farm efficiency in monetary terms.

To sum up, agricultural land itself forms part of the utilities that are valorized on the market or institutionally, if the intensity of farming is limited to a certain extent. However, this is conditioned by a certain level of "primary" capital accumulation, which features the economy at the stage of evolution where the society makes demand for the above-mentioned utilities.

As shown earlier, different stages of land rent valorization can be linked to the development of the market economy. At a certain stage of economic development, which is related to the evolution of social awareness, the market or/and institutions established for that purpose valorize intrinsic land utilities of a public goods nature and give them a monetary form. Thus, within the framework of the paradigm of sustainable agriculture, the reason for occurrence of land rent is the intrinsic utility of land, which in the commodity-money economy brings about a situation wherein the expected productivity of the capital factor in agriculture may be higher than in its market environment. The value of land rent is therefore determined by the positive difference between the expected productivity of capital in agriculture and in its market environment. Moreover, the agricultural land market realizes in prices the expectations of this surplus capital productivity in agriculture. This process of valuation, however, neglects the factor of farmer-owner's own work, since it has no real market value. Hence, income obtained by individual farms does not coincide with the values resulting from the land price (the mentioned rent gap).

The paradigm of sustainable agriculture is not so concerned with the principles of economic calculation as with the objective, scope and method of research. In the industrial model of development it was to maximize farm profit for farm needs. Herein, factors of production had a market price while others were considered free goods. In sustainable agriculture, there is an integrity of economic, social and environmental objectives, and the economic account must consider the benefits gained and lost, as well as the externalities.

There is a need for a new equilibrium, but it cannot be created solely by the commercial effects of agriculture. Food quality, carbon sequestration, water and soil protection, biodiversity, etc. also count. While economic benefits have specific addressees, negative externalities primarily affect all taxpayers, the environment and future generations.

Moving on to the implementation of the idea of sustainable agriculture, first of all, it is necessary to make an account of external costs and look at the sources of land rent in a different way. Secondly, the scale of using the environment must be determined by legal standards established by states and supranational institutions under administrative decisions on quality standards, fees, penalties and subsidies. The market mechanism should be institutionally supported, as only the activities of specific institutions can bring microeconomic and social criteria into line. Efficient supranational institutions can in fact better secure global common goods, provided that they have the support of nation states and legitimacy for their actions.

The paradigm shift from industrial to sustainable agricultural development is neither easy, nor quick. Agricultural economics shall treat agriculture not only as a business, but also as a way of life for millions of farmers. Economic theory should evolve towards an economics of sustainable development. This process has already begun, but it is creating a number of ongoing dilemmas. It is already clear, however, that agriculture must meet the demand for food products while reducing pressure on the environment, taking into account technical and biological progress, the need for food security and social and environmental rationality also at the global level. (Czyżewski, 2017).

Chapter 2
Integrating Three Dimensions of Agricultural Sustainability

2.1 Workable Approaches to Integrating the Environmental Dimension into the Concept of Sustainability

2.1.1 How to Deal with the Trade-Offs Implied by the Sustainable Agriculture Definition?

Eco-efficiency Concept

As can be seen from the previous sections, the crucial idea of sustainable agriculture consists in integrating the environmental and socio-economic dimensions. We summarized the workable approaches to build holistic sustainability measure in Fig. 2.1. Although there are many measures of so-called "environmental sustainability", considering them separately contradicts the logic of the concept of sustainability. It is not our intention to recall here the multiple approaches to measuring each separate dimension of sustainability, especially when it comes to "environmental sustainability". This thread can, however, be found in many works (Adenuga et al. 2018, 2019, 2020; Bartolini et al., 2012; Huang et al., 2016; Paracchini & Britz, 2010; Park et al., 2016; Povellato et al., 2007; Urdiales et al., 2016). Indeed, the concept of sustainability in itself implies balancing different dimensions—taking into account trade-offs between them. Therefore, the most promising concept seems to be eco-efficiency, which *ex definition,* confronts both the environmental and economic perspectives, and in an extended version, also includes the socio-economic one (integrated efficiency).

However, there are also some dilemmas encountered in using eco-efficiency as a measurement of agricultural sustainability. Firstly, we should find answers to the questions as to whether eco-efficiency is more of an economic or biophysical concept and, if so, whether outputs and inputs should be measured in economic or

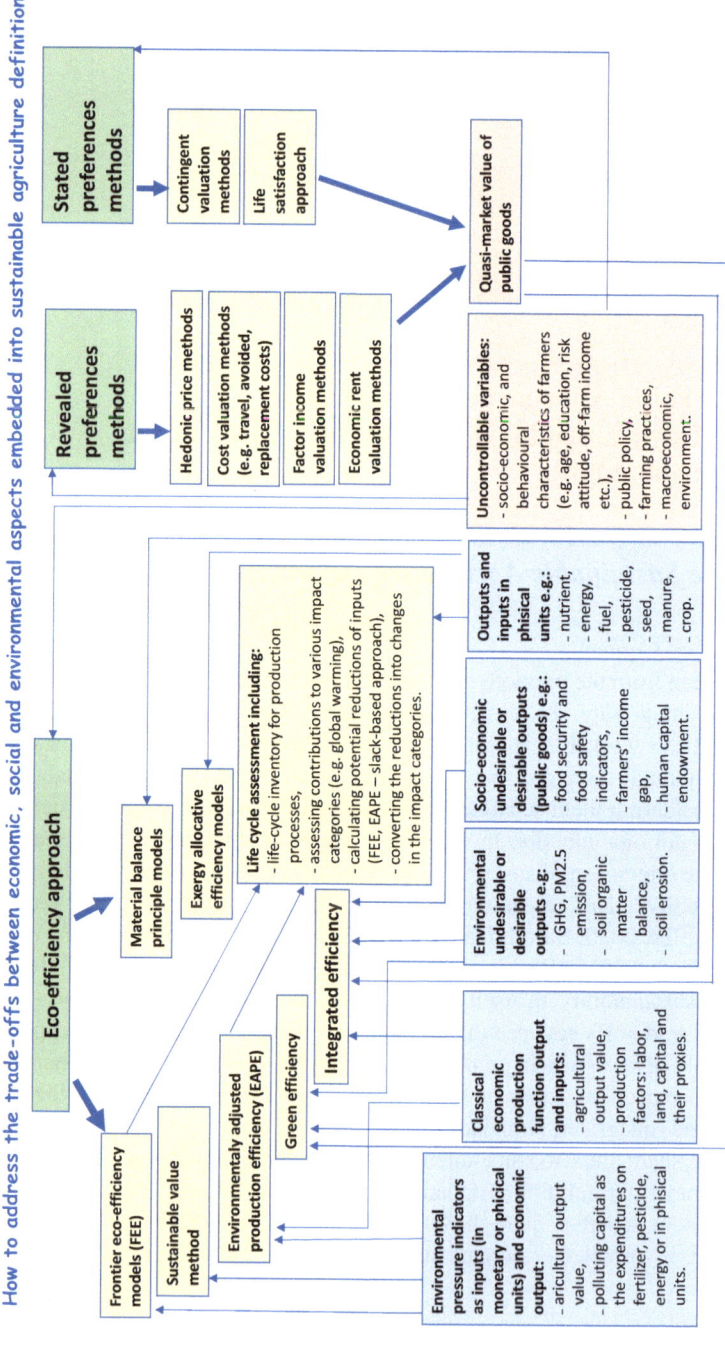

Fig. 2.1 Workable approaches to integrate economic, social and environmental dimension within sustainability concept

biophysical units. Secondly, we should discuss whether being eco-efficient does actually imply being sustainable.

Although the successful derivation of solutions to the above quandaries have not only been explicitly claimed very often, the great diversity of approaches in the selection of variables for effects and inputs in eco-efficiency analyses shows that this is still an unresolved issue. The origins of the term "eco-efficiency" itself shed some light on this discussion. It is important to note that the term has strict economic roots. In the 1990s, Schaltegger and Sturm (1990) coined the term "eco-efficiency" as the "business link to sustainable development". Then, it was disseminated so by the World Business Council for Sustainable Development (WBCSD, 2000), and in last decade, became frequently used with regard to agriculture (Bonfiglio et al., 2017a; Gadanakis et al., 2015; Pérez Urdiales et al., 2016; Stępień et al., 2021). OECD (1998) interpreted the notion of eco-efficiency as "the efficiency with which ecological resources are used to meet human needs" and thus operationalized it as a ratio of an economic output (i.e. the value of products sold by a firm, sector, or economy) and the input understood as the sum of environmental pressures generated by a production unit. For instance, since 2011, the European Commission has raised concerns that the growing demand for resources will inevitably lead to shortages and higher prices that will adversely affect the European economy: "Resources ought to be used more efficiently throughout their life cycle which means generating higher value with lower material costs" (EC, 2011).

The above definitions of eco-efficiency lack precision, so many approaches have been applied for measuring it: **environmentally adjusted production efficiency** (EAPE) (Dakpo et al., 2017; Huang et al., 2016; Song & Chen, 2019), **the frontier eco-efficiency model** (FEE) (Bonfiglio et al., 2017a; Gadanakis et al., 2015; Gómez-Limón et al., 2012; Pérez Urdiales et al. 2016; Picazo-Tadeo et al., 2011, 2012; Stępień et al., 2021), **the materials balance principle** (MBP) **adjusted approach** (Guesmi and Serra 2015; Hai and Speelman 2020; Hoang & Rao, 2010; Lauwers, 2009) **exergy allocative efficiency** (Liang et al., 2018; Mohammadzadeh et al., 2018), and **life cycle assessment** (Grados & Schrevens, 2019; Soheili-Fard et al., 2018), as well as **green efficiency** (also called "**Agricultural Green Total Factor Productivity**" (AGTFP)) (Ge et al., 2018; Han et al., 2018; Liu & Feng, 2019; Shen et al., 2018; Xie et al., 2018).

Does eco-efficient mean sustainable? A composite definition of sustainable agriculture provided at the very beginning implies the trade-off between: (1) satisfying human food and fibre needs, (2) being profitable and using on-farm resources efficiently to minimize adverse effects on the environment and people, (3) preserving the natural productivity and quality of soil, air and water, (4) sustaining vibrant rural communities. Theoretically, there is a risk that a higher environmental pressure on the inputs side can be compensated by an even faster growth of the production value, which ensures an increasing eco-efficiency without taking into account the depleting environment carrying capacity; protecting the latter is the crucial condition of sustainable development. Therefore, building eco-efficiency indicators that would be compliant with the idea of sustainability is a very topical research issue (Repar et al., 2017).

The above authors show that the insertion of additional environmental variables, such as greenhouse gas (GHG) emissions, water supply, phosphorus surplus etc. may have a significant impact on efficiency calculations. These additional variables can be introduced in both stochastic frontier analysis (SFA) and data envelopment analysis (DEA) frameworks. The main question that arises here is whether higher ambition regarding environmental protection must be in conflict with the economic objectives of the farm. This is especially important in promoting organic farming. Researches on this topic provide ambiguous results. Lakner and Breustedt (2017) found that organic farming is usually less technically efficient than conventional farming, while Beltrán-Esteve et al. (2017), based on investigation into Spanish citrus farming, claim that a shift from conventional to organic farming may reduce environmental impact significantly (by 80%) without any decline in economic performance.

Building upon on livestock grazing examples drawn from China, Huang et al. (2016) have shown that differences between technical efficiency and environmental efficiency may be significant. Herein, an average technical efficiency (TE) was estimated to be 0.837 (which is a high number) while environmental efficiency (EE) (after including total grassland net primary productivity capacity) was about 0.123. More interestingly, the authors concluded that there was a positive relationship between TE and EE for farms with smaller than average TE level, but when it comes to highly efficient farms (TE higher than 0.837), the relationship is an inverse U shape, meaning that the most technically efficient farms are sometimes environmentally less efficient.

Environmental efficiency is sometimes refereed to energy intensity. Ghali et al. (2016) have used the DEA framework and shown that highly technical efficient farms are often highly inefficient regarding energy use. There is a need, therefore, for a proper energy policy that is targeted mostly on farms with highest efficiency since they could significantly reduce their energy use.

Urdiales et al. (2016), in turn, used the DEA-based framework to calculate eco-efficiency for a sample of dairy farms in Spain, where environmental pressure was assessed by P, K, N surplus and CO_2 equivalent emissions. They found that environmental pressure may be decreased by as much as 37% while the current level of value added is maintained. Based on an Amazonian example, Peña et al. (2018) provide even more optimistic results. They claim that it is possible to increase the desired economic outputs by 19.5% with simultaneous reduction of inputs and undesirable output (degraded land) by 16.4%.

Optimistic results are also provided by Wettemann and Latacz-Lohmann (2017), based on a dairy industry example in Germany. The authors found that farms could reduce their variable costs and GHG emission by 10.5% without diminishing outputs. Moreover, when technical inefficiencies were eliminated and farms moved to a GHG-minimizing input mix, emission could then be decreased by 31.8%, on average. Similar results are also provided by Adenuga et al. (2018, 2019), who found that dairy farms in Ireland and Northern Ireland may reduce N surplus and increase production outputs simultaneously.

Guesmi and Serra (2015), on the example of Catalan arable farms, show that some desirable farming practices (such as efficient use of chemical inputs) may contribute to both bettered technical and environmental performance. Positive trade-offs between environmental and farm economic goals were also found by Hai and Speelman (2020). They saw that when farms use more appropriate input mixes, they became more cost efficient, and this outcome benefits the environment by default. What is more, from the other hand, efforts on improving environmental performance help to reduce production costs.

The situation in developing countries may be, however, different. Ullah et al. (2016) have shown that in the case of the majority of cotton farms in Pakistan, it is very hard to combine high economic performance with low environmental impact. In contrast, Soteriades et al. (2015) conclude that there is negative correlation between eco-efficiency and milk yield/cow based on a sample of specialized French dairy farms, but they also noted that this relationship depends on external circumstances and farming system. Generally, more self-sufficient farms are preferable.

The considerations set out above reveal that the question on the relationship between economic and environmental aspect of farm is still open. The ambiguity of results is related to the different econometric methods that were employed, but also to specific, local conditions. What is more, economic performance may be understood in different ways. If we equate performance with productivity, positive trade-offs are less likely to occur, as demonstrated by Soteriades et al. (2015). However, if we relate performance to profitability, this positive relationship is more probable since greater environmental commitment results not only in increasing levels of eco-efficiency, but it also helps to reduce production costs (e.g. lower use of fertilizers and pesticides).

Hoang and Coelli (2011) concluded that the EAPE models may be inconsistent with the materials balance principle (MBP), and thus both outputs and inputs in eco-efficiency analysis should be expressed in physical units (e.g. energy equivalent). Although the aforementioned approach seems logical from the MBP perspective, it stands in some contradiction with the definition of production efficiency and economic theory of value. A larger effect in physical units is not necessarily reflected in a higher market value of products that is derived from utility of a good to consumers. The utility translates into prices and market equilibrium quantities, as well as into an ability to meet society needs. However, inconsistency of expressing inputs in biophysical units and outputs in economic units implies that MPB-compliant measurement may have limited application for production economics and economic policy.

The concern expressed in the previous paragraphs is reflected in many works, particularly where the authors have attempted to supplement the classical production function with additional polluting inputs (Dakpo et al., 2017; Huang et al., 2016; Song & Chen, 2019) or included undesirable outputs on the effects side (Peña et al., 2018). The ecosystem carrying capacity can be addressed in different ways. The first possibility is to include excess pressures on the inputs side. For instance, Park et al. (2016) employed the renewability index which consists of the ratio of renewable resources used, to the total ecological resources, and by this means, represented the depletion of renewable resources. Other studies have applied different types of nutrient surpluses, e.g. P, K, or N surpluses and also grassland net productivity capacity (Adenuga et al., 2019, 2020; Huang et al., 2016; Urdiales et al., 2016). The second option is to treat excess pressures as undesirable outputs (i.e. N and P surpluses, as put forth by Adenuga et al. (2018)). This approach seems to be a reasonable compromise resolving the inconsistency between MBP and value theory. It applies the concept of an economic production function in which undesirable outputs on effect side (expressed as excess pressures beyond ecosystem carrying capacity) are linked with standard economic inputs and outputs .

In this book, we would also like to contribute to this methodological discussion, which is looking for a form of eco-efficiency measurement that is compliant with the concept of sustainable agriculture. We propose using slack-based measure approaches (SBM), including undesirable and desirable outputs that reflect three dimensions of sustainability, i.e. social, economic and environmental. The model that we argue is the most compliant with the sustainable agriculture definition. It assumes strong (free) disposability of bad outputs, which means that bad outputs are separable from good outputs (Halkos & Petrou, 2019). In other words, it is possible to reduce negative environmental outcome (such as GHG emission) without depleting the value of agricultural production. It is also possible to expand production without increasing the pressure on the environment. Such an approach follows the multi-objective logic of the sustainable agriculture definition and thus brings eco-efficiency closer to sustainability—as it is more reliable when technological progress in farming is taken into account. Thus, we called our approach **"integrated efficiency" (IE).**

IE also supposes the occurrence of so-called "slacks". This means, on the one hand, that there is reduction potential in a particular input and/or undesirable output and/or expansion potential of desirable output without general changes in farm technology and affecting the other inputs/outputs (i.e. Pareto-like improvement). Moreover, the desirable outputs may be increased and the bad outputs and inputs may be reduced simultaneously. In contrast to standard radial DEA models, individual inputs and outputs do not have to be expanded or reduced proportionally. This assumption seems to be reliable since there is evidence that the above issue factually concerns farmers in many regions of the world. For instance, in Europe, they can use fewer pesticides without a negative effect on the obtained value added (Chèze et al., 2020). Similarly, European small farms may relatively easily convert into organic farming or implement agri-environmental schemes (AES), while simultaneously increasing their incomes (EC, 2019). It seems also feasible to reduce several inputs simultaneously but each of them to a different extent (not only proportionally as classical DEA assumed).

Case of Organic Farming in CEECs
The case of organic production in Central-Eastern Europe also confirms the occurrence of slacks: Since 2013, contrary to expectations, organic food production has been stagnating or regressing in Central and Eastern European countries (i.e. in Poland and Romania—the biggest players in this part of Europe), and the problem

mainly touches small-scale farming. For example, in Poland, year by year, fewer and fewer farmers are engaged in organic production. In 2013, there were more than 26,500 thousand organic farms, while at the time of this writing (2021) there are slightly more than 19,000, which contributes to the decrease in the area of organic crops, which currently amounts to 484,000 ha—for comparison, in 2017 there were 10,000 ha more. Paradoxically, 2019 brought a record increase in demand for organic products in Poland and throughout Europe (EC, 2019). Initial estimates indicate that sales in Poland increased by over 20% to over PLN 1.2 bln PLN. The discrepancy between the dynamics of domestic supply and demand for organic products shows that the problem is growing. The easiest answer why this has happened is that organic farming is simply not profitable in Poland. . . but it might be, as EC (2019) data clearly indicates.

Even excluding subsidies, net market income per annual work unit (AWU) was higher in organic crop farms in France, Austria, Spain and Italy. In the milk sector, the net market margins per cow are higher for organic farms in every country. Moreover, subsidies compensate for potential income loss—says the quoted report. In this case, the compensation comes mainly from the Common Agricultural Policy (CAP) agri-environmental payments. These payments may reflect the institutional value of public goods delivered. However, the adoption of environmental schemes among small-scale farmers in Eastern European countries is significantly lower than in other Western European countries.

Many researchers and European institutions point out the need to uncover lock-ins and find levers to encourage farmers to move to and stay in sustainable, climate-neutral and biodiversity friendly food production systems. Hence, a number of authors have explored farmers' attitudes regarding voluntary conservation schemes (Christensen et al., 2011; Ducos et al., 2009; Ruto & Garrod, 2009; Schulz et al., 2014).

In several studies high transaction costs and bounded rationality (Ducos et al., 2009) were identified as significant barriers to farmers' interest in AES. Other authors focused on how risk influences farmers' decisions (Chèze et al., 2020). In small-scale farming, the situation is, however, more complex due to the predominant role of behavioral factors in the adoption of environmentally sustainable practices (Dessart et al., 2019). It is also likely that perceptual limitations play preeminent roles in the decision-making processes of small-scale farmers—as based on the theory of planned behavior framework (TPB) and as explained in the recent literature review by Dessart et al. (2019) and Czyzewski et al. (2021).

Perceptual limitations explain the difference between optimal beliefs and choices (concerning i.a. perceived cost and benefits, perceived control, perceived risk) from the beliefs and choices that people have. This idea may be especially interesting in the context of small-scale farm holders, but has not yet been empirically explored with regard to them. The European Commission has high hopes for small-scale farms as providers of public goods, but these expectations may turn out to be futile if those farmers remain reluctant to adopt sustainable practices. This is the case mainly in Central and Eastern European countries such as Poland and Romania (EC, 2019), but also in several EU associated countries (e.g. Moldova, Serbia).

There are over six million small-scale farms in EU-27 countries (Davidova et al., 2013) for which the concept of eco-efficiency and slacks reduction seems to be very promising. This is also a dilemma for the new CAP perspective, 2021–2027. Small-scale farms are mainly located in European countries with the most fragmented agriculture, such as Poland, Romania, Serbia and Moldova in which the agricultural sector continues to play a relatively important role for maintaining jobs. The subject of agricultural development in the mentioned countries has received relatively little attention from international research. This has happened for numerous reasons, the most important among which is the lack of data availability and poor infrastructure in the countryside.

Market-Based and Non-market Value of the Environment
Another approach to valuing environmental or social public goods (PG) is that with regard to resources and production factors and the incomes they generate. The value of PGs provided by agriculture and rural areas is usually assessed using environmental and social variables describing cultural amenities, farmland biodiversity, water quality and availability, air quality, soil quality, climate stability, resilience to fire and resilience to flooding (Delbecq et al., 2014; Santos et al. (2016). This set of goods was taken by the cited Santos et al. as the "landscape". In such an approach, we may refer to multiple studies on landscape valuation, such as that of: Scarpa et al. (2007), Chiueh and Chen (2008) and Borresch et al. (2009).

In a narrow perspective, pure PGs are considered to meet two conditions: "non-rivalry" and "non-excludability". In practice, however, such goods are scarce in the economy (examples include national services, national defense, order and security). We, therefore, consider quasi public goods, including goods characterized by "rivalry" and "non-excludability", also called "common goods" or "merit goods". Some of these might be private in terms of their physical traits, but as a result of social doctrine and the social policy implemented by public authorities, they are provided to citizens even without their acceptance.

A specific feature of PGs related to agriculture and rural areas is the fact that they may be the external effect of "regular" agricultural production or of abstinence from specific activity, a purpose-specific effect or a common resource in society's possession (Geoghegan et al. 1997; Hvid, 2015; Plantinga & Miller, 2001; Shi et al., 1997).

Market-based value methods are the most frequent manner of measuring the value of landscape amenities in rural areas. In practice, indirect attempts at assessing the influence of public goods on the income are very rare, except for the "institutional" valuation of public goods by CAP (e.g. agri-environmental payments, support for least favored areas, set-aside payments, and subsidies to rural area development; Czyżewski & Matuszczak, 2016a, 2018). The **hedonic price method (HPM)** is used most frequently with reference to the influence of public goods in rural areas on the prices of agricultural land, house prices, or prices of agrotourism services. Indirect valuing of externalities such as the **cost/factor income valuation** method (i.e. travel costs, avoided costs, replacement costs) is applied less often

(Groot et al. 2002). The mentioned family of market-based valuing methods is also called **revealed preferences methods**. The theoretical foundation of HPM was developed in the work of Rosen (1974), the base concepts of whom have been built upon by many authors, each contributing to the framework of this method in its application to environmental notions. For instance, Delbecq et al. (2014) reveal that farmland values are only partially explained by agricultural returns. These authors identified multiple non-agricultural attributes of farmland, which include biodiversity, climate regulation, rural culture and open space, as well as features that indirectly impact food quality and human health. A more recent study, that by Bilbao-Terol et al. (2017), tested the hypothesis that positive externalities from agriculture, namely, "maintaining and preserving an attractive landscape", have a beneficial impact on the prices of rural accommodation provided by farmers or other rural citizens. The aim of the research was to identify which agricultural activities affect the profitability of rural tourism. The latter, in turn, explores the impacts of environmental amenities associated with agricultural and silvicultural land use on the price of self-catering cottages.

Some authors point out the weaknesses related to problems of individual perception, subjectivity, continuity, aversion behavior, market segmentation, and the assumption of equilibrium, omissions of important characteristics and the debaTable mathematical specification of the model (Vanslembrouck et al., 2005). We would like to draw attention to another problem of HPM that is related to the interpretation of the situation in which public goods in rural areas turn out to have a negative sign next to the regression coefficient. This means that they negatively affect prices (of land, houses or services). The cited above authors (Bilbao-Terol et al., 2017; Vanslembrouck et al., 2005) tested, for instance, the hypotheses about the positive influence of various public goods on prices, as well as their marginal effects. So, what if it turns out that a feature universally recognized as a public good has a negative influence on market prices? Does it stop being a public good then, or does it have a negative value?

The second answer is contradictory in itself, because a negative value of a public good is at odds with the definition of a good, which becomes a "public ill". In Vanslembrouck's study, the negative sign accompanies the "forest" variable, which is positive in the study by Bilbao-Terol.

Schläpfer et al. (2015) in turn examined how land use and environmental amenities and disamenities affect rental prices across urban, suburban, periurban, and affluent communities in Switzerland and obtained ambiguous results (negative signs contrary to expectations) for public goods such as "forest", "open space", and "cultural objects". Similarly, Garrod et al. (1992) also noticed an unexpected result in their studies, i.e. the negative effect attributed to the possession of woodland. In our opinion, we cannot talk about a negative value of a public good or leave it to the market to decide what is and what is not a public good and classify amenities and disamenities based on the sign next to the regression coefficients. Negative signs may simply indicate that the given public good in the given area was not taken into consideration by the market for various reasons, because the market is unreliable when it comes to the valuation of public goods. Thus, this does not mean that that good does not provide certain amenities to the consumer. This is why we recommend the eco-efficiency approach that separates negative and positive externalities (i.e. bad outputs and public goods).

The problem of a negative valuation, in particular, when it comes to environmental public goods, is not rare in the literature. According to our previous findings, in Poland, farmland rich in public goods is cheaper (Czyżewski & Matuszczak, 2016b, 2018). Moreover, Nilsson and Johansson (2013) noted that agricultural environmental payments in Sweden have a negative influence on land prices. Rutherford et al. (1990) also reached similar conclusions. This is not because these goods have no economic value at all, or even have negative value, but because individual consumers do not realize this value, which is, however, noticeable from the point of view of social well-being. So they realize consumer rent, even though they may not know it. For this reason, a search for the value of public goods should be considered, both from the perspective of producer (seller) rent and consumer (buyer) rent.

To address the latter concern, an alternative approach has been proposed, i.e. the **economic rent valuation method** (ERV). A factor income valuation method, it advocates the adoption of the variance component model to estimate the changes of income or revenues (in monetary values) resulting from the occurrence of specific public goods in rural areas (Czyżewski et al., 2020). In this approach, consumer rent is defined as a situation in which the consumer pays less for the good than they would be willing to pay based on its utility function.

Let us add, though, that in the case of public goods, we also need to consider amenities that are not conscious or explicit at individual level, but they exist in public awareness. For instance, if the prices of houses located near cultural facilities are lower, it means that in the studied sample, consumers evaluated the utility of these facilities incorrectly in their preferences concerning house location. Regardless of that, they have the opportunity to take advantage of these cultural goods, even though they have not paid for them, therefore, this is consumer rent—although perceived not at an individual, but at a collective level. Hence, in the case of public goods, the market criterion saying that a good is worth as much as the customer is willing to pay for it should not be used, because otherwise it may turn out that culture funding or supporting nature is unnecessary, because it has negative market value (Czyżewski et al., 2020).

Another approach to valuing public goods is by means of assessing individual's willingness to pay (WTP). However, it is commonly known that individuals have no incentive to disclose their true demand for non-excludable goods, or we do not know whether they are aware of public goods' utility. Therefore, some economists are very pessimistic as to whether it is possible to assess people's preferences and willingness to pay for public goods (Frey et al., 2009).

Essentially, two approaches have been pursued: **stated preference methods** (e.g. the contingent valuation method), and **the life satisfaction approach**—a method of valuing the psychological costs of public ills (Kahneman et al., 1997; Kahneman & Thaler, 2006). In the first strand of research, a choice-modeling questionnaire has been developed which makes it possible to estimate multinomial logit and random parameters models for valuing alternative scenarios (estimated coefficients stand for the rate at which respondents are willing to trade-off one attribute for another). For example, a study of Santos et al. (2016) represents the most recent and comprehensive attempt at building an empirically-based framework to value multiple agricultural or rural public goods. Herein, Santos et al. (2016) use context-rich valuation scenarios at broad supranational scale, and employs the typology of Macro-Regional Agri-Environmental Problems associated with 13 clusters of European agrarian structures.

In contrast, the life satisfaction approach correlates the degree of public goods with individuals' reported subjective well-being, and evaluates them directly in terms of life satisfaction. This reported subjective well-being can serve as an approximation for individually experienced welfare (Levinson, 2012; Luechinger, 2009). For example, A. Levinson (2012) has combined air quality data with individuals' self-reported levels of "happiness", as a function of their demographic and economic characteristics and the current air quality. The estimated function is employed to calculate the average marginal rate of substitution between annual household income and air quality that makes respondents equally happy. The stated preference methods are quite similar, since they employ utility functions for pairs of chosen goods, one of which is a public good. However, all these methods have substantial weaknesses in that individuals have no incentive to disclose their true preferences as regards public goods and they are not aware of PGs utility from the point of view of social benefits.

To sum up, there are four potential approaches to deal with trade-offs between different dimensions of sustainability in agriculture: **(1) eco-efficiency (including Integrated Efficiency approach), (2) hedonic price methods, (3) factor/cost income valuation, and (4) stated preferences method**. The methods (2)–(4) allow assessing the part of income that a consumer is willing to pay or already paid/should pay (in case of ERV) for a public good or the minimum monetary amount that a producer is willing to accept to provide a public good. This is, therefore, a direct trade-off between economic and environmental or social well-being. The Integrated Efficiency (IE) concept tends to estimate an optimal (fully efficient) frontier within a given technology that comprises all possible combinations of economic effects, inputs and desirable and undesirable externalities. Moreover, IE reveals possibility to optimize each inputs and outputs separately within a given issue. Hence, IE inclusive with undesirable outputs and free disposability seems to be the most comprehensive attempt to operationalize the trade-offs that appear in sustainable agriculture definition.

2.1.2   Major Approaches to Computing Eco-efficiency

Since the 1990s, investigators of eco-efficiency measurement have been engaged in three strands of research. The first concerns assessing environmental policy influence in terms of its effectiveness, and considers national expenditures (Adam & Tsarsitalidou, 2019) or transnational schemes (Arata & Sckokai, 2016; Gocht et al., 2017). The second strand explores corporate performance with regard to environment (Li et al., 2017). The third strand is anchored particularly into the practice of sustainable agriculture and indicates the trade-offs between its ecological and economic aspects (Bonfiglio et al., 2017a; Caiado et al., 2017; Gadanakis et al., 2015; Picazo-Tadeo et al., 2012; Staniszewski, 2018). Within this thread, the notion of **"sustainable intensification"** (SI) has been coined. This postulates the specific way of measuring eco-efficiency, i.e. by supplementing the classical factor-based production function with polluting inputs or their intensity (EAPE).

Various studies have, however, reached similar conclusions—that the understanding of SI differs accordingly to the economic level of country development. For instance, for highly developed Member States of the EU (i.e.EU-15), SI means improving environmental farming performance without decreasing economic productivity, while in less developed Central and Eastern European Member States, SI is more concentrated on building economic productivity without depleting the natural capacity of the environment (Moutinho et al., 2018; Staniszewski, 2018).

Generally speaking, eco-efficiency measurement can be computed by applying **frontier methods**, i.e. radial or non-radial DEA (including directional distance functions—DDF, free disposable hull DDF) or SFA. Indeed, many studies have used radial frontier methods to estimate the eco-efficiency for DMUs in farming. Those studies focus on environmental pressure as the input of agricultural production (e.g., Maia et al., 2016; Xing et al., 2018), but the outcome of such an approach cannot be directly used in regression of policy instruments on eco-efficiency. It needs "bootstrapping" or other technics (see in Chap. 5) to be considered as outcome variable in a regression model.

There have been multiple attempts to adopt the double-bootstrapped truncated regression of the eco-efficiency outcome (Bonfiglio et al., 2017a; Gadanakis et al., 2015; Pérez Urdiales et al., 2016; Picazo-Tadeo et al., 2011, 2012). These analyses are methodologically similar to ours, but they advocate an input-oriented approach that makes it vulnerable to the criticism of being incompliant with MBP, and which also does not address the possibility of providing public goods as desirable outputs or optimizing both outputs and inputs at the same time. Thus, we propose SBM with undesirable outputs assuming full disposability (see Sect. 3.3.1).

The second strand of methods makes reference to the value-based approach and is known in subject literature as the **sustainable value method (SV)**. This originates from the works of Figge and Hahn (2004, 2005) and has quite rarely been used (Czyzewski et al., 2020; Grzelak et al., 2019; van Passel et al., 2009). Probably this lack of application came about because applying an SV estimator based on linear production function that has been strongly criticized by Kuosmanen and Kuosmanen (2009) due to unrealistic assumptions. However, one should bear in mind that SV is a kind of second-best solution when we deal with aggregated (average) data, and a non-linear production is inapplicable (Felipe & McCombie, 2012a, 2012b; Salois et al., 2006).

Originally, SV was defined as follows:

$$SV_i = \frac{1}{m} \sum_{j=1}^{m} x_{ij} \left(\frac{y_i}{x_{ij}} - \frac{yf_i}{xf_{ij}} \right) \tag{2.1}$$

where: SV_i is the sustainable value; x_{ij} and y_i represent the inputs used of type-j and the DMU i and the return of the resources (e.g. agricultural output) of the DMU I, respectively and yf_i related to xf_{ij} represents the opportunity cost; $i = 1...n$ can be the region or country; and $j = 1...m$ is the type of analyzed inputs.

We agree that the criticism by Kuosmanen and Kuosmanen (2009) is legitimate in reference to microeconomic production function, as the eco-efficiency gap between the DMU and benchmark $\left(\frac{y_i}{x_{ij}} - \frac{yf_i}{xf_{ij}} \right)$ was assumed in the above equation to be a function of x_{ij} inputs—which is rather not linear as Figge and Hahn (2004, 2005) had believed. Obviously, product elasticities cannot be equal with the factor shares which provide variable returns to scale. This reasoning is not, however, applicable for aggregated data. Quoting Felipe and MCcombie (2012a): "the best statistical fit given by estimating putative regional aggregate production functions must give estimates of constant returns to scale with the output elasticities equal to their factor shares... Regressions that find increasing returns to scale and any differences between the values of the output elasticities and the factor shares do so by virtue of being misspecified".

We, however, disagree with Figge and Hahn's notion that opportunity costs equal market average. This assumption does not work with regard to environmental performance and the trade-offs that occur in sustainable agriculture. Firstly, the market mechanism does not take into account public goods (i.e., social and environmental), yet these are affected by production inputs. Therefore, the market average will always be underestimated as it relates to potential opportunities to reduce pressure factors or to use assets in a more sustainable way. Indeed, only farm leaders can implement managerial or production solutions that will successfully trigger their willingness to pay for public goods (e.g., ecological food, agritourism). Secondly, the principle that "the average equals opportunity cost" may not be suitable for agriculture because of the existence of the market treadmill to which especially small and medium farms are subjected.

The treadmill theory asserts (as discussed in Chap. 5) that farm income does not grow parallel to increase of the productivity. Although farmers invest in new technologies, these efforts will not provide the expected increase of income, as the growing supply meets flexible prices for raw agricultural materials. As a result, the marginal revenue from agricultural output decreases, thereby forcing producers to continuously invest in ever-more-efficient technologies. Only those farmers who are the first to introduce these new technologies gain the deserved benefits, and a majority will not notice any income growth. For these reasons, average productivity does not reflect the real opportunity cost in agriculture.

Hence, looking for a consensus, we argue that it is possible to adopt the original SV approach, albeit, using non-parametric frontier measurement for assessing the benchmark, and we can maintain the application of linear return to scale of using specific input if we employ region or country-level aggregated data. Therefore, we accept the concept of "eco-efficiency gap", but changing Figge and Hahn's approach to opportunity cost as the benchmark.

It should be noted that SV assesses the distance of a given region or country to a specified eco-efficiency frontier. The result is expressed in monetary units with regard to each input and can be totaled. The frontier DMU sample can be identified, for example, by solving a typical linear programming problem (i.e. the CCR multiplier model as developed by Zhu, 2008, as well as by Czyzewski & Guth, 2021), wherein:

$$\max \theta = \sum_{i=1}^{m} \mu_r y_{ro} \tag{2.2}$$

and is subject to:

$$\sum_{i=1}^{z} v_i x_{io} = 1$$

$$\sum_{r=1}^{s} \mu_r y_{rj} - \sum_{i=1}^{z} v_i x_{ij} \leq 0$$

$$\mu_r, v_i \geq 0(\varepsilon)$$

where: x_{io} is the input i used by object $o(i = 1 \ldots z)$; y_{ro} is the output r used by object $o(r = 1 \ldots s)$; ε is the infinitesimal constant; and v_i and μ_r are called multipliers.

In theory, SV takes positive or negative value in monetary terms. A minus sign reflects a value of so-called 'clean production' (zero-inputs growth) that might be provided by a DMU (e.g. farm) that manages to reach a 'frontier' eco-efficiency within a given technology. Moreover, the relatively rare situation of a positive SV can be interpreted as the 'clean product' that has been obtained without additional inputs, as compared with an average frontier DMU (Czyzewski & Guth, 2021).

2.1.3 Regressing Eco-efficiency and Its Potential Drivers

It is often the case that some variables cannot be directly included on either the outputs or inputs side of the eco-efficiency analysis. We call such variables "uncontrollable variables". This is particularly the case if the use of a certain input or production technique is beyond the decision of the DMU and is, for example, imposed by agricultural policy in the form of a specific support scheme. In considering environmentally adjusted production efficiency, such variables definitely should not be treated as inputs for frontier estimation. However, we are usually interested in investigating the impact of policy on eco-efficiency and therefore we are on the look out for ways to regress the uncontrollable against the eco-efficiency score. This issue will also be discussed in Chap. 5.

The simplest way to accomplish the aforementioned would be to estimate a linear model for eco-efficiency, but such a model will be likely biased, firstly, due to the fully efficient DMUs for which the score equals 1. Therefore, some researchers have applied censored regression (i.e. the so-called "two-stage approach"—where in the first stage, one performs basic DEA measurement). This solution was, however, criticized by econometrists because of serial correlation of the scores and its artificial data generating processes (see Chap. 5).

A better solution in this situation is to employ the so-called "super efficiency model" and, subsequently, to estimate a bootstrapped regression using uncontrollable variables. This approach was originally proposed by Andersen and Petersen (1993) and later applied by Long et al. (2018). It enables comparison of even fully efficient DMUs that are laying on the frontier, as their scores differ (since they are greater than 1) and the greater the value, the better positioned is the unit. We will described this solution in detail in Chap. 3 and consider it to be a quite optimal approach in the case of using EAPE with undesirable outputs.

Estimating determinants of eco-efficiency measurement obtained through frontier-based methods has been debatable for several technical reasons, especially since Simar and Wilson (2007) criticized a commonly used approach in which censored or truncated regression is applied (Stanton, 2002). Firstly, Simar and Wilson argued that one should not treat DEA scores as if they were independent observations due to serial correlation, and for this reason, a truncated rather than censored regression model is more appropriate. Secondly, the quoted authors emphasized the fact that fully efficient DMUs are an artifact of the finite sample bias inherent in DEA, but they do not reflect the true underlying data generating process. This is why the mentioned authors proposed a parametric bootstrapping procedure that tends to be consistent with the assumed data generating process (Badunenko & Tauchmann, 2018).

This procedure is called **"double bootstrapped truncated regression"**, as both the DEA scores in the first stage and then truncated regression are bootstrapped to control for potential bias. In many studies, this procedure has been applied in order to estimate regression models for the eco-efficiency score computed under the FEE input-oriented model (Bonfiglio et al., 2017a; Gadanakis et al., 2015; Pérez Urdiales et al., 2016; Picazo-Tadeo et al., 2012).

There are also alternative methods that consist of three or four steps, and therefore, they are called **"four/three-stage models for incorporating uncontrollable variables"**. The first stages of these models are the same, and assume the initial performance evaluation for each DMU using typical DEA non-parametric techniques. In the second stage, SFA may be engaged to decompose the input stage-one slacks, or alternatively, the total slacks (radial plus non-radial) are regressed against a set of uncontrollable variables using Tobit regression. The initial input values are then adjusted based on the estimated regression model coefficients and statistical noise. This operation can be considered as a separate stage. The last stage recalculates the DEA score using the adjusted input data.

What are potential drivers of eco-efficiency (see Fig. 2.1)? If we agree that becoming more eco-efficient can be seen as a farm management goal, the question arises as to what factors stimulate this process. There are many possible determinants of eco-efficiency. These include: socio-economic characteristic of farmers, farm practices, macroeconomic reasons, and policy and economic environment impact. The most popular methods for studying the impact of determinants on eco-efficiency are SFA-based models in which possible factors are introduced in the model to explain inefficiencies (e.g. Adenuga et al., 2020; Song & Chen, 2019) or different types of DEA-based models wherein double truncated regression is applied (Long et al., 2018; Urdiales et al., 2016).

Regarding the socio-economic characteristic of the farmer, age and education are most often studied. Interestingly, the results are contradictory. Building upon on a Northern Ireland example, Adenuga et al. (2019, 2020) found that older farmers are more efficient and they explain this by greater experience in farm management revealed by older farmers. Urdiales et al. (2016), in turn, found that greater eco-efficiency is positively correlated with lower farmer age. What is more, farmers that are more eco-efficient are those who plan to continue their operation in the next 5-year period, and are in favor of restrictive environmental policy. Perhaps, the most important finding of the author is that farmers who claim themselves as environmentally conscious are in fact more eco-efficient.

Guesmi and Serra (2015), in turn, conclude that a farmer's age and education level has no significant impact on environmental efficiency, but eco-efficiency is positively impacted by the family size. In addition, focusing on farming only rather than conducting non-farming activities also has a positive impact on eco-efficiency.

Study after study has concluded that the level of farm eco-efficiency is directly dependent on farming practices. Based on a sheep meat example in France, Dakpo et al. (2017) have shown that just by modifying the polluting inputs mix, farmers could decrease inefficiency by 35%. What is more, by applying better farm level decision practices, farmers could reduce the amount of physical inputs without reducing production levels (Ullah et al., 2019). This means that better farm management can help in reducing costs and in improving eco-efficiency.

The negative impact on eco-efficiency is often linked to intensification processes. Production intensity could be understood differently, but is usually related to the use of fertilizers (Long et al., 2018), feed concentrates (Adenuga et al., 2020) or chemical inputs (Guesmi & Serra, 2015). The results of all these studies confirmed that increasing intensity has negative impact on eco-efficiency, but on the other hand, more efficient use of these inputs improves farm performance. Another farming practice that can be important is proper management of machinery use (Khoshroo et al., 2018). This could help in reducing farm energy intensity.

Progress towards more eco-efficiency farming is supported by national and paranational agricultural and environmental policy, so the effectiveness of such policies is of a major importance. Song and Chen (2019) found that the proportion of government expenditures set aside for environmental protection has a positive impact on grain production eco-efficiency in China. What is more, the GDP per capita in China's provinces was also seen to have positive impacts on eco-efficiency. In another example, in Brazil (Amazonia), Peña et al. (2018) discovered that the eco-efficiency of a municipality improves when the share of farms receiving support is higher. Moreover, Guesmi and Serra (2015) demonstrated that CAP subsidies positively impacted environmental efficiency in Catalonia (Spain). However, Adenuga et al. (2020) did not find any significant impact of environmental subsidies on Northern Ireland farming practices. Other studies, such as that of Dong et al. (2018) do not analyze policy impact formally, but rather claim that low carbon agriculture should be supported.

To sum up, we may say that appropriate studies of eco-efficiency are very important, especially since they may provide the basis for policy recommendations. However, until now, there are too many discrepancies in the results. Based on this short review, we cannot unequivocally say whether characteristics such as age, or education contribute to higher eco-efficiency. The impact of policy is also unclear. Without any doubt, we may only conclude that farm-level management practices are very important. Studies of this subject often conclude that by changing some farming practices or the input mix, it is possible to improve eco-efficiency or decrease the environmental burden. However, their authors often do not provide any specific examples of what exactly should be done.

2.2 How to Integrate the Socio-economic Dimension into the Eco-efficiency Concept?

2.2.1 Economic Dimension of Sustainability

Although the term "economic sustainability" is commonly encountered, we would like to recall that the term "sustainability" should be reserved for the simultaneous balance of three aspects: social, economic and environmental (see Fig. 2.1). None of these alone constitutes sustainability in the proper sense of the definition, and therefore when we say "economic sustainability" or "social sustainability", we rather mean "resilience".

Economic sustainability (sometimes called "economic resilience") is very often defined as the income gap between the agricultural and non-agricultural sectors. There is evidence from various countries (Argentina, Austria, Bulgaria, and developing countries, including the Third World) that industrial agriculture generates this income gap since it has negative impact on the rural population over the long-term, while impoverishing farmers and depriving them of opportunities for development (Bachev, 2017; Berlan, 2013; Chang, 2009; Gizicki-Neundlinger & Güldner, 2017; Guth et al., 2020; Hill & Bradley, 2015; Zawalińska et al., 2016).

Practicing sustainable agriculture generates the opposite effect, as Bacon et al. (2012) and Hediger (2008) argue, in that it diminishes the social costs of industrial farming, provides higher socio-economic stability and brings about willingness to accept pro-ecological schemes (Bock, 2012). The relationships between relative farmer deprivation and its drivers have been less documented. There are, however, several recent studies focusing on this issue (Cai & Pandey, 2015; Czyżewski & Poczta-Wajda, 2017). The agricultural policy role in increasing employment in rural areas has been also emphasized as a remedy for depopulation and income gap (Subić et al., 2013; Weingaertner & Moberg, 2011).

However, agricultural policy may turn out to be futile in reducing income gap. As an example, we can recall the EU CAP. One of its objectives has always been "ensuring fair standard of living for the agricultural community, in particular, by increasing the individual earnings of persons engaged in agriculture" (European Commission, 2015). This means decreasing the difference between earnings in agriculture and in non-agricultural activities. Despite the significant improvement of the economic conditions of farmers in the EU, the average entrepreneurial income in agriculture per non-salaried annual work unit has reached only 40% of average wage in the economy as a whole per full-time equivalent—as the EC's quoted in above report reveal. It is worth noting here that people's well-being also depends on the comparison of their relative standard of living with the others (i.e. with employees of different sectors). Seeing that the people working outside the agriculture are usually richer and better situated, farmers may have the impression of being subjected to social injustice. This phenomenon forms the basis for the notion of "relative deprivation" and is used in sociology to understand why people advocate social change.

The concept of relative deprivation was firstly defined by Davis (1959), and further disseminated by Runciman (1966). The crucial assumption of this theory is that people usually judge their situation by comparing it with the chosen reference group. Struggling with the relative deprivation at the collective level consists in participation in social activities, i.e. protests or lobbying, which tend to change the position of the relatively deprived group (Grant et al., 2015). Recently, the theory of relative deprivation has been engaged in studies on the perception of differences in quality of life (Chen & Ravallion, 2013; Jayanta & Dipti, 2013) and also has been used to improve understanding of population migration (Hyll & Schneider, 2014; Stark & Fan, 2011). In agricultural economics, the concept has been employed to explain inequalities in the possession of production factors (Fałkowski, 2013) and to find the links between farmers' income gap and macroeconomic and policy factors affecting this phenomena (Czyżewski & Poczta-Wajda, 2017). Cai and Pandey (2015) also applied a similar concept to European agriculture, for comparing the productivity gap, defined as the difference in value added in the agriculture and non-agricultural sectors.

Despite numerous studies focusing on agricultural incomes (e.g. Hill & Bradley, 2015; Zawalińska et al., 2016), the application of relative deprivation is less documented, and to best of our knowledge, has not been applied in eco-efficiency studies and holistic sustainability measurement.

The thread of economic sustainability in literature is very broad and, of course, goes beyond the concept of relative deprivation. Many different measures of economic sustainability in agriculture can be listed. It has been stressed that sustainable development in terms of economic standard of living is inseparably linked with the social sphere (Guth et al., 2020; Sebaldt, 2002) including social cohesion, equalizing opportunities, counteracting marginalization and providing access to education, culture and healthcare. We focus on these issues in the next section. However, the most frequently applied measures of economic sustainability in farming are the quantitative indicators of agricultural output in physical or monetary units, but also income per person within a household, farm expenses, less often—the wage level.

For measuring economic efficiency, the following indicators have been employed: professional activity indicators, labor force productivity, capital and energy intensity, the level of investment, expenditures on research and development (Eurostat, 2019; National Statistical Office, 2011), but also the indices of cost-effectiveness, liquidity and stability (Latruffe, 2010). Moreover, economic sustainability (or resilience) has been applied to assessing farm viability, wherein farm viability means capability to last in competitive market surroundings for the long term (Latruffe et al., 2016).

On the macroeconomic level, gross domestic or national product and national income, as well value-added of agricultural sector are used, but also the issue of its distribution is frequently engaged with the Gini coefficient based on the Lorenz curve that is usually adopted in this case (Peacock et al., 1988). However, Sen proposed a different measure of well-being, taking into account that a high level of income per capita is not always synonymous with a high level of socio-economic development (Sen, 1992). A last but not least example is the Index of Sustainable Economic Welfare (ISEW) developed by H.E. Daly and J.B. Cobb, in which individual expenses on consumption are adjusted for losses derived from unequal distribution of income (Daly & Cobb, 1989).

2.2.2 Social Dimension of Sustainability

Farm environmental performance does not, however, rely only on production factors viewed in a classical way (amount of labor, capital and land), but on human capital (HC) as well—this involves farmer's knowledge, experience and qualifications—which, in turn, influences the farmer's decision. In many studies, human capital has been found to be predominant for sustainable farming (Czyzewski et al., 2021; Elstrand, 1969; Fielding et al., 2008; Menozzi et al., 2015; Parman, 2012). Let us cite what Parman (2012, p. 317) said: "Human capital plays an important role in modern agriculture, helping farmers use a chosen set of inputs efficiently and improving a farmer's ability to choose between different sets of inputs, outputs and technologies". The other authors also agree that difference in assets productivity stands only for about a half of total economic performance of farms, and the rest depends on the capabilities of human capital (Elstrand, 1969; Fielding et al., 2008; Menozzi et al., 2015). The role of human factor has been studied on different aggregation levels and in various systemic conditions. For example, at a sector level, we can cite the work of Foster and Rosenzweig (1995), at farm level—Bingen et al. (2003), Pindado et al. (2018), or Kijek et al. (2016), and in developing countries—Caicedo (2019).

HC in farming has been recently analyzed in the context of the Theory of Planned Behavior (TPB) (Ajzen, 1991), focusing on behavioral factors that influence farmers' attitudes towards sustainable practices (Czyzewski & Sapa, 2021; Martínez-García et al., 2013; Menozzi et al., 2015; Power et al., 2013; Wauters et al., 2010). According to the TPB, behavior is guided by: (1) personal attitude towards the behavior; (2) perceived social pressure (i.e. how other people and authorities judge or may influence the behavior; (3) perceived ability to carry out the behavior—called "perceived behavioral control" (PBC); and (4) perceived moral obligation (added by Fielding et al. (2008) to the TPB model). Herein, the more favorable the attitude and subjective norm and the stronger the perceived control, the more likely that a given behavior will occur (Ajzen, 1991).

We argue that perceptual limitations play a predominant role in decision-making processes, especially in small-scale farming—as put forth by the recent literature review by Dessart et al. (2019). Perceptual limitations explain the difference between optimal beliefs and choices (concerning i.a. perceived cost and benefits, perceived control, perceived risk) from the beliefs and choices that people have. As we mentioned, the recent literature review by Dessart et al. (2019) supports the hypothesis about the crucial role of perceptual behavioral factors for adoption of sustainable farming practices. The cited author revealed that cognitive factors have direct influence on farmer's decision-making, and among them, perceptual behavioral factors dominate. Among these are: perceived cost and benefits, PBC and perceived risks. For example, despite the commonly known fact that any integration (both vertical and horizontal) is advantageous for farmers, the level of integration of small and medium farms in Poland, but also in other Central and Eastern European Countries has remained very low. This has been explained by scarce awareness of the benefits that may result from the integration, but also the reluctance to cooperate due to perceived risk of being tricked by the others (Livingston et al., 2010; Stępień & Polcyn, 2019).

TPB has been several times used to address behaviors related to sustainable and environmental farming practices. Examples include the work of Martínez-García et al. (2013), who studied improvements in grass-land management, Wauters et al. (2010), who analyzed the adoption of soil erosion control practices, Power et al. (2013), who focused on the determinants of adoption of environmentally oriented behaviour and Menozzi et al. (2015), who devoted their research to participation in sustainability programs.

It is the farmer who decides how sustainable his/her farming practices are or will be, and his/her behavior is determined by human capital endowment. Human capital is mainly forged by education, training and participation in social or cultural events. The farmer's education contributes directly to his/her stock of knowledge, but the indirect impact of education is also important. More educated farmers are more also likely to continue to improve their skills and seek out information about new techniques and their potential implementation. They also perceive knowledge as something valuable and understand that it may bring substantial benefits (Bindlish & Evenson, 1997; Wozniak, 1993). Moreover, a well-educated farmer is capable of implementing better combinations of inputs and technology and hence to improve productivity (Abdulai & Huffman, 2005; Lin, 1991; Wozniak, 1993). The likeness to adopt a new technology is particularly crucial in the context of climate change. Schultz (1975) argued that education enabled an individual to perform an action even under uncertain circumstances. If this is true, one should seem education as an indicator of what potentially a farm environmental performance might be.

Martey et al. (2014) concluded that education improves farmers' ability to cooperate and participate in group activities. Mojo et al. (2017) add that more educated and self-aware farmers are more aware of advantages of vertical integration. For example, Chagwiza et al. (2016) revealed a positive relationship between the level of education and the decision to join a cooperative. Jitmun et al. (2020) and Zhang et al. (2017) also confirm this finding.

However, it is difficult to build human capital if the more basic needs from Maslov's pyramid are not met. In particular, we mean the satisfaction of food needs and a decent standard of living. Research by Poczta-Wajda et al. (2020) has demonstrated that even in countries regarded as highly developed (i.e. Poland), half of the small farms declare some kind of food security problems. Food and nutrition security problems in small-scale farming are also reported in different Central and Eastern European countries. In such a situation, it is difficult to build human capital conducive to sustainable practices. Existential problems may strongly determine the aforementioned perception of behavioral control, which is why it is important to focus on them first when studying the social aspect of sustainable agriculture.

At the same time, it is worth being aware of the feedback that exists between social and economic aspects. For example, cooperatives play an important role in facilitating access to education—sometimes they provide funds and school infrastructure (Wanyama, 2014). Similarly, lower levels of relative deprivation of farmers income can build stronger perceived behavioral control and a sense of greater self-efficacy. Moreover, a narrowing of the income gap enables deeper social participation. Therefore, in the analyses in the following chapters we will focus on food security and the income gap as meta-factors that determine the social and economic dimensions of sustainability (see also Fig. 2.1).

Chapter 3
How Sustainable Is Agriculture Worldwide?

3.1 Agricultural Sustainability as a Multi-dimensional System

As stated in Chaps. 1 and 2, agricultural sustainability is a broad concept and may be analysed from many different perspectives. However, there is a consensus that it should cover the economic, social and environmental aspects of farming. Before we move on to policy issues, a detailed description of the current state of sustainability in different parts of the world should be carried out. In the final point of this chapter, we will study sustainability in an integrated efficiency framework showing how efficient is a country in achieving different goals. Therefore, the integrated efficiency concept can be very informative itself, but it also gives us the possibility to separately analyse several indicators that constitute 'inputs' or 'outputs' of the agricultural operation. Based on the previous literature, we will study several different dimensions of sustainability in the following subsections. Regarding the data availability, some aspects can be studied only for selected countries. We have collected the data from several sources: the World Bank database, FAOSTAT—the database of the Food and Agriculture Organization of the United Nations, the Penn World Table, and ILOSTAT—the database of International Labour Organisation. Data is provided for 2006–2018, unless otherwise noted.

We analysed the data and indicators for several groups of countries with a special focus on developing countries, following the classification proposed by the FAOSTAT. These groups are: the least developed countries (LDC), land-locked developing countries (L-LDC), small island developing states (SIDS), low-income and food deficit countries (LIFDC), net food importing developing countries (NFIDC), developing countries (DC) which covers all previous sub-categories, the

The research in this chapter was partly funded by the National Science Centre, Poland under the project No. UMO-2021/41/B/HS4/02433.

B. Czyżewski, Łu. Kryszak, *Sustainable Agriculture Policies for Human Well-Being*, Human Well-Being Research and Policy Making, https://doi.org/10.1007/978-3-031-09796-6_3

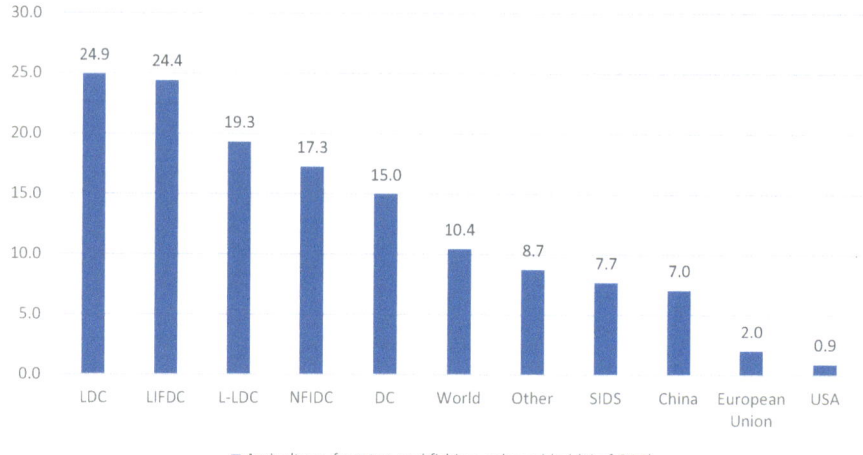

Fig. 3.1 The role of agriculture in the total economy and gap between gross value added in agriculture and other sectors. *Note*: *LDC* stands for least developed countries, *L-LDC* are land-locked developing countries, *SIDS* are small islands developing states, *LIFDC* are low-income food deficit countries, *NFIDC* are net food importing developing countries, *DC* are developing countries, *and EU* stands for the European Union

European Union (EU) and others. Regarding their size and share in the global economy, China and the USA are treated separately in most of the analyses.

3.1.1 Economic Dimension of Agricultural Sustainability

The agricultural sector is highly diversified worldwide, regarding its economic dimension and role in the total economy (cf. Fig. 3.1). The agriculture, forestry and fishing sectors comprised 10.44% of domestic gross value added (GVA) in 2018, on average, but the situation clearly differed in different parts of the world. The natural resources sector (NRS) constituted only 2.01% of GVA among EU Member States and 0.9% in the USA. In China it was 7.04%, so it indicates that NRS was still relatively more important in China in comparison to the Western world. On the other hand, among developing countries, this share amounted to 14.98% but for the least developed countries it was as much as 24.92% and 17.26% for the net food importing developing countries and 24.44% for the low-income and food deficit countries. The only one group of developing countries with clearly smaller share of NRS in the total economy was the group of small developing island states.[1] It means that for low-income countries, the natural resources sector played a relatively more important role. However, it does not mean that the sector was effective.

---

[1] The detailed list of developing countries belonging to particular groups can be found in Table A.1 in Appendix.

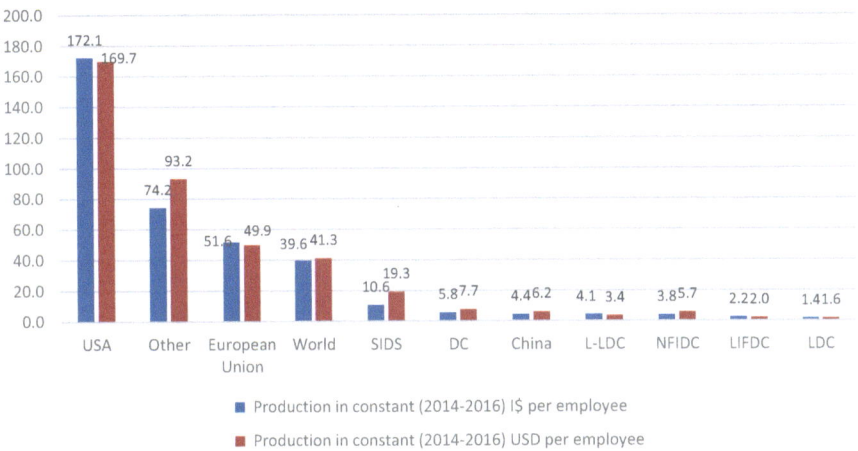

Fig. 3.2 Labour productivity in agriculture in constant thousands of USD and constant thousands of international dollars. *Note*: Abbreviations are as in Fig. 3.1

The average **labour productivity** between different groups of countries differs to a large extent (Fig. 3.2). The value of this indicator is related to the level of agricultural development but also the agrarian structure. Normally, in countries where large-scale agriculture predominates, it is easier to obtain a high productivity level because in these countries more hectares of land are available per one employee. Therefore, the production calculated per one employee is also higher.

Regarding countries and the country groups analysed in this chapter, it is clear that the highest labour productivity was noted in the USA. It exceeded $172,000 when calculated in international dollars[2] (I$) and it was around $170,000 using USD. In the European Union, average labour productivity was more than three times lower and amounted to $51,600 and $49,900, respectively. In developing countries (in general), the labour productivity was much smaller and it did not exceed $8000, but the situation differed between particular subgroups.

In the small island developing states labour productivity was higher—when calculated in USD it amounted to $19,300 and in I$ it was 10,600. On the other hand, among the least developed countries, labour productivity was very low—$1400 and $1600 in I$ and USD, respectively.

The differences in labour productivity between the measurement in constant I$ and USD could be calculated for 147 countries (values in both constant USD and

---

[2]According to FAO "international dollars", are calculated using a Geary-Khamis formula that attributes a single "price" to each agricultural commodity. For example, one metric ton of corn obtains the same price regardless of the country where it was sold. Hence, the fluctuations of exchange rate for national currencies and the differences in local prices of agricultural products have no influence on the production value expressed in international dollars referring to the law of one price. However, the latter might apply only if there were no barriers to international trade which is quite unrealistic assumption.

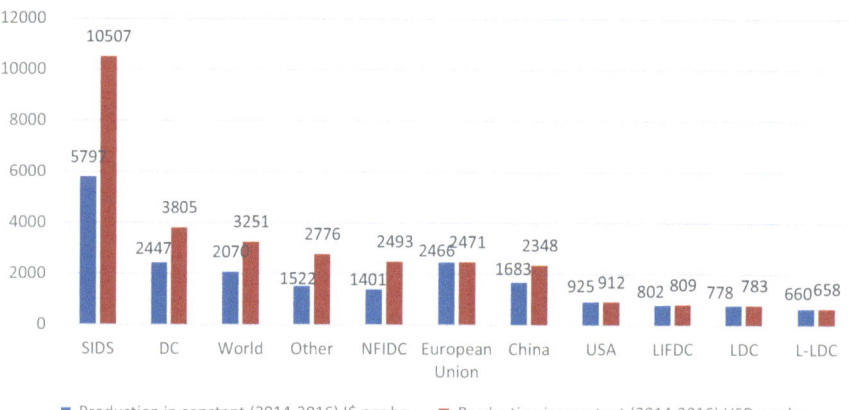

Fig. 3.3.   Land productivity in agriculture in constant USD and constant international dollars. *Note*: Abbreviations are as in Fig. 3.1

constant I$ are available) and the average difference amounted to $5235. It would mean that, on average, values in I$ were higher, but when the median difference is taken into account, it turns out that it was around 0. In the case of 75 countries, the values in I$ are lower and sometimes the differences are significant. For example, in Congo the average agricultural production in constant I$ per worker was $1013, while production in constant USD was equal to $5226. This means that the difference constituted 81% of the of value of production in USD. A big negative difference was also recorded, among others, in countries such as Mauritius (80%), Eritrea (73%), or Japan (70%).

On the other hand, there are some countries for which values for labour productivity in I$ are much higher than values in USD. This was mostly the case of developing countries where the price levels, (including) agricultural prices, were rather low. There were five countries in which values in I$ are more than doubled in comparison to values in the USD, namely Botswana (574%), Oman (294%), Kuwait (155%), Guinea (143%) and the Central Africa Republic (102%). It seems that the use of international dollars is more suitable for between-countries comparisons, especially when the sample studied is highly diversified. It would be especially appropriate if the world agricultural market worked effectively and the law of one price really held. However, the clear differences between calculations in international dollars and standard US dollars show that agricultural prices do not converge to the same level worldwide. Therefore, in the further parts of the book we will focus on the constant USD while calculations in I$ will serve as a background for comparison purposes and will be attached in Appendix.

The situation looks different when **land productivity** is analysed (Fig. 3.3) as farming systems differ around the world. Both intensive and extensive farming systems exist in both developing and developed countries. Average land productivity in developing countries is similar (even slightly higher) compared to all countries—$3805 vs $3251 per ha, respectively. However, the results for developing

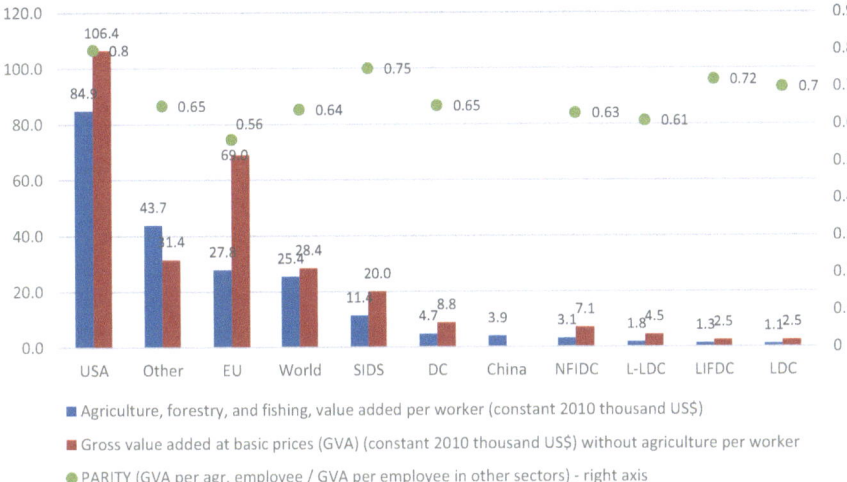

Fig. 3.4 Gross value added per worker in agriculture and other sectors in constant 2010 thousands of USD. *Note*: Abbreviations are as in Fig. 3.1. Average parity is calculated as the mean of parity levels of all countries belonging to the given group

countries is inflated by the small island developing states where land is rather a scarce factor and intensive farming systems predominate. In LDC and LIFDC, the land productivity was below $1000. Interestingly, land productivity was also rather low in the USA. It was even lower than in China and around three times lower in comparison to the EU. This is because US farming is rather extensive and large-scale agriculture prevails.

When it comes to **value added** (which may serve as an approximation of income) (Fig. 3.4) we have used data from the World Bank, so the agriculture is understood here broadly as the natural resources sector, including forestry and fishing as well. We use here values in constant USD from 2010 and focus on the problem when there was an income (dis)parity between the NRS sector and other sectors of the economy. The NRS value added per worker was equal to $25,380 in all the studied countries, on average, but in the USA it was much higher—$84,870—while in China it was $3930 only (cf. Fig. 3.4). For the developing countries, the average was $4740 but it was inflated by SIDS, for which it was $11,370. For the least developed countries, the average GVA per employee was only $1100 and it was only slightly more for land-locked developing and low income and food deficit countries. It seems clear that in the poorest countries the agricultural sectors are relatively large but they does not produce sufficient value added, which may result in severe food security problems.

From the socio-economic point of view, the so-called **income gap** remains an important issue worldwide. It can be understood as the relation of income in agriculture to incomes in other sectors. However, the notion of agricultural income is vague and it can be differently defined in different countries. Therefore, for the sake of international comparisons of the income gap problem, we present the indicator defined as the relation between GVA in NRS per agricultural worker to GVA in other sectors calculated per person employed outside the NRS sector. The income gap in all world countries for which data is available was 0.64 meaning that the GVA per worker in the NRS was, on average, 36% less than the GVA per worker in other sectors. In developing countries the income gap is estimated at 0.65 and for less developed countries—0.7. Interestingly, in the highly developed European Union, the income gap remains significant—the GVA per worker in agriculture was 44% smaller than in the other sectors, but the situation differed a lot between Member States. This is due to the different agrarian structures. For example, the income gap was significant in countries with a predominance of small-scale farming (e.g. Greece, Poland or Romania), whereas in countries with a very modern farming sector (e.g. The Netherlands or Sweden) the problem has almost disappeared and the value of our income gap indicator exceeded 0.9. It means that in these cases, agricultural incomes were more or less on the same levels as in other sectors. In the USA, large-scale farming predominates so the gap between the GVA per worker for the agricultural and non-agricultural sectors was relatively small, but is equal to 0.8. These calculations show that the problem of the income gap is apparent in both developing and developed parts of the world. Lower levels of value added often translate into lower levels of the so-called entrepreneurial income. This can be seen as the justification for income support by agricultural policy.

The above analyses have shown that the economic situation of agriculture is complex and different indicators should be used to get a more detailed picture of the field (Alston & Pardey, 2014). It is important which farming system is developed—intensive or extensive. However, following Hayami and Ruttan (1970a, 1970b), it depends on the resources' availability and, therefore, their relative prices. The economic and income aspects of farming remain the core issues of the sustainability problem. Without solving the economic dilemmas, especially in developing countries, the agricultural sector cannot become fully sustainable.

3.1.2 Social Resilience

One of the basic challenges in the contemporary food sector is related to the food security problem. Obviously, it is especially important in developing countries but also in developed countries the food quality is of a great importance. In this subchapter, we analyse the social dimension of agricultural sustainability using two indicators—**the average dietary energy supply adequacy** (in %) and **the hunger index** (cf. Fig. 3.5). However, the latter is not calculated for developed countries, including the EU. The average dietary energy supply in all sample countries equates to 118.78 which means that, on average, dietary needs are met on a world scale. However, the situation between countries differs a lot. In the USA, the value of this indicator is 148, meaning that much more food than needed is available in the country. In the EU and in China, the average dietary supply is also far above 100%, so the food shortage problem should not exist if the food distribution is

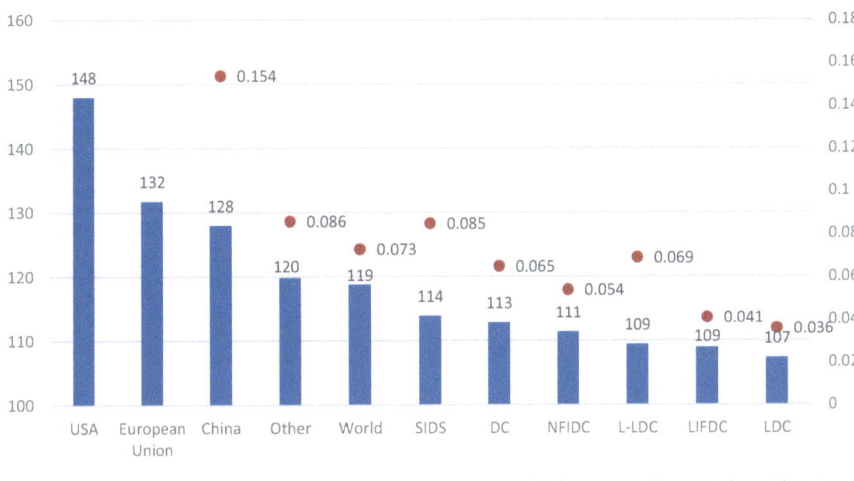

Fig. 3.5 Social dimension of agricultural sustainability of agriculture—food availability and the problem of hunger. *Note*: Abbreviations same as in Fig. 3.1

appropriate. Interestingly, in the group of all developing countries the value of the indicator exceeds 100, even for low-income and food deficit countries. The average dietary supply for these countries is equal to 108.91. Of course, there are some countries in the sample that have problems with their food supply—in 27 countries in 2018 the value of the indicator was below 100 and the average for this group was 91.93%. A particularly difficult situation concerned Somalia and the Central African Republic—the dietary energy supply in these two countries was 76% and 79%, respectively.

The consideration above can be enriched by the analysis of inverse of the hunger index. This index is constructed using the proportion of undernourished in the population, the prevalence of wasting in children under 5 years, the prevalence of stunting in children under 5 years, and the under-five mortality rate. The interpretation of the index score is not straightforward, but we can compare it between different groups of countries.

There is a large difference between China (0.154) and developing countries (0.065). The situation in LDC is, however, much worse. The value of the indicator is equal to 0.036 only. For example, in the Central African Republic the proportion of undernourished people in the population (2016–2018 average) was 59.6%. What is more, it increased from 32% in 2009–2011. The prevalence of stunting in children under 5 years in this country was 47.4% and the under five mortality rate was 12.2%. Our analysis demonstrates that food production on the world scale is still sufficient but there are some parts of the world that face severe food problems. It is therefore really important for these countries that economic development of the sector should translate into improved nutrition.

3.1.3 Environmental Aspects

The third pillar of sustainability is related to the natural environment. It is clear that from developing countries point of view, ensuring a fair standard of living in rural areas and food safety is of a great importance, but environmental challenges should be taken into account not only in high-income countries but also in developing parts of the world. There are several different indicators of environmental conditions, regarding air, land and water quality, but probably the most popular indicator is the **greenhouse gas emissions (GHG)**. It is estimated that agriculture is responsible for 25–30% of GHG emissions (see Chap. 4 for details). These emissions depend, among others, on the model of agriculture in a given area (Czyżewski & Kryszak, 2018; Hamuda & Patkó 2010). However, it is important whether the GHG emissions are calculated per agricultural production or per hectare of agricultural land. In the intensive agriculture, GHG emissions may be large when calculated per hectare, but if the production value from the hectare is significant, then the emission per production unit may be relatively low. As there is no clear answer which of these two indicators should be used, in this part we analyse both (cf. Fig. 3.6).

Regarding the whole sample, the average GHG emission per hectare is 2.29 tonnes (CO_2 equivalent) per hectare. In the EU, it amounted to 2.84 tonnes, but interestingly in China and in the USA it is much lower—1.21 and 0.888 tonnes, respectively. It may come as a surprise that the GHG emission per unit of land in developing countries is lower in comparison to the EU only by about 17.3% and equates to 2.35 tonnes. However, this result is inflated to some extent by small island developing states with their average emissions 40% higher than in the EU. In the other subgroups of developing countries, the GHG emissions are significantly lower—only 1.32 tonnes per ha in L-LDC and 1.66 in LIFDC. In the land locked countries, the emissions are therefore more than 50% lower in comparisons to the EU Member States.

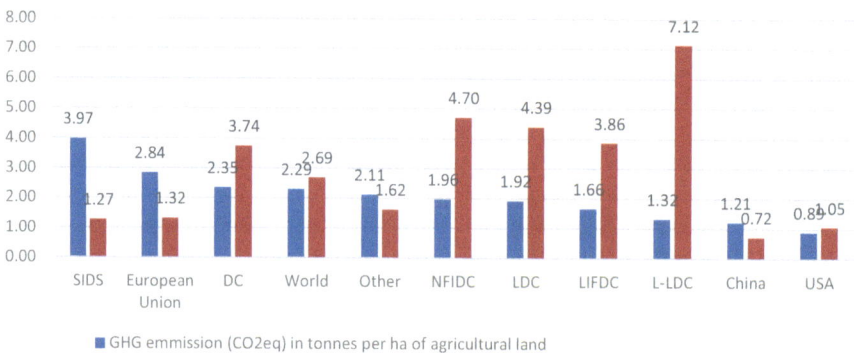

Fig. 3.6 The greenhouse gas emissions per agricultural sector in different groups of countries (2018). *Note*: Abbreviations same as in Fig. 3.1

When the GHG emission is calculated per unit of production, we see that this is also an important problem in developing countries. The agricultural production in these countries is inefficient from the emissions perspective. The average of GHG emissions per $1000 in developing countries is 3.74 tonnes, while in the EU it is only 1.32 and in China and the USA—0.72 and 1.055 respectively. What is more, the GHG emissions per unit of production is even larger in some subgroups of DCs. It is 4.7 tonnes for net food developing countries and 7.12 for land locked developing countries. These calculations show that the GHG emissions are also important issue in these countries. Furthermore, the fact that economic productivity is much lower while the emissions are significant, makes the issue of the GHG emissions even more severe.

3.1.4 Inputs of Agricultural Production

The efficiency of the agricultural sector lies in the relations of its outputs (good and undesirable) to inputs. In this part we will analyse basic agricultural inputs. These are the labour and land factors, the number of livestock units (LSU), energy use, nitrogen, phosphate and potash nutrients, and pesticides. Since the aim of this part is to provide a comparison between different parts of the world, we will present inputs in the relative form. We will show the average number of hectares of land per agricultural worker and the share of agricultural land in the total area of the country (Fig. 3.7), nutrients use per ha of cropland (Table 3.1), as well as the LSU and pesticides per ha of total agricultural land (Fig. 3.8).

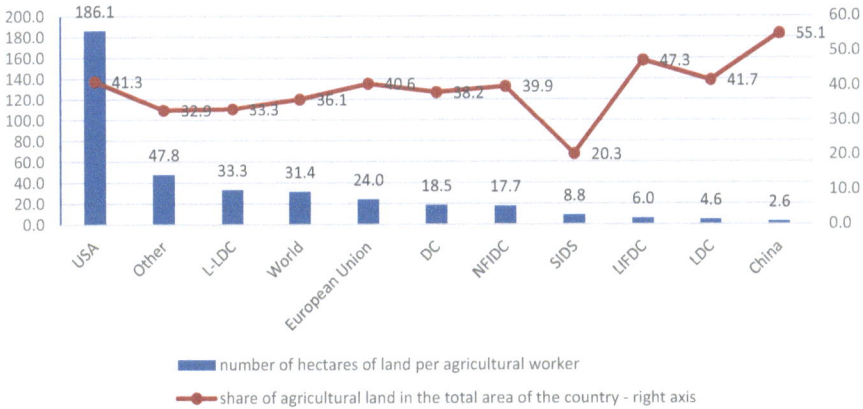

Fig. 3.7 The number of hectares per agricultural worker and the share of agricultural land in the total country area (%). *Note*: Abbreviations same as in Fig. 3.1. The number of hectares of land per worker for "Other" and "World" are calculated without Argentina which is an outlier (the value for this country is 8396 ha per worker)

Table 3.1   The use of nutrients (nitrogen, phosphate and potash) in kilogrammes per ha of cropland

Country group	Nutrient nitrogen N (total) kg/ha of cropland	Nutrient phosphate P_2O_5 (total) kg/ha cropland	Nutrient potash K_2O (total) kg/ha cropland
China	208.53	58.04	79.61
European Union	95.86	21.82	24.88
USA	72.58	25.44	28.58
Other	64.54	26.04	33.65
World	58.40	18.85	21.61
SIDS	45.09	15.52	19.71
DC	38.36	11.94	10.50
NFIDC	37.46	10.94	9.99
L-LDC	29.23	6.99	4.55
LIFDC	21.33	9.32	6.53
LDC	18.76	7.29	4.61

Note: Abbreviations same as in Fig. 3.1

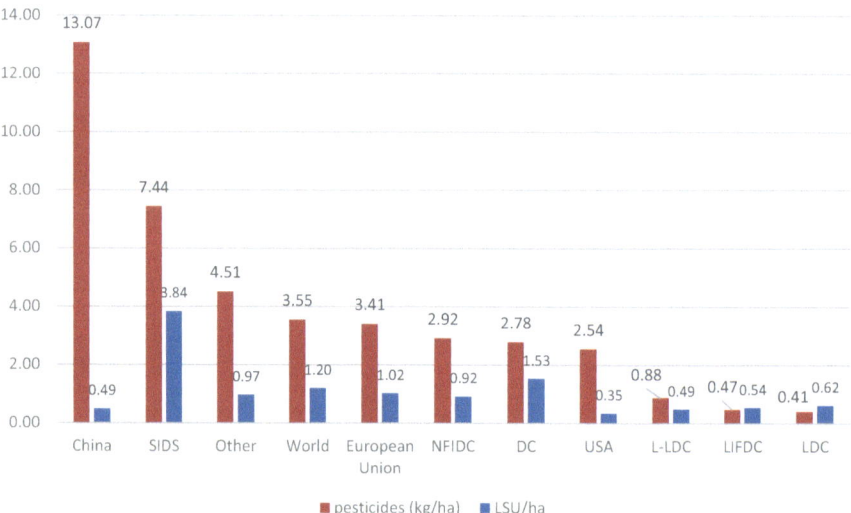

Fig. 3.8   The number of LSU and pesticides use per ha of total agricultural land. *Note*: Abbreviations same as in Fig. 3.1

The big differences in the methods of agricultural production in different parts of the world are related to the use of **fertilisers** (cf. Table 3.1). For example, the average nitrogen use in all sampled countries was 58.4 kg/ha of cropland but in the EU it was 95.86, in the USA −72.58 and in China −208.53 kg (excluding nitrogen from organic manure). On the other hand, in developing countries the average is 38.36 kg/ha and for the least developed countries 18.76 kg/ha. Similarly, the big differences are related to phosphate nutrient. The world average is 18.85 kg/ha, in the EU it is 21.82 and in USA 25.44 kg/ha. Interestingly, in China the phosphate use is much larger and the value is 58.04 kg/ha. In developing countries the phosphate use is 37% lower when compared to the all countries average and it is equal to 11.94 kg/ha. In land locked developing countries the phosphate use is only 6.99 kg/ha, which constitutes only 37% of the sample average.

The **land structure** is highly diversified in different parts of the world. For example in China, small-scale farming predominates. The average number of hectares per one worker is only 2.6. In developing countries, the average is 18.5 ha/worker, but among the least developed countries it is equal to 4.6 ha only. The size of the farms in the EU is bigger but 24 ha/worker is not a large number when compared to the situation in the USA (186 ha) or in some other countries where the extensive type of agriculture is developed. The share of agricultural land in the total area of the country is not such a differentiating factor. For example, the share of agricultural land is, on average, similar in developing countries and in the EU, or the USA. Only small island developing states have a clearly smaller share of agricultural land. Obviously, the problematic issue is land productivity rather than the amount of land available.

The average number of **LSU** for the all studied countries is 1.2 LSU/ha, but the indicator differs between world regions (cf. Fig. 3.8). In the EU it is similar to the world average and equals 1.022 per ha. The animal production in the EU is much more intensive when compared to China or the USA where the indicator values are 0.49 and 0.35, respectively. In developing countries the average livestock density is 1.531 but, once again, the value is inflated by the small island developing states where the value of indicator exceeds 3.8 LSU/ha. In other groups of developing countries the intensity is much lower. For example, for land locked developing countries it is 0.49 only and for low-income and food deficit countries it is 0.54.

Pesticides use in the EU Member States is 3.41 kg/ha which is less than in the World category. It may be linked to the EU common agricultural policy. One of its main aim is to make agriculture more sustainable and in some programs (especially under second pillar) pesticides use reduction is required. However, pesticides use in the EU is still higher when compared to the USA (2.54 kg/ha). In developing countries the average is 2.78 kg/ha. Interestingly in net food importing developing countries it is 2.92 and in small island developing states it is very high—7.44. On the contrary, the pesticides use in the least developed countries is marginal and amounts to 0.407 kg/ha only. In China, in turn, pesticides use is extremely high—13.07 kg/ha, similarly to the nutrient use.

3.1.5 Institutional Background

Integrated efficiency is the result of the relationship between outputs and inputs that were studied above. However, there are some exogenous, institutional factors that influence the process of transforming inputs into outputs. In the standard microeconomic approach, several socio-economic characteristics of farmers and their farms are usually used to explain the integrated efficiency level. Furthermore, policy

Table 3.2 The exogenous determinants of efficiency in agriculture

Country groups	Political stability and absence of violence/terrorism (index) <−2.5;2.5>	Human capital index (PWT) <0;5>	Foreign Direct Investments as % of GDP
LDC	−0.71	1.84	2.90
L-LDC	−0.57	2.37	3.10
SIDS	0.50	2.87	7.00
LIFDC	−0.95	2.09	3.70
NFIDC	−0.36	2.22	3.70
DC	−0.31	2.40	4.50
European Union	0.67	3.11	3.10
China	N/A	2.67	3.10
USA	0.48	3.74	1.20
Other	−0.08	2.73	2.60
World	−0.07	2.66	3.60

Note: Abbreviations same as in Fig. 3.1

measures or subsidies are also introduced. In this research we follow the macroeconomic approach, so we analyse some general indicators, i.e. the impact of the political environment (the index of political stability and absence of violence/terrorism) and the human capital index (downloaded from the Penn World Table) (cf. Table 3.2). Further, we also use foreign direct investments (FDI) (% of GDP) as a proxy of economy openness and its ability to absorb new technologies.

The value of the **political stability indicator** falls within the range −2.5 and 2.5 and the higher the index, the more politically stable is the country. The values for the USA and the EU are obviously larger than for developing countries but there is also a difference between the former two. The level of political stability in the EU (0.671) is larger than in the USA (0.48). The average value for developing countries is −0.312. However, we may say that small island developing states are quite stable in political terms since the indicator value is, on average, even larger than in the EU. In turn, in the other developing country groups, the situation is clearly different. In low-income and food deficit countries and the least developed countries, the value of the indicator is −0.948 and −0.708, respectively.

The very important factor that may influence integrated efficiency level is **human capital**. There is no single indicator for this, but the advantage of the index provided by the Penn World Table is that it is based on the average years of schooling from Barro and Lee (2013) and is available for a large number of countries. The average value of the indicator in China is similar to the all sample average and is equal to 2.674. In the EU the human capital index is higher—3.111 but it is clearly lower than in the USA—3.774. In the developing states the average level of human capital is smaller and amounts to 2.398. However, in the least developed countries it is only 1.836. The low level of human capital remains one of the most important problems in LDC and it creates significant barriers not only in the agricultural sector.

Foreign direct investments are usually higher (in relation to the economy's size) in developing countries. Indeed, in developing countries in our sample FDI constitute, on average, 4.5% of GDP. In small developing countries it is, however, even 7% since these countries are usually more open and have better infrastructure to receive investments. In the EU countries and in China, FDI are relatively less important and their share in the economy is 3.1%. In the USA this share is even lower and is equal to 1.2% only.

The analysis of the above figures and indicators has shown that the agricultural sector is highly diversified and it is at very different stages of development in different parts of the world. Different agricultural policies therefore will be needed to deal with the specific local problems; however, the integrated efficiency indicator seems to be relatively objective. For example, if the agricultural production of the country is relatively low but the inputs use and bad environmental outputs are also on the rather low level, the integrated efficiency level may be high. But if the agricultural production in a food deficit country is low, we should not praise the country in international comparisons even if it is efficient due to its low inputs use. This is why, in our view, the integrated efficiency calculation should be preceded by the detailed description of its different elements and accompanying factors.

3.2 Determining Clusters of Countries with Similar Agricultural Models

In the analyses so far, we have focused on the typology of developing countries proposed by the FAO. Furthermore, we have treated the EU, China, and the USA separately and the remaining of the countries as a one group. However, for the following analysis, this division would be insufficient. We aim not only to compare integrated efficiency levels, but also to propose solutions for agricultural policy. Therefore, we need a smaller number of groups but these groups should be maximally different from each other but internally consistent. This is because policy should be tailored to the specific needs. However, we cannot analyse a large number of small and very coherent groups because of the data availability. Calculations of the DEA-based efficiency require a sufficient number of observations.

In the following analysis, we have treated the EU as a separate group, especially since these countries are subjected to a unified agricultural policy. To obtain further groups, we have performed a cluster analysis for 79 countries in our dataset (for which all the necessary data is available), based on the data for 2018, as we wanted to know the most current state of challenges and problems in the countries studied.

There are different possibilities of establishing dominant agricultural models in a given area. However, in this book we adopt the approach proposed, among others, by Kryszak and Herzfeld (2021) who claim that the relations between basic agricultural inputs and outputs may be a good indicator of the agricultural model, especially in cross-country or cross-regional studies.

Table 3.3 Descriptive statistics of variables used for the clustering procedure

	Cluster 1		Cluster 2		Cluster 3 (EU)	
Productivity of:	Mean	SD	Mean	SD	Mean	SD
LSU	4.02	2.01	1.19	0.68	2.95	1.03
Energy	7030	29,980	12,095	49,954	404	229
Employment	78,619	245,152	11,569	23,754	52,455	41,352
Land	3607	6283	852	1389	2600	2365
Nitrogen	100	91	135	295	39	24
Pesticides	7758	22,450	18,856	55,069	1461	835

Table 3.4 Multivariate test of means for variables used for clustering procedure

Test	Statistic		F(df1,	df2)	F	Prob > F
Wilks'	Lambda	0.4162	12	198	9.08	0.000
Pillai's	Trace	0.5978	12	200	7.11	0.000
Lawley-Hotelling	Trace	1.3692	12	196	11.18	0.000
Roy's	Largest	1.3442	6	100	22.4	0.000

The following cluster analysis is based on six variables, i.e. partial productivities. Basically, we calculated the following relations:

- Gross Production Value (current thousand US$) / LSU (number)
- Gross Production Value (current thousand US$) / energy consumption in tj
- Gross Production Value (current thousand US$) / employment in agriculture (1000 persons)
- Gross Production Value (current thousand US$) / land in 1000 ha
- Gross Production Value (current thousand US$) / nitrogen use in tonnes (number)
- Gross Production Value (current thousand US$) / pesticides use in tonnes (number)

We have performed several different clustering procedures. Based on the Harabasz-Calinski criterion, we decided that the best division would be that achieved by the k-mean clustering procedure with the segment option. This specifies that k nearly equal partitions be formed from the data. Approximately the first N/k observations are assigned to the first group, the second N/k observations are assigned to the second group, and so on. The group means or medians from these k groups are to be used as the starting group centres. The procedure produced two groups—the first group consists of 29 countries, and the second has 50 countries. The descriptive statistics for the three groups (including the EU as a separate cluster) are provided in Table 3.3. In Table 3.4 we have performed a multivariate test of means to compare the mean levels of variables used for clustering. This will help to assess whether our clustering procedure was effective, namely whether our clusters really differ between each other.

The agricultural sectors in the countries of the first cluster are represented by a higher level of production intensity in comparison to the cluster 2. The first cluster is represented by a particularly high level of LSU, employment and land productivity. The average level of these productivities are 2 times, 6.5 times and 4.2 times higher in comparison to the cluster 2. Obviously, a higher level of land and labour productivity translates into higher incomes. Therefore, from the farmers point of view, their socio-economic situation is usually better in this group. However, the agricultural model dominating in cluster 1 can generate some pressure on the natural environment. Nitrogen and especially pesticides productivity is low, which means that there is much more fertilisers and pesticides used to obtain a given amount of production. When it comes to the EU, it manifests an average level of LSU, land and employment productivity. But energy, pesticides and nitrogen productivity is low. This can be caused by the relatively high level of nutrient and pesticides use, as manifested in Table 3.1 and Fig. 3.8.

Multivariate tests of means have shown that our clustering procedure was effective—all tests clearly indicate that the three clusters are statistically different. This is of a great importance because further calculations will be run for these three clusters, and policy recommendation will also be based on this. From the technical point of view, we can assume that available technology is different among these clusters, which justifies separate integrated efficiency calculations for each cluster.

3.3   How to Improve the Sustainability of Agriculture in Different Parts of the World?

3.3.1   Methodology of Computing Integrated Efficiency Scores

Efficiency in the agricultural sector is often studied using the non-parametric DEA (data envelopment analysis) framework. In this approach, the relative efficiency of a given decision making unit (DMU) is assessed by measuring the distance of this DMU to the efficient frontier which is created by efficient DMUs (Charnes et al., 1978). In contrast to the basic productivity or efficiency measures, DEA allows to include multiple inputs and outputs in the model. Traditional DEA is a radial model which means that all inputs (outputs) should be reduced (expanded) proportionally based on the efficiency score in the input (output) oriented model (Ullah et al., 2019). For example, if the efficiency score of a given DMU in the input-oriented model is equal to 0.7, then it means that all inputs of this DMU should be reduced by 30% for the DMU to become efficient. However, it is clear that in the real world situation, the potential reduction of inputs (or expansion of outputs) is not always equal (Chen & Jia, 2017). This leads us to apply the hybrid DEA model (Tone, 2004). This model allows us to divide our inputs and outputs into radial ones (proportionally changes) and non-radial. The latter variables are treated as in the slack-based measure (**SBM**) approach, firstly proposed by Tone (2001). The advantage of treating some variables as non-radial is that in this approach it is possible to directly estimate the so called "slacks" which inform about the potential reduction (expansion) of each individual input (output) in the sense of the Pareto improvement. The idea of slacks will be explained in details in the next point (Sect. 3.3.2). When it comes to the radial model, one needs to distinguish between proportionate movement and slack movement, which in this context could be treated as a possibility of improvement in the Pareto sense. In other words, occurring of slack means that there are possibilities to improve the DMU's performance within existing technology without a need to proportionally change the values of other inputs/outputs.

When using the hybrid DEA approach described above, a few question may arise: First, we need to decide whether we should use an input or output oriented model. On the one hand, farmers have more control on the inputs of agricultural production while the outputs are subjected to fluctuations caused by the factors such as weather. On the other, we analyse sectors (not individual farms), and the outputs increase is especially important from the perspective of developing countries. Therefore, we employ the non-oriented (input-output) model.

Second, agriculture as a typical variable returns to scale activity since the increase in inputs will not always lead to an equivalent increase in output. The analysis has shown that when integrated efficiency is under consideration on the sectoral level, then agriculture usually cannot benefit from the increasing scale effect. Moreover, we have to take into account the issue of self-accelerating degradation of natural environment. For example, agricultural lands subject to erosion and negative organic matter balance require increasing fertilization to maintain productivity levels. Biodiversity loss, e.g. in the number of invertebrates in the soil, or the number of insects or birds in the rural ecosystem also translates into lower yields of crops, fruit trees and bushes. In summary, the progressive degradation of the environment requires increased inputs of fertilizers, energy, and pesticides, which only serves to maintain the current level of yield and do not translate into increased production. Thus, it is a case of diminishing marginal changes in production. This is why we prefer to employ non-increasing returns to the scale model (NIRS) rather than variable returns to the scale model (VRS) or constant returns to scale (CRS). This issue will be elaborated in detail in Sect. 3.3.2.

Third, traditional efficiency analysis was focused on the analysis of desirable economic outputs which were produced from the inputs typically used for agricultural production. Nowadays, however, we cannot ignore bad outputs that are inherently related to agricultural production. There are four main ways of dealing with bad outputs in efficiency analysis (Halkos & Petrou, 2019). The first option is to simply ignore them, but as stated above, this option shouldn't be taken into account. Efficiency analysis without the inclusion of bad outputs may only be used for comparison purposes as a baseline. Comparing calculations "with" and "without" bad outputs may generate a valuable insight on how these bad outputs are changing the view on the functioning of the agricultural sector.

The second option would be to introduce undesirable outputs on the input side of the model. In this approach, we assume that both bad outputs and standard inputs should be decreased. The third option is to apply some transformation on the bad outputs values and treat them as a regular outputs in the efficiency calculation. For example, a researcher may multiply the value of the bad output by –1 or employ the multiplicative inverse, i.e. dividing 1 by the value of the bad output.

Perhaps, the most widely used approach nowadays is to treat undesirable outputs as outputs in the production function in their actual format. This can be done using the directional distance function (DDF) proposed by Chung et al. (1997) or in the SBM framework (Tone 2004). In these approaches, the desirable outputs may be increased and the bad outputs may be reduced simultaneously. Therefore, these models assume a strong (free) disposability of bad outputs in contrast to the weak disposability model in which bad outputs may be decreased only if good outputs are also reduced (bad outputs are inseparable from good outputs). The weak disposability model is found to be consistent with physical laws and it shows the opportunity cost of reducing bad outputs (Färe & Grosskopf, 2003), but on the other side this approach "leaves the impact of undesirable outputs on efficiency undetermined"(Hailu & Veeman, 2001). What is more, the weak disposability model assumes that all units apply the same abatement factor, which is not really likely (Kuosmanen, 2005). Many researchers, therefore, employ additive SBM with bad outputs (Dong et al., 2018; Le et al., 2019; Ullah et al., 2019).

The fourth issue is comparing efficient DMUs which may be of a great importance in the analysis where the DMU number is rather low. According to traditional approaches to efficiency calculations, DMUs that lie within production possibility sets are inefficient (values of efficiency between 0 and 1) while DMUs that lie on the production frontier are efficient and have the value 1. The common problem of this approach is that usually many DMUs are found to be efficient (especially under the variable returns to scale—VRS—assumption) and it is impossible to compare DMUs on the frontier (Long et al., 2018; Yang et al., 2015). To overcome this issue, the so called super-efficiency model proposed by Andersen and Petersen (1993) may be used. This model enables to compare efficient DMUs since the efficiency values for these DMUs differ (they are greater than 1) and the bigger the value, the better positioned is the unit. When we assess super efficiency, we need to exclude the kth DMU and designate frontier without this DMU. The model was then further extended to the SBM framework.

Fifth, there is a problem how to distinguish between radial and non-radial variables in hybrid models. Chiu et al. (2013) and Wang et al. (2019) made a decision based on the correlation between indices of variables. If the correlations were high, then they treat those variables as radial since they change in a similar manner. In this work, we choose radial variables based on the theoretical and policy-based criteria rather than simply correlations.

In this analysis we will use the following outputs and inputs:

- **Good economic outputs**: gross production value in total agriculture (constant thousands of USD)
- **Good social output**: average dietary energy supply adequacy (percent)
- **Bad environmental output**: GHG emission (CO_2 equivalent, gigagrams)

- **Inputs**: the number of livestock units (LSU), energy consumption in terajoule (tj), employment in agriculture (in 1000 of employees), land (in 1000 of ha), nitrogen use (in tonnes) and pesticide use (in tonnes)

On the input side in the efficiency analysis of the agricultural sector some further inputs (machinery use, intermediate consumption) are sometimes used, but we were limited by data availability for a large country set. Furthermore, we need to exclude different nutrients (potash and phosphate) to be in line with the minimum number of DMUs to be used regarding the number of inputs and outputs in the model (Cooper et al., 2007). However, these two nutrients were highly correlated with the nitrogen use.

In Cluster 1 and 2 we treat three inputs (land, LSU, energy) and undesirable output (GHG) as radial variables. In addition to this, in the EU (Cluster 3) we treat production value and nitrogen as radial variables as well.

Therefore, from the technical side our approach is a hybrid super-efficiency DEA model with undesirable output. In the analysis in the following subchapters we will focus mostly on slacks values rather than integrated efficiency levels. This is because our aim is to find problematic areas in the functioning of the agricultural sector. We want to see which aspects should and can be improved to increase integrated efficiency level, which provides our measure of agricultural sustainability. The technical side of our model is described below and it correspond to the original work of Tone (2004) and the application paper of Wang et al. (2019).

Let the observed data matrix be $X \in R_+^{m \times n}$, where n and m are the numbers of DMUs and inputs, respectively. This input data matrix can be decomposed into radial and non-radial parts—$X^R \in R_+^{m_1 \times n}$ and $X^{NR} \in R_+^{m_2 \times n}$. The total number of inputs is equal to $m = m_1 + m_2$. Similarly, we have the good (desirable) output and the bad (undesirable) output data matrix: $Y^g \in R_+^{s \times n}$ and $Y^b \in R_+^{k \times n}$, where s and k are the numbers of good outputs and bad outputs, respectively. They can be both decomposed into radial and non-radial parts: for good outputs $Y^{gR} \in R_+^{s_1 \times n}$ and $Y^{gNR} \in R_+^{s_2 \times n}$; for bad outputs $Y^{bR} \in R_+^{k_1 \times n}$ and $Y^{bNR} \in R_+^{k_2 \times n}$.

For the specific DMU $(x_0, y_0) = \left(x_0^R, x_0^{NR}, y_0^{gR}, y_0^{gNR}, y_0^{bR}, y_0^{bNR}\right) \in P$ and the hybrid super-effiency DEA model is described as follows:

$$\min \frac{1 - \frac{m_1}{m}(1 - \theta) - \frac{1}{m}\sum_{i=1}^{m_2} \frac{s_i^{NR-}}{x_{i0}^{NR}}}{1 + \frac{s_1}{s}(\phi - 1) + \frac{1}{s}\sum_{r=1}^{S_2} \frac{s_r^{NR+}}{y_{r0}^{gNR}} + \frac{k_1}{k}(\omega - 1) + \frac{1}{k}\sum_{t=1}^{k_2} \frac{s_t^{NR-}}{y_{t0}^{bNR}}}$$

s.t.

$$\theta x_{i0}^R \geq \sum_{j=1, \neq 0}^{n} x_{ij}^R \lambda_j, \; i = 1, 2, ..., m_1$$

$$x_{i0}^{NR} \geq \sum_{j=1, \neq 0}^{n} x_{ij}^{NR} \lambda_j - s_i^{NR-}, \; i = 1, 2, ..., m_2$$

$$\phi y_{r0}^{gR} \leq \sum_{j=1, \neq 0}^{n} y_{rj}^{gR} \lambda_j, \; r = 1, 2, ..., s_1$$

$$y_{r0}^{gNR} \leq \sum_{j=1, \neq 0}^{n} y_{rj}^{gNR} \lambda_j + s_r^{NR+}, \; r = 1, 2, ..., s_2$$

$$\omega y_{t0}^{bR} \geq \sum_{j=1, \neq 0}^{n} y_{tj}^{bR} \lambda_j, \; t = 1, 2, .., k_1$$

$$y_{t0}^{bNR} \geq \sum_{j=1, \neq 0}^{n} y_{tj}^{bNR} \lambda_j - s_t^{NR-}, \; t = 1, 2, ..., k_2$$

$\sum \lambda_j \leq 1, s_i^{NR-} \geq 0, s_r^{NR+} \geq 0, s_t^{NR-} \geq 0, \theta \leq 1, \phi \geq 1, \omega \geq 1$, where $\sum \lambda_j \leq 1$, means that we assume non-increasing returns to scale and $s_i^{NR-} \geq 0, s_r^{NR+} \geq 0, s_t^{NR-} \geq 0$ are the slack values for inputs, good outputs and bad outputs, respectively.

To evaluate the dynamics of integrated efficiencies, namely changes in the efficiency of a DMU between two periods, we employ the Malmquist index (MI). This index may be decomposed into the catch-up effect (often called efficiency change—EC—in productivity literature) and technological change (TC) which represent the frontier-shift effect. Let the two time periods be t_1 and t_2. Then, the catch-up effect is:

$$EC = \frac{\delta^{t_2}\left((x_0, y_0)^{t_2}\right)}{\delta^{t_1}\left((x_0, y_0)^{t_1}\right)}$$

and the frontier-shift effect is:

$$TC = \left[\frac{\delta^{t_1}\left((x_0, y_0)^{t_1}\right)}{\delta^{t_2}\left((x_0, y_0)^{t_1}\right)} \times \frac{\delta^{t_1}\left((x_0, y_0)^{t_2}\right)}{\delta^{t_2}\left((x_0, y_0)^{t_2}\right)}\right]^{\frac{1}{2}}$$

The Malmquist index is a product of two variables:

$$MI = \left[\frac{\delta^{t_1}\left((x_0, y_0)^{t_2}\right)}{\delta^{t_1}\left((x_0, y_0)^{t_1}\right)} \times \frac{\delta^{t_2}\left((x_0, y_0)^{t_2}\right)}{\delta^{t_2}\left((x_0, y_0)^{t_1}\right)}\right]^{\frac{1}{2}}$$

In the standard approach for MI calculations (contemporaneous index), the TC component classifies each change in the productivity of frontiers' DMUs as a technical change. This means that it covers forward and backward shifts of the frontier (Shestalova, 2003). In our work, we employ a modified sequential MI index. In this approach we assume that past production techniques are always available for current periods (Alene, 2010). Therefore, technical regress is not possible and the TC component will only show technical progress. All deteriorations in performance are therefore attributed to the EC component. Shestalova (2003) has shown that the differences between MI under contemporaneous and sequential technology settings are rather small, but decomposition of index may differ a lot. In contrast, based on the example of the agricultural sector in African countries, Alene (2010) have shown that a difference in the productivity calculation under two methods may be signif-icant. They found that under conventional settings, an average productivity growth in the African agriculture rate was 0.3% while under sequential technology it was 1.8%. In EC, TC and MI, the terms $\delta^{t_1}\left((x_0, y_0)^{t_1}\right)$, $\delta^{t_2}\left((x_0, y_0)^{t_2}\right)$, $\delta^{t_1}\left((x_0, y_0)^{t_2}\right)$, $\delta^{t_2}\left((x_0, y_0)^{t_1}\right)$ are the within inter-temporal efficiencies calculated by the super-efficiency DEA model. The model for $\delta^{t_1}\left((x_0, y_0)^{t_1}\right)$ is as follows:

$$\delta^{t_1}\left((x_0, y_0)^{t_2}\right) = \min \frac{1 - \frac{m_1}{m}(1-\theta) - \frac{1}{m}\sum_{i=1}^{m_2}\frac{s_i^{NR-}}{x_{i0}^{NR}}}{1 + \frac{s_1}{s}(\phi - 1) + \frac{1}{s}\sum_{r=1}^{S_2}\frac{s_r^{NR+}}{y_{r0}^{gNR}} + \frac{k_1}{k}(\omega - 1) + \frac{1}{k}\sum_{t=1}^{k_2}\frac{s_t^{NR-}}{y_{t0}^{bNR}}}$$

s.t.

$$\theta x_{i0}^{Rt_2} \geq \sum_{t'=1}^{t}\sum_{j=1, \neq 0}^{n} x_{ij}^{Rt''}\lambda_j^{t''}$$

$$x_{i0}^{NRt_2} \geq \sum_{t'=1}^{t}\sum_{j=1, \neq 0}^{n} x_{ij}^{NRt''}\lambda_j^{t''} - s_i^{NR-}$$

$$\phi y_{r0}^{gRt_2} \leq \sum_{t'=1}^{t}\sum_{j=1, \neq 0}^{n} y_{rj}^{gR}\lambda_j^{t''}$$

$$y_{r0}^{gNRt_2} \leq \sum_{t'=1}^{t}\sum_{j=1, \neq 0}^{n} y_{rj}^{gNR}\lambda_j^{t''} + s_r^{NR+}$$

$$\omega y_{t0}^{bRt_2} \geq \sum_{t'=1}^{t}\sum_{j=1, \neq 0}^{n} y_{tj}^{bR}\lambda_j^{t''}$$

$$y_{t0}^{bNRt_2} \geq \sum_{t'=1}^{t}\sum_{j=1, \neq 0}^{n} y_{tj}^{bNR}\lambda_j^{t''} - s_t^{NR-}$$

$$\sum \lambda_j \leq 1, s_i^{NR-} \geq 0, s_r^{NR+} \geq 0, s_t^{NR-} \geq 0, \theta \leq 1, \phi \geq 1, \omega \geq 1$$

The next parts of the text will be organised as follows:

First, we will focus on the average slack values in the whole analysed period (2005–2018). We will analyse the values calculated for production in constant USD with reference to calculations using constant international USD (see Tables in Appendix). This will help us to see how the calculations of integrated efficiency would change if the law of one price truly held and agricultural prices were the same across the world. Thank to this approach we can also see to what extent is the integrated efficiency impacted by the market failure.

Second, we will analyse how the slack values have changed. More specifically, we will analyse the slack changes between the 2005–2007 and 2016–2018 sub-periods. Third, we will move on to the dynamic analysis. We will study integrated efficiency changes calculated by the MI index. Fourth, we will sum up our slack analysis from the global perspective to see what the global profits would be if countries eliminated slacks.

3.3.2 Facing Decreasing Returns to Scale

In this section we would like to address two issues: First, to draw attention to the problem of the decreasing return to scale in the Clusters 1–3 of the countries analyzed. Second, to show in practice how Pareto improvement slacks, which are the basic criterion of our analysis, may operate.

In some countries (e.g. China), production and related food security is considered the leading criterion for sustainability in agriculture. Consequently, the question of compensating for the increase in environmental pressure due to e.g. higher fertilization by a more than proportional increase in production that results from increasing return to scale (IRS) becomes crucial. In frontier-basis analysis, the realization of IRS implies a parallel shift along a section of the technological frontier characterized by a higher slope coefficient than the constant return to scale (CRS) function, as illustrated in the graphs in the lower right corner of Figs. 3.9, 3.10, and 3.11. However, the position of the DMU relative to the technology frontier set in a given group of countries does not always allow the IRS to be realized (see the areas shaded in red in the illustrative graph in Figs. 3.9, 3.10, and 3.11). In some cases, regardless of the dominant optimization orientation (input- or output-oriented), the agricultural sector faces a decreasing return to scale (DRS), as we have shown in Figs. 3.9, 3.10, and 3.11 referenced to above.

The IRS/DRS problem can be best illustrated in two-dimensional frontier plots, where one type of output and input is assumed. Hence, such frontier plots were drawn for each cluster analyzed, showing the transformation of each input type into agricultural production at a given technological frontier. The agricultural output was expressed in 2018 international dollars to focus on a pure technical efficiency and resulting economies of scale. The countries in the Figs. 3.9, 3.10, and 3.11 plots above the red dotted line encounter DRS regardless of the adopted production orientation. This situation in practice means that increases in environmental pressure are less than proportionately offset by higher production. Therefore, improving

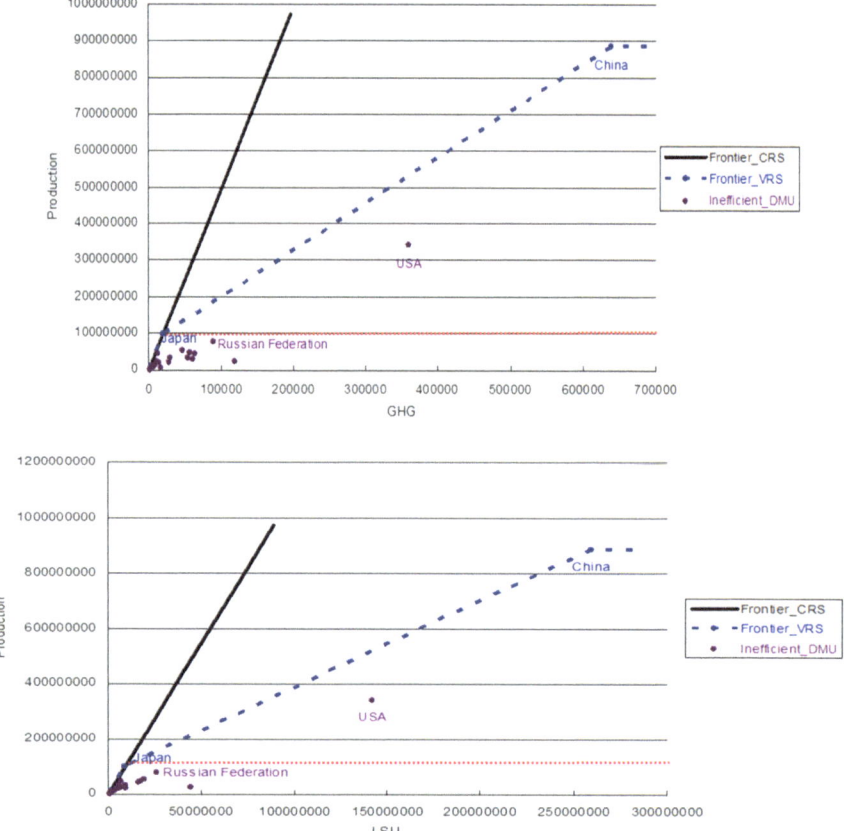

Fig. 3.9 Production frontiers, decreasing return to scale and potential slacks in Cluster 1 (two-dimensional frontier plots, radial DEA model, constant international dollars, 2018). *Note:* red arrows in frontier plots depict potential slacks; DRS/IRS means a decreasing and increasing return to scale, respectively; DMUs above red dotted line face DRS

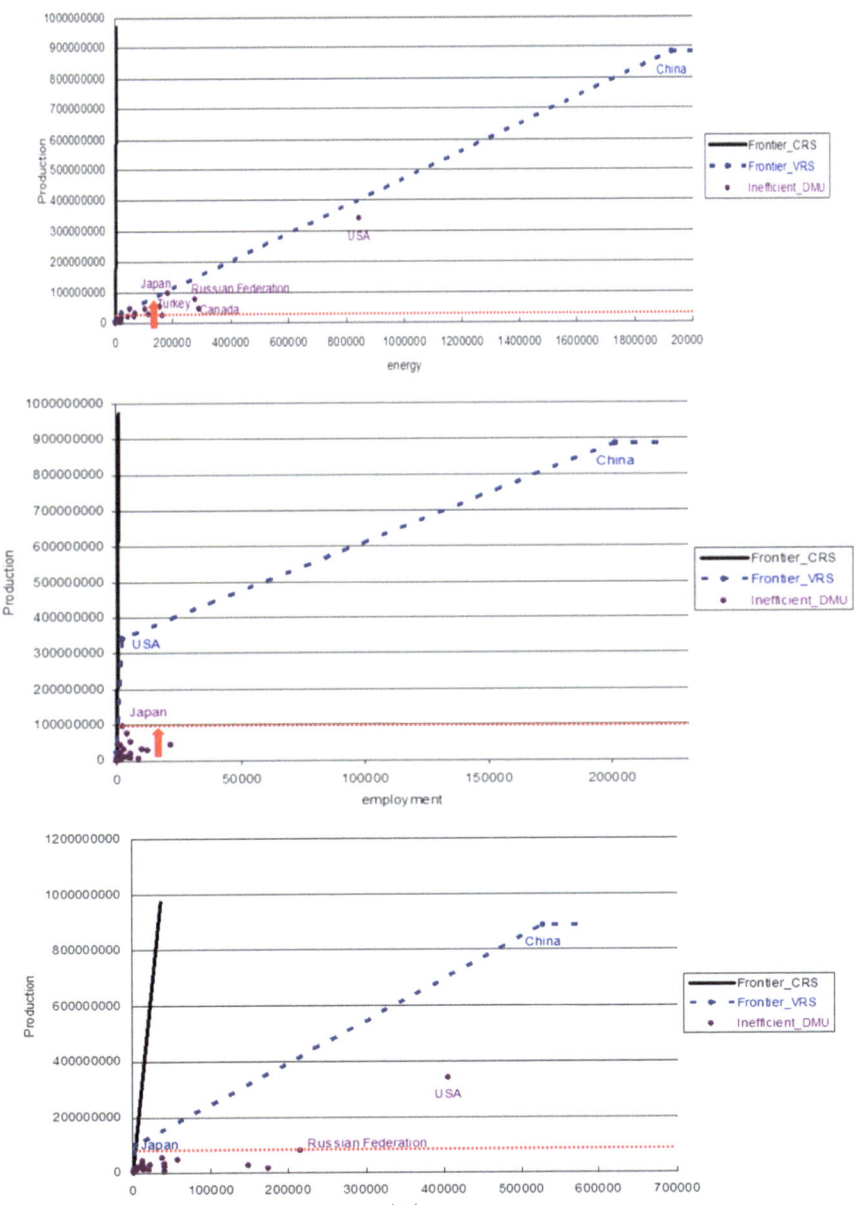

Fig. 3.9 (continued)

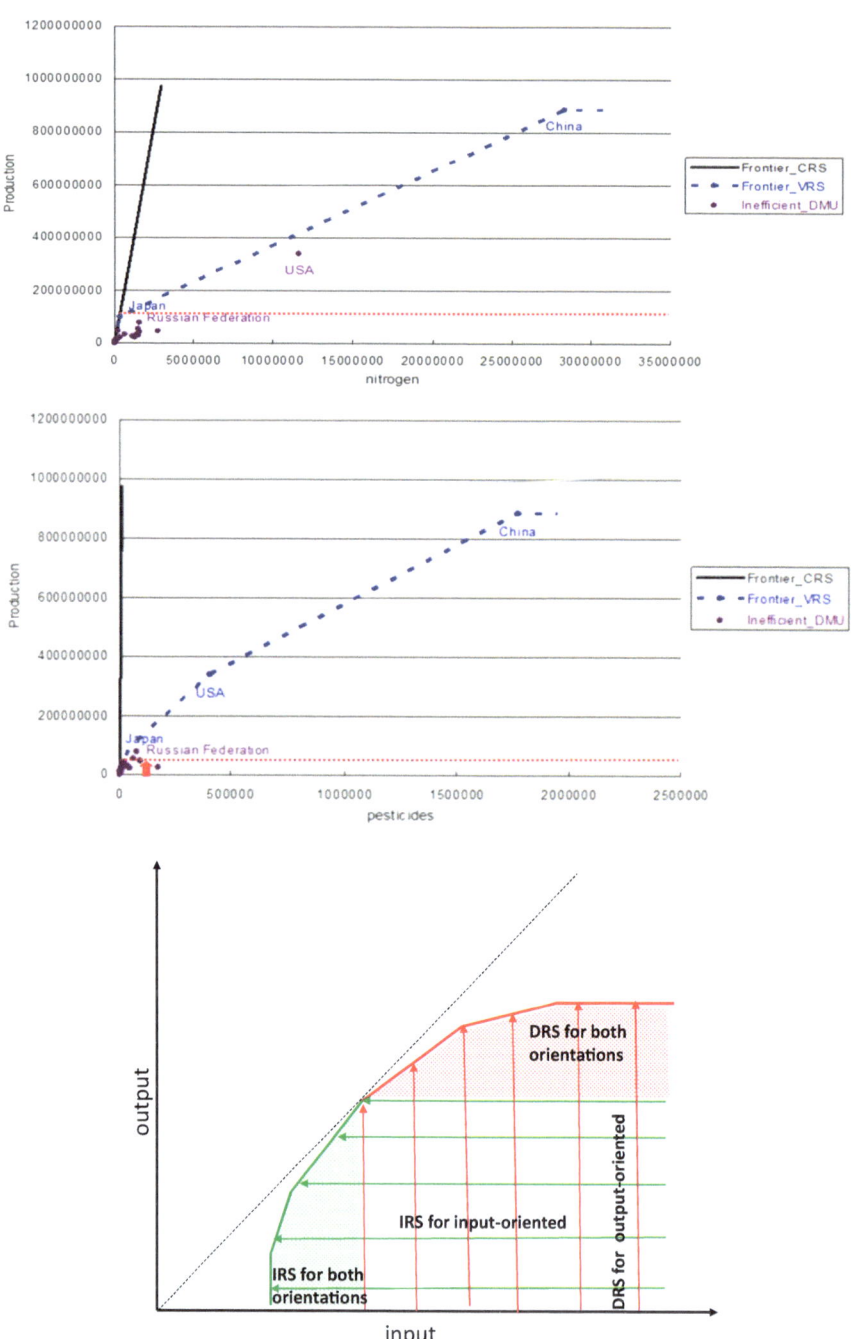

Fig. 3.9 (continued)

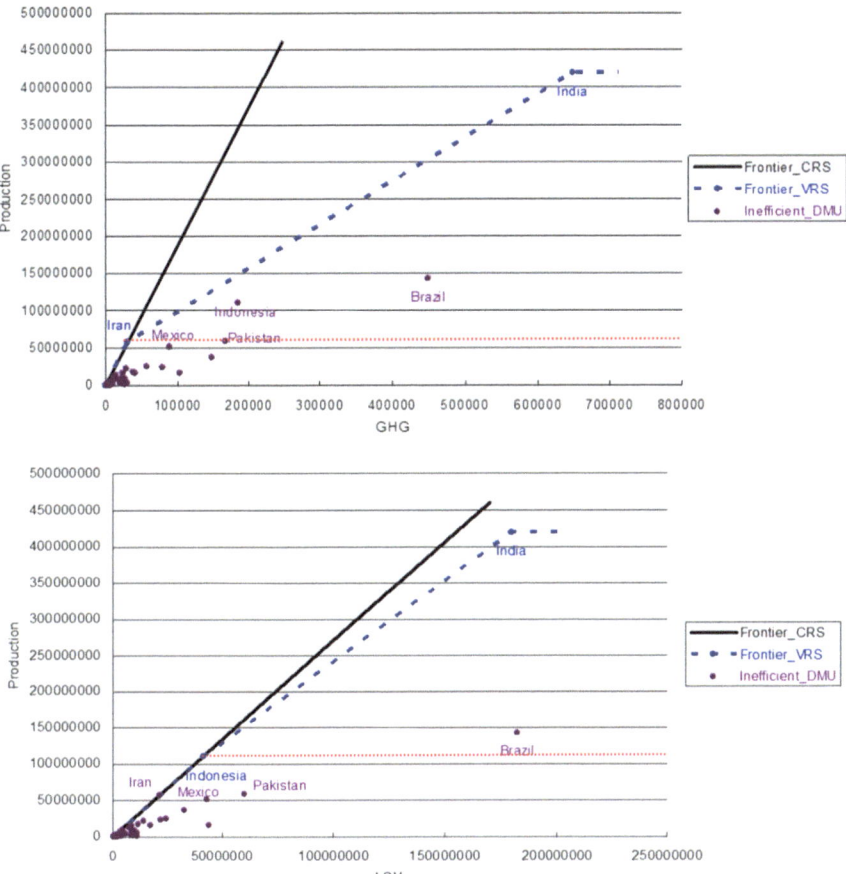

Fig. 3.10 Production frontiers, decreasing return to scale and potential slacks in Cluster 2 (two-dimensional frontier plots, radial DEA model, constant international dollars, 2018). *Note*: red arrows in frontier plots depict potential slacks; DRS/IRS means a decreasing and increasing return to scale, respectively; DMUs above red dotted line face DRS

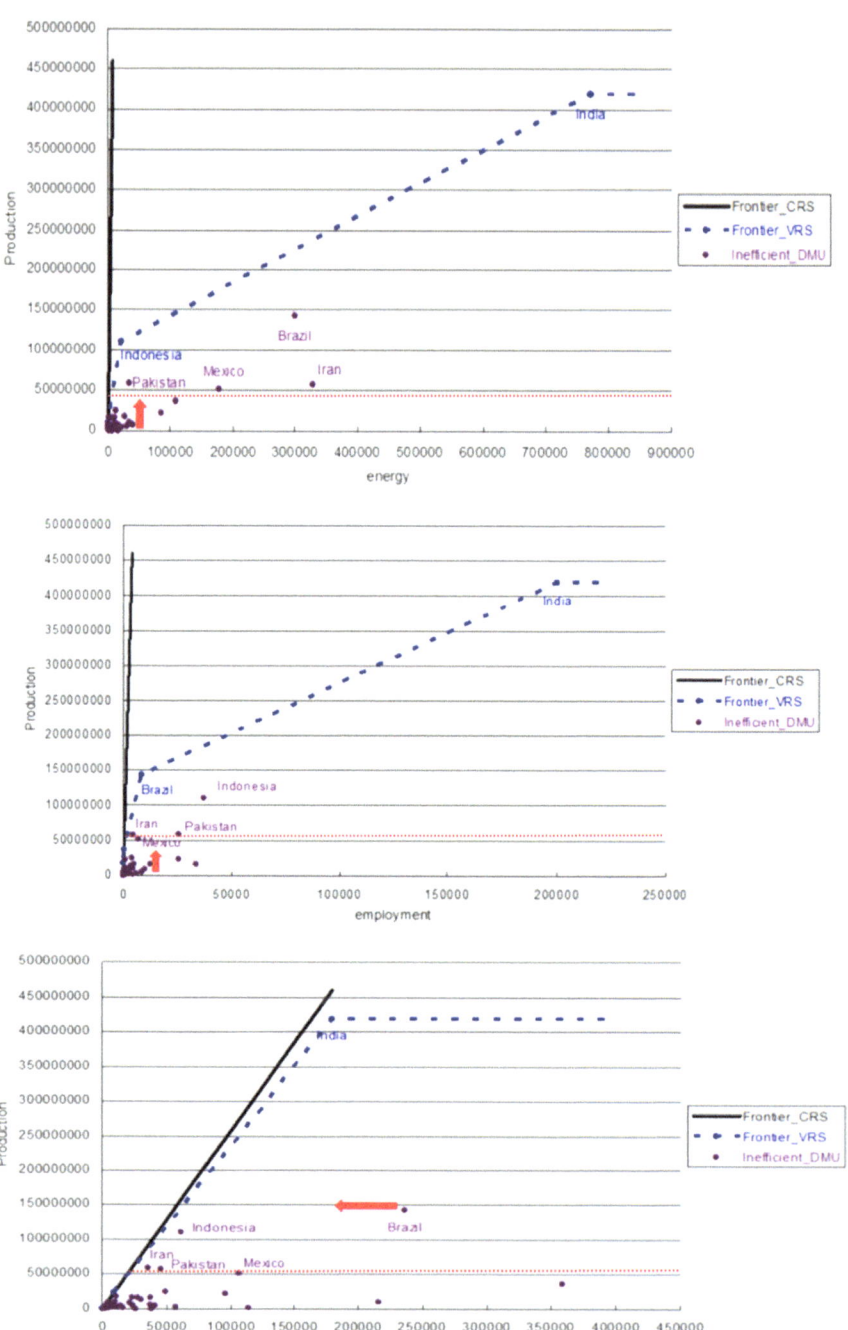

Fig. 3.10 (continued)

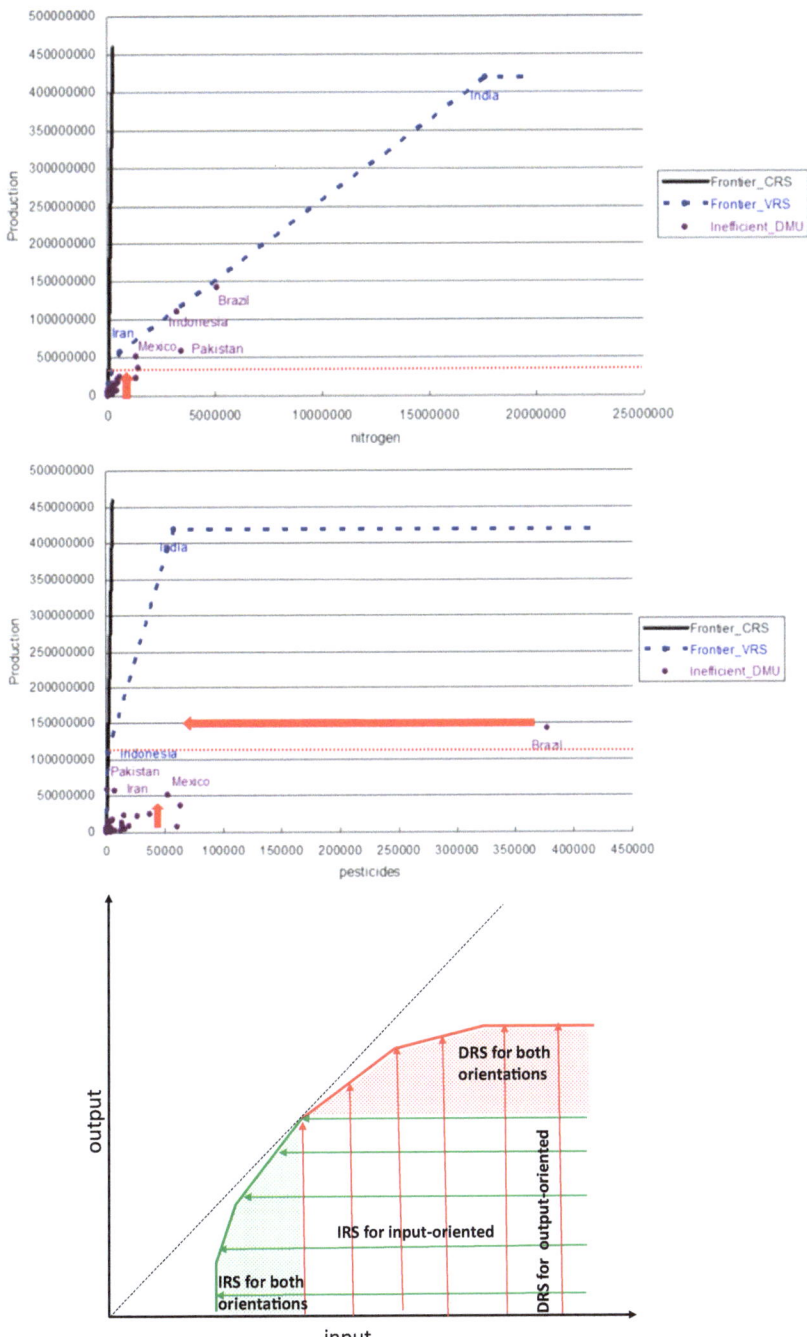

Fig. 3.10 (continued)

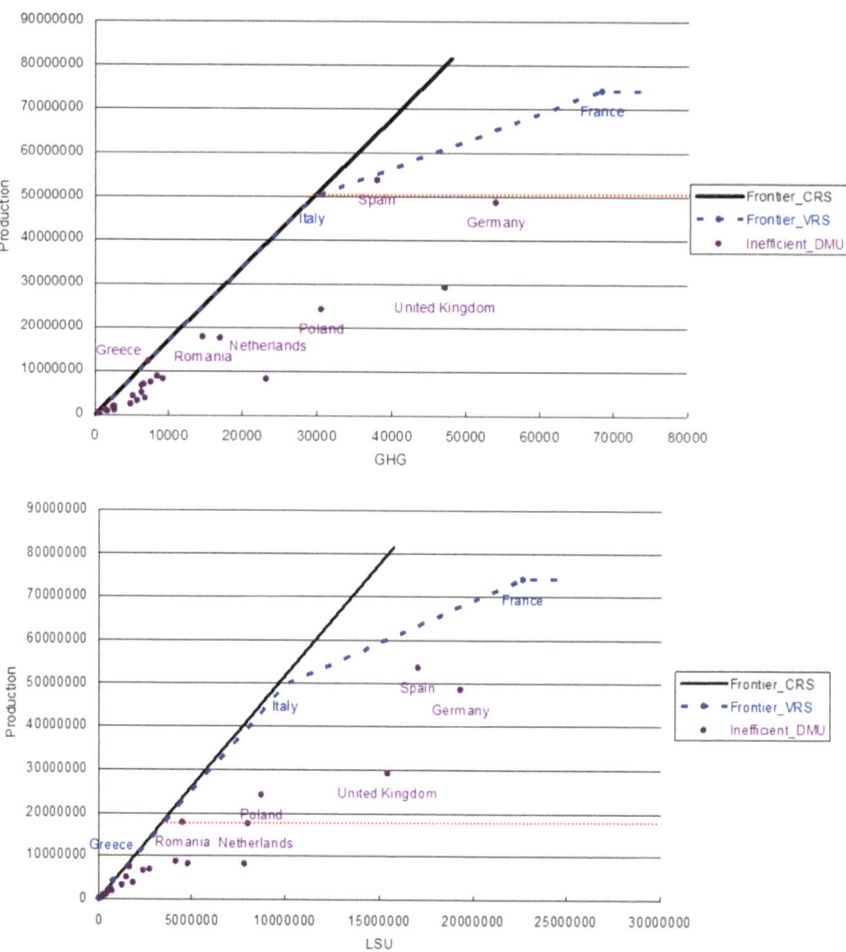

Fig. 3.11 Production frontiers, decreasing return to scale and potential slacks in Cluster 2 (two-dimensional frontier plots, radial DEA model, constant international dollars, 2018). *Note*: red arrows in frontier plots depict potential slacks; DRS/IRS means a decreasing and increasing return to scale, respectively; DMUs above red dotted line face DRS

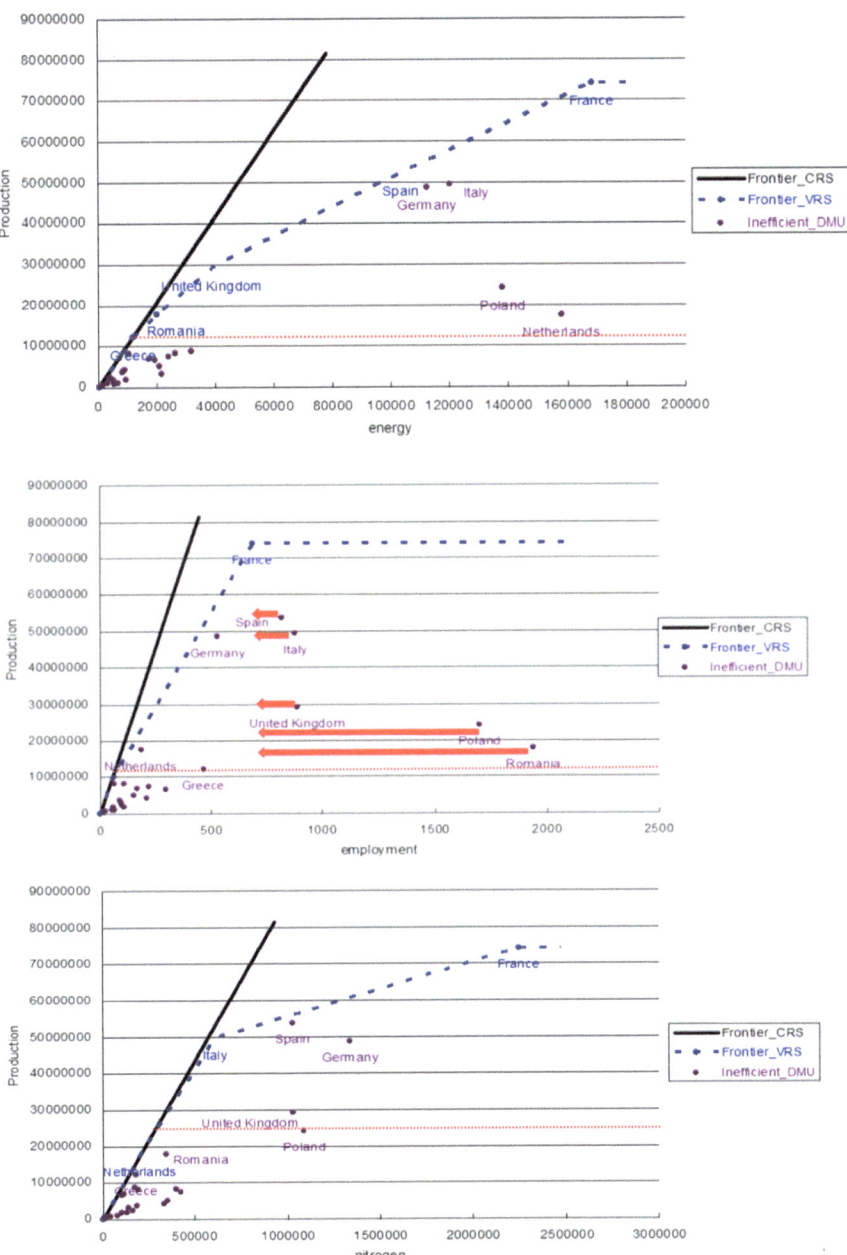

Fig. 3.11 (continued)

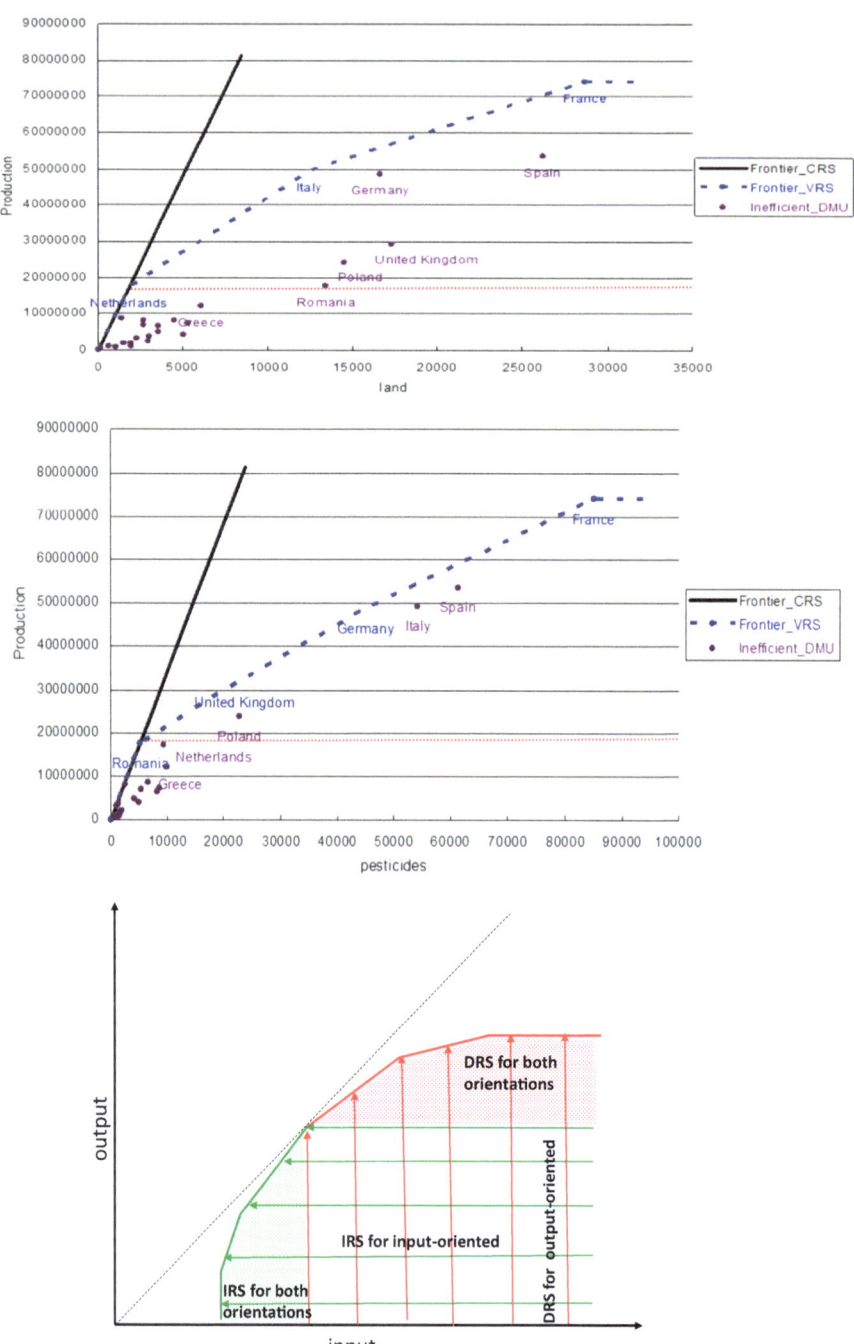

Fig. 3.11 (continued)

economic performance induces relatively large trade-offs in the remaining areas of agricultural sustainability and is ever less rational.

DRS are becoming an increasingly common problem under conditions of self-accelerating degradation of the environment as explained in the previous point (Sect. 3.3.1). In such a situation, it is necessary to look for another way to increase the overall level of agricultural sustainability than an input-intensive stimulation of production.

Figures 3.9, 3.10, and 3.11 also show the idea of slacks related to Pareto improvement. Potential slacks in terms of the inputs management are shown with red arrows. In principle, a slack is a DMU movement along a vertical or horizontal section of the VRS frontier. Of course, it can only be drawn for production functions with one explanatory variable (or possibly two). Potential room for improvement assuming nine dimensions (as our IE analysis assumes) may be much more complex and is impossible to be graphically presented. Moreover, it is likely that many more slacks will be identified through optimization of the multidimensional frontier model. Nevertheless, approximating the slacks problem on two-dimensional frontier plots allows for a better understanding of the problem and points to potential directions to look for opportunities to optimize resource management in agriculture across countries.

In **cluster 1,** most countries face DRS for energy consumption. With respect to other inputs, China, the USA, Japan and Russia are generally in the area of DRS (see the countries above the red dotted line in Fig. 3.9). Among the countries facing DRS, the USA and Russia are clear outliers in terms of inefficiency in input management, except for employment and pesticide use.

Potential production slacks can be observed for countries beyond those listed above in terms of: energy use, pesticides and agricultural employment. These findings will be checked in the multidimensional analysis. This type of slacks means that there may be opportunities to increase production while maintaining the current level of the given input. The shape of the isoquants of production drawn indicates the correctness of the assumption of a non-increasing return to scale (NIRS) adopted later in the multiscale IE analysis.

In **cluster 2** there are many more countries encountering DRS, usually: India, Brazil, Indonesia, Mexico, Pakistan, and Iran for all inputs except pesticide use (see Fig. 3.10). There are also many more areas with potential slacks. First, it is worth noting the input slacks in land management and pesticides that Brazil is likely to face. This implies that the consumption of these inputs can be reduced without affecting production change. Second, with respect to energy consumption, nitrogen, pesticides, and employment, the problem of output slacks may apply to all countries except those listed above. It means that in this cluster there are great opportunities to increase efficiency in the use of the above inputs, which is understandable, knowing that we are dealing with less developed countries.

In **cluster 3,** the situation is the most diverse (Fig. 3.11). In terms of fertilizer management, GHG emissions and LSU, only a few countries face DRS: France, Spain, Germany and Italy. For the remaining inputs, the DRS threatens, in addition to the countries mentioned above, the UK, Poland, the Netherlands and Romania (and Greece in terms of employment in agriculture).

The issue of employment in the EU's agriculture deserves special attention. It is the only area where slacks accumulate, which means that employment can be significantly reduced without affecting production levels. This applies to Spain, Italy, the UK, Poland and Romania. Hidden unemployment in agriculture is in the case of Poland and Romania a remnant of the transition from a socialist to a capitalist economy in the beginning of the 1990s. It can also be assumed that to some extent it is an effect of direct payments of the CAP, which often take the role of social support.

3.3.3 Potential for Optimization in the Use of Inputs and the Production of Outputs

To produce good agricultural output, several different inputs are needed. Simultaneously, undesirable by-products (e.g. greenhouse gases) are also produced. In this part we analyse which inputs and bad outputs should be decreased and which desirable outputs should be increased. The total number of countries in Clusters 1 and 2 is 72 (instead of 79) since some variables values were not avaliable in the whole 2006–2018 period. In Table 3.5 the results for Cluster 1 are provided. Since we use the super-efficiency approach, some slack values for inputs and bad outputs may be positive and for good outputs they may be negative. For example, a positive value of slack in employment input for Israel (2.25%) means that this country could even increase employment in agriculture by 2.25% and it still would be efficient. As stated in the methodology section, we use the hybrid approach, meaning that we treat some variables as radial while others as non-radial. Since slack values are calculated differently under these assumptions, we recalculated the slack values for radial variables to obtain "real" slack values which are comparable with their non-radial counterparts and remain in the line with the assumption that a slack movement has no impact on the change of any other input or output. The formula for the real slack is as follows:

$$Sr = (Sp \times CL)/(Pm + CL)$$

where Sr stands for 'real slack' movement for radial variable; Sp stands for slack movement associated to radial variable before transformation; CL indicates the current level of input, bad output, or good output; Pm indicates proportionate movement in %.

Finally, we present real slack values for all variables. As may be noted in Tables 3.5, 3.6, and 3.7, real slacks for radial variables are usually lower in comparison to non-radial variables. This is because in the radial approach some progress can be made without changes in any other input or output but another part

Table 3.5 The average value of proportionate movement (Pm) and slack (Sr) in the whole analysed period (in %) in countries of cluster 1

DMU	Pm land	Pm LSU	Pm energy	Pm GHG	Sr Land	Sr LSU	Sr Employment	Sr Energy	Sr Nitrogen	Sr Pesticides	Sr Production	Sr Dietary	Sr GHG
Albania	-0.04	-0.04	-0.05	-0.04	0.00	-7.05	-8.92	-5.88	-6.70	-3.69	1.25	-3.18	-16.40
Algeria	0.00	0.00	0.00	0.00	-11.09	-11.10	0.87	-3.06	0.45	1.97	-12.85	-4.34	-9.98
Argentina	0.00	0.00	0.00	0.00	-17.90	-16.46	9.50	-12.62	1.35	0.28	-10.34	-0.38	-18.77
Armenia	-8.73	-9.09	-7.85	-9.32	-26.70	-11.95	-43.05	-25.37	-64.09	-1.01	-0.24	-3.24	0.00
Canada	-21.28	-21.08	-21.30	-21.36	-16.30	-3.51	0.00	-44.47	-48.36	-36.53	0.00	9.90	-14.82
Chile	-4.01	-3.98	-8.53	-3.66	-0.71	-5.43	-7.49	0.00	-20.73	-10.30	-2.89	3.28	-7.40
China	0.00	0.00	0.00	0.00	-2.83	-4.70	1.23	-5.49	0.04	0.00	-8.06	-0.55	-4.69
Côte d'Ivoire	-20.35	-20.26	-20.55	-19.86	-38.31	0.00	-71.30	-1.60	-67.06	-15.66	-1.13	3.32	-29.77
Egypt	-7.86	-7.69	-9.20	-7.70	0.00	-12.98	-15.29	-8.76	-14.83	0.00	-4.96	-1.07	-13.79
Ghana	-4.84	-5.69	-6.35	-5.30	-7.53	-5.62	-29.79	-2.92	-20.32	-60.19	-7.47	-1.38	-26.45
Israel	0.00	0.00	0.00	0.00	-2.03	-9.72	2.25	-7.06	2.58	0.14	-7.43	-4.93	-4.67
Japan	-0.04	-0.04	-0.03	-0.03	-1.18	-7.18	0.02	-10.48	1.06	-0.18	-9.21	0.12	-5.47
Jordan	-1.46	-1.57	-1.61	-1.61	-4.33	-9.13	-1.12	-10.12	-6.66	6.42	-2.96	-4.48	-6.52
Malawi	0.00	0.00	0.00	0.00	-0.77	-18.69	0.00	0.00	1.43	0.53	-12.55	-2.11	-9.88
Malaysia	-6.91	-7.02	-8.45	-6.83	-3.07	-0.82	-40.00	0.00	-49.74	-51.62	-3.57	7.56	-21.83
Mozambique	-18.28	-19.14	-35.73	-18.00	-46.08	-7.77	-35.78	-58.23	-22.84	-30.13	-4.09	-0.32	-37.98
North Macedonia	0.00	0.00	0.00	0.00	-4.02	-5.04	0.21	-6.82	1.07	3.88	-5.74	-13.45	-8.77
Norway	-0.17	-0.18	-0.11	-0.17	-0.05	-7.13	7.17	-9.85	-0.31	1.06	-1.40	-4.14	-9.53
Philippines	-1.42	-1.41	-1.67	-1.41	-0.53	-6.83	-1.19	0.00	-7.93	-1.89	-6.26	0.01	-10.32
Russian Federation	-37.79	-37.83	-37.79	-37.77	-79.10	-20.36	-40.91	0.00	-31.20	-13.23	0.00	0.00	-45.99
Saudi Arabia	-25.44	-25.44	-25.52	-26.71	-67.17	-16.61	-31.91	-54.56	-45.16	-0.18	-1.02	2.01	0.00
Sri Lanka	-6.29	-6.45	-6.44	-6.37	0.00	0.00	-38.51	0.00	-68.86	-37.91	-0.49	6.19	-51.18
Switzerland	-1.06	-1.06	-1.04	-1.05	-0.74	-7.65	-0.34	-0.15	-1.41	-0.34	-6.66	-0.31	-6.20

(continued)

Table 3.5 (continued)

DMU	Pm land	Pm LSU	Pm energy	Pm GHG	Sr Land	Sr LSU	Sr Employment	Sr Energy	Sr Nitrogen	Sr Pesticides	Sr Production	Sr Dietary	Sr GHG
Thailand	-40.08	-40.46	-40.18	-40.20	-13.13	0.00	-82.37	-3.19	-79.38	-57.93	-2.08	14.76	-61.90
Turkey	-0.22	-0.20	-0.31	-0.20	-10.74	-9.86	-0.50	-9.18	-0.36	-1.36	-1.94	-3.52	-10.52
Ukraine	-39.99	-40.55	-40.12	-40.18	-26.23	0.00	-75.15	0.00	-87.55	-76.06	0.00	11.28	-44.02
USA	-0.16	-0.16	-0.17	-0.16	-8.58	-9.07	-0.05	-8.31	-0.62	-0.30	-8.97	-0.05	-8.82
Viet Nam	0.00	0.00	0.00	0.00	-0.06	-8.72	0.00	0.00	0.00	0.18	-4.05	-0.47	-6.23
Average	-8.80	-8.90	-9.75	-8.86	-13.90	-7.98	-17.94	-10.29	-22.72	-13.72	-4.47	0.38	-17.57

of the move to the frontier needs to be done proportionally for all radial variables. Therefore, for the radial variables we also present proportionate movements (PM). The percentage of PMs for all radial variables for a given country in a given period are equal by definition. However, in Tables 3.5, 3.6, and 3.7 we present the average values for 2005–2018 that are calculated using the following formula:

$$average\ proportionate\ movement(\%)$$
$$= \frac{average(2005 - 2018)\ proportionate\ movement}{average(2005 - 2018)initial\ value\ of variable}$$

Therefore, the PMs for different radial variables are slightly different.

The biggest average slack for countries in Cluster 1 (cf. Table 3.5) was recorded for nitrogen input and then for employment and the GHG emissions. On average, nitrogen use should be reduced by 22.72%, employment by 17.94% and GHG by 17.57%. Furthermore, GHG should be reduced by a further 8.86% resulting from proportionate movement. However, these reductions would be more difficult to implement in practice because they would require changes in existing technology. The results above are not a surprise since cluster 1 contains countries with more input-intensive agriculture. Therefore, there is relatively large room to decrease the inputs use. Interestingly, the average slack on production is negative. It is also negative for a great majority of countries. It means that agricultural production in this cluster could be even decreased without any harm on integrated efficiency. Obviously, the situation clearly differs between countries. There are some countries with a significant slack on nitrogen (above 50%). It doesn't necessarily mean that nitrogen use in these countries is large in absolute values. It means that nitrogen use is too large in comparison to the output values. The results in Table 3.5 show the importance of multi-dimensional analysis. In two-dimensional analysis, Russia and the USA were found as outliers laying far from the frontier, but multi-dimensional analysis shows that while Russia exhibits large relative slacks in most of the variables, the USA's slacks are below average for all inputs, except for LSU.

The average slack values in countries of Cluster 2 are larger in the case of all inputs except for nitrogen use. An exceptionally large difference is related to pesticides—countries in cluster 2 should reduce pesticides use by 32.24%, on average. Obviously, the situation is different across the cluster. There are some countries without any slack on pesticides (e.g. Brazil, Cameroon, Indonesia), but there are also countries (e.g. Belarus, Guyana, and Senegal) where the slack on pesticides is very large. Once again, we should recall that it may just mean that pesticides use is too high in relation to the input/output mix used and not in absolute terms. Although we have shown that the dominant agricultural model in Cluster 2 is not that intensive, the slacks on GHG emissions and energy are significant, especially if proportionate movements are also taken into consideration. The PMs for energy and GHG are –20.85% and –19.99%, respectively, while in Cluster 1 they are –9.75% and –8.86%. Interestingly, in Cluster 2 one may find countries with relatively large positive slacks on dietary energy supply adequacy (e.g. Iraq,

Table 3.6 The average value of proportionate movement (Pm) and slack (Sr) in the whole analysed period (in %) in countries of cluster 2

DMU	PM land	PM LSU	PM energy	PM GHG	Sr Land	Sr LSU	Sr Employment	Sr Energy	Sr Nitrogen	Sr Pesticides	Sr Production	Sr Dietary	Sr GHG
Australia	-2.09	-2.06	-2.08	-2.13	-13.24	-8.44	-0.04	-9.26	-1.08	-2.19	-8.81	-1.29	-10.74
Azerbaijan	-10.63	-11.14	-12.57	-11.39	-0.61	-6.78	-15.79	-9.80	-10.33	-23.26	-6.08	-0.79	-4.64
Bangladesh	-1.19	-1.08	-0.98	-1.09	0.00	-54.59	-37.60	-48.81	-30.31	-27.45	-1.37	5.83	-51.76
Belarus	-7.06	-6.98	-7.07	-7.00	-2.06	-0.56	-39.54	-6.05	-59.32	-69.90	0.00	0.61	-59.15
Bolivia	-3.73	-3.78	-3.42	-3.83	-10.50	-15.38	1.29	-0.70	0.00	-7.31	-6.51	0.14	-16.46
Brazil	0.00	0.00	0.00	0.00	-7.84	-9.03	2.23	-5.68	0.59	0.00	-8.87	-0.56	-9.04
Brunei Darussalam	0.00	0.00	0.00	0.00	-6.33	-14.38	1.21	-24.01	4.49	-0.29	-14.50	-7.97	-9.18
Burkina Faso	-46.46	-47.84	-55.73	-47.71	0.00	-36.45	-23.70	0.00	-18.12	-66.27	-1.03	0.46	-31.08
Cameroon	0.00	0.00	0.00	0.00	-1.74	-4.59	0.19	-0.93	0.39	0.00	-11.04	-0.92	-3.66
Colombia	-35.47	-34.92	-38.02	-35.03	-31.93	-10.57	-15.76	0.00	-3.32	-66.15	0.00	0.72	-3.74
Costa Rica	0.00	0.00	0.00	0.00	-3.07	-5.28	0.41	-1.71	0.80	0.01	-10.78	-0.52	-5.24
Ecuador	-35.83	-36.15	-36.82	-36.09	0.00	-21.44	-34.53	-33.25	-10.11	-56.37	0.00	11.03	-7.15
El Salvador	-26.94	-27.45	-26.46	-27.40	-5.02	-8.84	-30.65	-6.53	-42.78	-18.61	0.00	8.88	-9.01
Ethiopia	-28.40	-30.70	-37.55	-30.75	-8.04	-37.24	-40.62	0.00	-0.69	-29.13	-2.29	5.61	-31.13
Georgia	-45.75	-45.25	-52.93	-45.03	-33.33	-0.92	-76.79	-31.58	-43.92	-44.29	0.00	2.58	-13.47
Guyana	-54.14	-53.93	-54.18	-53.71	-73.32	0.00	-91.99	-60.37	-91.36	-95.17	0.00	0.41	-70.76
Honduras	-57.41	-57.55	-58.21	-57.47	-3.51	-12.27	-59.75	0.00	-42.64	-20.68	0.00	8.93	-2.73
Iceland	0.00	0.00	0.00	0.00	-18.87	-2.56	0.22	-16.85	0.67	3.35	-10.22	-2.26	-15.99
India	0.00	0.00	0.00	0.00	-8.98	-11.74	0.04	-15.42	0.00	0.00	-11.86	-0.41	-11.34
Indonesia	0.00	0.00	0.00	0.00	-13.89	-16.93	0.94	-23.71	0.33	0.00	-17.82	-1.18	-14.62
Iran	-0.71	-0.71	-0.84	-0.60	-9.11	-9.34	-0.67	-16.63	3.81	0.87	-11.31	-0.07	-3.80
Iraq	-34.28	-35.24	-33.69	-34.59	-11.37	-0.96	-60.41	-40.03	-39.55	-35.67	0.00	24.90	-1.05
Kazakhstan	-3.91	-3.90	-3.69	-3.99	-30.04	-4.85	-1.10	-5.52	1.53	-8.40	-7.21	-2.44	-15.81
Kenya	-10.40	-11.11	-9.54	-11.00	-3.31	-20.64	-14.94	-1.08	0.00	-1.49	-4.55	6.35	-15.54

Kyrgyzstan	−5.49	−5.75	−5.23	−5.80	−32.60	−2.60	−2.76	−9.27	0.00	−24.24	−3.47	3.26	−18.15
Madagascar	−6.43	−6.72	−8.68	−6.08	−33.88	0.00	−26.96	0.00	−65.77	−56.87	−2.26	11.97	−21.90
Mexico	−25.89	−26.49	−27.07	−26.97	−25.98	−11.69	−5.52	−2.75	−15.65	−51.79	0.00	−0.51	−1.74
Mongolia	−57.64	−59.49	−60.66	−59.05	−72.74	−57.45	−3.60	0.00	−34.87	−74.62	0.00	17.18	−59.85
Morocco	−8.11	−8.08	−8.52	−8.10	−31.67	−18.33	−31.07	−3.15	−6.56	−43.48	0.00	−0.74	−1.83
Namibia	−85.77	−85.72	−85.76	−85.86	−32.40	0.00	−71.06	−14.05	−92.18	−81.37	0.00	23.63	−25.76
Nepal	−0.32	−0.35	−0.53	−0.34	−1.14	−15.78	−3.79	−18.79	0.06	−7.15	−8.74	−0.95	−14.65
New Zealand	0.00	0.00	0.00	0.00	−2.22	−8.65	0.05	−5.05	0.68	0.00	−8.65	−0.22	−9.25
Nicaragua	−12.66	−14.36	−15.56	−14.21	−36.12	−50.19	0.08	0.00	−32.39	−43.08	−5.29	0.00	−48.83
Niger	0.00	0.00	0.00	0.00	−15.66	−20.50	0.00	−7.77	0.32	1.78	−17.11	−1.78	−20.16
Pakistan	−13.47	−12.69	−13.62	−13.03	−1.49	−43.22	−12.73	−20.39	−34.35	−57.09	−2.28	6.85	−29.03
Paraguay	−42.49	−43.48	−43.08	−43.37	−38.72	−33.21	−10.51	0.00	−2.69	−25.54	−0.43	2.73	−35.61
Peru	−25.54	−25.54	−24.27	−25.61	−29.96	−17.88	−3.50	0.00	−2.32	−69.51	0.00	3.51	−2.18
Senegal	−73.11	−73.54	−74.51	−73.44	−9.82	−4.02	−45.96	0.00	−49.77	−79.90	0.00	14.14	−11.89
South Africa	−35.72	−35.71	−38.41	−35.60	−70.51	−12.37	−18.70	0.00	−12.01	−75.98	0.00	2.13	−1.42
Tajikistan	−20.38	−19.88	−22.06	−19.94	−7.21	−1.16	−27.14	−41.04	−8.60	−46.40	−0.92	29.67	−26.54
Tunisia	0.00	0.00	0.00	0.00	−8.07	−4.07	0.00	−2.92	0.00	0.00	−0.01	−2.88	−0.66
Uruguay	−8.12	−8.31	−8.35	−8.23	−56.71	−48.26	−9.27	−0.97	−10.25	−27.12	0.00	−0.30	−44.88
Yemen	0.00	0.00	0.00	0.00	−20.59	−17.71	−0.43	−32.80	−1.66	0.00	−22.72	0.35	−7.83
Zambia	−44.94	−46.72	−47.14	−45.06	−72.40	0.00	−55.35	0.00	−51.19	−87.59	0.00	39.37	−54.38
Average	−19.78	−20.06	−20.85	−19.99	−20.36	−15.47	−19.67	−11.75	−18.18	−32.24	−8.81	4.67	−19.40

Table 3.7 The average value of proportionate movement (Pm)and slack (Rs) in the whole analysed period (in %) in countries of cluster 3 (EU)

DMU	Pm land	Pm LSU	Pm energy	Pm nitrogen	Pm production	Pm GHG	Sr Land	Sr LSU	Sr Employment	Sr Energy	Sr Nitrogen	Sr Pesticides	Sr Production	Sr Dietary	Sr GHG
Austria	-1.65	-1.66	-1.66	-1.89	1.68	-1.69	-1.54	-4.98	-3.26	-0.17	-6.07	-1.15	0.74	-1.05	-5.25
Belgium	1.31	1.32	1.26	1.35	-1.33	1.32	-1.41	-7.37	19.91	-6.58	-6.24	3.49	0.00	-2.13	-6.83
Bulgaria	-49.23	-49.89	-50.22	-48.30	48.57	-49.06	-60.82	-0.73	-19.41	0.00	-74.43	-50.79	0.00	1.84	-42.06
Croatia	-0.13	-0.13	-0.09	-0.11	0.13	-0.13	-15.69	-16.77	4.16	-18.43	-18.55	27.32	0.00	-2.78	-13.26
Cyprus	-58.87	-58.90	-58.80	-58.71	58.74	-58.80	-46.14	0.00	0.00	-33.24	-49.43	-56.51	0.00	10.80	-16.97
Czechia	11.89	10.22	11.89	9.49	-10.43	10.14	-5.82	-13.90	0.58	-15.71	-10.92	-0.29	1.08	-5.00	-9.03
Denmark	-0.11	-0.11	-0.10	-0.13	0.11	-0.11	-11.28	-4.51	13.07	-1.28	-7.20	10.05	0.00	1.20	-1.24
Estonia	-28.89	-28.91	-28.77	-28.70	28.78	-28.90	-79.91	0.00	-4.77	-61.23	-44.49	-60.24	0.00	1.27	-53.28
Finland	-28.72	-28.70	-28.69	-28.85	28.68	-28.72	-34.74	0.00	-0.26	-49.31	-62.13	-35.23	0.00	0.00	-43.74
France	1.36	1.33	1.40	1.34	-1.41	1.36	-6.10	-6.66	1.15	-3.85	-10.94	0.18	0.00	0.16	-6.60
Germany	-0.74	-0.73	-0.72	-0.68	0.71	-0.71	0.00	-12.10	0.22	-1.08	-11.34	2.84	0.00	0.53	-10.11
Greece	3.55	3.36	3.95	2.97	-3.69	3.45	-13.68	-5.48	-1.59	-7.95	-10.91	9.22	0.00	-3.79	-5.88
Hungary	-26.57	-26.55	-26.54	-26.92	26.47	-26.65	-49.30	0.00	-1.30	-11.92	-65.96	-21.93	0.00	0.09	-14.81
Ireland	-0.05	-0.05	-0.06	-0.05	0.05	-0.05	-15.62	-22.16	1.33	0.00	-20.15	0.00	0.00	-3.82	-24.85
Italy	2.98	2.79	3.07	3.23	-3.12	2.85	-3.21	-4.47	-3.78	-6.92	-4.46	1.00	0.00	1.12	-5.78
Latvia	-49.63	-49.70	-49.58	-49.56	49.63	-49.65	-76.01	0.00	-43.10	-46.14	-55.18	-76.87	0.00	0.00	-50.65
Lithuania	-16.50	-16.04	-16.15	-16.83	16.76	-16.29	-52.11	-0.60	-36.06	0.00	-64.11	-49.28	0.00	0.00	-49.18
Luxembourg	-0.10	-0.10	-0.10	-0.10	0.10	-0.10	-3.66	-5.33	5.93	-2.26	-6.05	1.39	0.52	-9.84	-6.90
Netherlands	3.21	3.13	3.26	3.37	-3.08	3.15	0.00	-9.80	3.61	-11.77	-8.15	4.73	0.00	-0.41	-9.40
Poland	-28.55	-28.51	-28.49	-28.68	28.65	-28.61	-1.32	0.00	-67.66	-42.39	-47.16	0.00	0.00	4.19	-11.76
Portugal	-16.61	-16.79	-16.23	-17.44	16.76	-16.72	-5.90	-8.65	-50.91	-2.72	0.00	-65.15	0.00	-0.09	-13.92
Romania	2.35	2.64	0.48	1.98	-2.44	2.56	-22.75	-15.01	-15.83	0.00	-23.14	2.36	0.00	-2.25	-15.17
Slovakia	-50.61	-50.30	-50.42	-51.13	50.20	-50.62	-68.33	0.00	-0.01	-10.05	-60.11	-66.93	0.00	7.85	-30.86

Slovenia	−55.67	−55.77	−55.80	−55.74	55.79	−55.75	−43.61	−0.17	−45.15	−27.90	−3.98	−77.13	0.00	0.00	−22.03
Spain	−4.39	−4.42	−4.48	−4.60	4.42	−4.38	−37.31	−22.68	−8.46	−3.18	−16.71	−1.40	0.00	7.79	−4.13
Sweden	−19.84	−19.80	−19.68	−19.92	19.80	−19.84	−64.56	0.00	0.00	−46.78	−53.93	−33.98	0.00	0.22	−40.41
United Kingdom	2.35	2.48	2.31	2.47	−2.32	2.51	−8.87	−9.90	1.49	0.00	−8.03	3.40	0.00	−0.01	−9.04
Average	−15.11	−15.18	−15.15	−15.26	15.12	−15.16	−27.03	−6.34	−9.26	−15.22	−27.77	−19.66	0.09	0.22	−19.38

Namibia, Mongolia). At the same time, these countries do not exhibit a negative slack on production (in contrast to the majority of countries). It indicates that an agricultural output deficit may be a problem for these countries. Average slacks and proportionate movements on basic agricultural inputs such as land and LSU are also higher in comparison to Cluster 1. In particular, this is the case of Latin America countries. This may be linked to the more extensive type of agricultural production in these countries. Under this model, large resources of land and LSU are used, but the production (in relations to inputs) is not on a very high level, which translates into a low level of eco-inefficiencies. This is an important conclusion of a more general nature. It turns out, that from the perspective of integrated efficiency, the more intensive agriculture may be beneficial since the relations between inputs and outputs is more rational. It can be interesting that two-dimensional analysis has identified Brazil as a country with large slacks in land and pesticides use. The multi-dimensional analysis proved the slack problem only with respect to land input. In percentage terms, also this slack is not particularly high, but one should take into account the size of the country. From the global perspective, any improvements in major agricultural producers have a large impact for the general eco-system.

When analysing average slack values in the EU countries, one conclusion stands out. Namely, there are clear differences between countries, in particular (but not exclusively) between so called 'old' and 'new' Member States. For example, the average real slack value on land is the highest in Cluster 3, but this is mainly caused by the high slack values in the Central and Eastern EU countries. In Slovakia, Sweden, or Bulgaria, the slack on land exceeds 60%. In addition, these countries exhibit large room for improvement as manifested by proportionate movement values. This means that in some parts of the EU, the resources of land for agricultural production should be lower or, from the other perspective, output from this amount of land should be higher.

Employment in agriculture in most of the EU countries is very low, therefore we are not surprised that the average slack (−9.26%) for this variable is low. However, this average can be obtained thanks to Western EU countries—in some Member States (e.g. Denmark or the Netherlands) employment could be even increased. At the opposite end, in some countries (such as Poland, Lithuania, Latvia, Slovenia, but also in Portugal) the slack on employment is very high. This is because the agricultural sectors in these countries are in a transformation process and the level of employment is still relatively high. In this context, the initial results from the two-dimensional analysis can be proved.

When it comes to energy, one may say that the average slack (−15.22%) and average proportionate movement (−15.15%) is quite big but similar to other vari-ables; this value is inflated by the large slack in some 'new' member states and, in this case, Sweden. Another problem that we identified in the EU is the excessive use of nitrogen—the average real slack value is the highest. From the integrated effi-ciency point of view, it could be decreased in all countries, except for Portugal. However, particularly large slack values are recorded among the 'new' MS and in Nordic countries, such as Finland and Sweden.

When compared to other clusters, the mean slack on pesticides (−19.66%) is on an average level. Interestingly, in some MS, the pesticides use could even increase, but the negative slacks in some other countries (such as Estonia, Latvia, Slovenia) is very high. GHG emissions should be reduced in all countries in the cluster and the average slack is equal to −19.38%. The proportionate movement is −15.16%. Similarly to inputs use, the reduction of this bad output should be especially high in 'new' member states and in Sweden and Finland. The average slacks on production and dietary adequacy are close to 0, meaning that EU agriculture provides a sufficient amount of production and supply of food. Inefficiencies in some countries result rather from excessive and irrational use of certain resources.

The comparison of slack calculations when using constant USD (Tables 3.5, 3.6, and 3.7) and constant international dollars (Tables A.2, A.3, and A.4 in Appendix) show that market imperfections have a very serious impact on the calculated level of IE and slacks. In particular, this is the case of Clusters 1 and 2. The average slack values in Cluster 2 are lower by ca. one-third when using international prices, while in Cluster 1 slacks in the modified approach often constitute less than half of their counterparts when using the constant USD. In other words, a half of slack values could be diminished if the world agricultural market worked optimally and agricultural prices would converge to one equilibrium level. When it comes to cluster 3, slack values are smaller when dealing with international dollars, but the difference is not that apparent. This is because the EU constitutes a single market and agricultural prices are more similar (though not the same, of course).

The main message from the average slacks calculations in 2005–2018 for our sample of countries would be that modern and relatively intensive agriculture still pays off. Modern agriculture sparingly uses certain basic resources and at the same time achieves very good production results. Of course, GHG emissions remain a concern and should be reduced, but from an efficiency point of view, the potential for reducing emissions in countries with modern agriculture is moderate. It is also important to remember that emissions per unit of production in these countries are often low. At the same time, resources such as land and labour can be shifted to other uses. Large slack values (low integrated efficiency levels) were observed in some developing countries with low input use but with very poor production results. However, this was also the case in countries where good outputs are produced on a sufficient level but the use of inputs is relatively too high.

3.3.4 How Are the Slacks Values Changing?

In the previous subchapter we were analysing the average value of slacks in the 2005–2018 time period. It showed that significant saving on inputs and bad outputs could have been achieved in that period if inefficiencies were eliminated. However, we need to remember that a 14 years period is relatively long and important changes could occur, in particular in countries in the transition process. This means that some problems could have diminished while some others may have even escalated. To provide a more comprehensive picture of the field, in this subchapter we will analyse changes in relative average slack values between the beginning (2005–2007) and the end (2016–2018) of the research period. In Tables 3.8, 3.9, and 3.10 we present the results of these calculations for designated clusters. Positive values in the Tables mean that situation worsened, i.e., the average slack has increased between two sub-periods. When it comes to inputs in Cluster 1 (Table 3.8), the situation improved regarding land, LSU, and energy. The average slacks for land and energy decreased by ca. 5 pp. and slack on LSU decreased by 14.19 pp. Although there were few countries in which slacks on these variables increased and remained stable (e.g. Saudi Arabia and Israel for the land factor, Armenia for LSU, Canada and Mozambique for energy), in most of the countries changes were negative, i.e. the slacks decreased. These changes are somehow offset by the increases in PMs but this impact is limited, regarding the values.

On average, countries of Cluster 1 faced bigger problems with the excessive use of labour and pesticides (see Table 3.5) and when it comes to these inputs, the slacks even increased, on average by 4.83 and 4.32 pp. However, in these cases the tendencies are not straightforward. There are some countries where slacks on these variables decreased significantly (Cote d'Ivoire and Russia for employment, and Sri Lanka or Switzerland for pesticides), but on the other hand, in some countries the slacks increased dramatically. For example, the slack on employment in Ghana increased by 81.66 pp. and the slack on pesticides in Malaysia increased by 44.99 pp.

Previous analysis has shown that GHG emissions are an important problem in all the clusters. In this context, it is a good sign that slacks on GHG increased only in six countries. In Saudi Arabia and Armenia they remained unchanged while in most of the cluster, these slacks decreased. The average decrease was 10.25 pp. while in big agricultural producers, such as Argentina or the USA, the slacks on GHG decreased by more than 30 pp.

Interestingly, the basic trends in Cluster 2 are similar to Cluster 1. The average slacks on employment, nitrogen and pesticides increased but these trends differ to a large extent for different countries. When it comes to land and LSU, the slacks decreased (on average by 13.18 and 12.20 pp., respectively) more than in cluster 1 and the values range was rather small, meaning that most of the countries optimised the use of these basic inputs. In the case of Cluster 2, the changes in slacks for land, LSU and energy are reduced to a greater extent by proportionate movements. Most of the countries managed to reduce slacks on GHG and the average change (decrease by 11.13%) is similar to Cluster 1. It is a piece of good news that the slacks on GHG were reduced also for big agricultural producers such as India, for example.

The average changes in slacks in EU countries are usually lower in comparison to Cluster 1 and Cluster 2 and do not exceed 5 pp. except for GHG for which the average decrease in the slack is 6 pp. (cf. Table 3.10). When it comes to emissions, there are some countries that experienced significant slack reduction (e.g. Croatia or the United Kingdom), but in some Member States, the slacks have even increased. Particularly high growth in slack values were recorded in some 'new' members, such as Bulgaria (19.8 pp.), Hungary (14.5 pp.) or Slovakia (9.8 pp.). When it comes to

Table 3.8 The change in relative proportionate movement (Pm) and slack (Sr) values between 2005–2007 and 2016–2018 sub-periods in Cluster 1 (in pp.)

DMU	Pm land	Pm LSU	Pm energy	PmGHG	Sr Land	Sr LSU	Sr Employment	Sr Energy	Sr Nitrogen	Sr Pesticides	Sr Production	Sr Dietary	Sr GHG
Albania	0.20	0.20	0.20	0.20	0.00	−24.72	−19.56	−16.42	12.77	7.66	−3.49	−3.62	−27.32
Algeria	0.00	0.00	0.00	0.00	−31.15	−30.30	0.22	1.34	−4.03	−2.97	−22.15	−11.91	−21.71
Argentina	0.00	0.00	0.00	0.00	−40.51	−37.31	58.04	−27.16	0.00	1.31	−14.48	0.85	−37.94
Armenia	23.17	23.15	23.21	23.16	22.75	5.17	45.12	−47.95	39.44	3.02	0.71	−15.07	0.00
Canada	0.86	1.12	0.29	0.58	−5.58	−14.78	0.00	24.14	28.75	44.91	0.00	5.12	−4.43
Chile	15.35	15.54	19.34	15.43	0.00	−12.47	29.20	0.00	40.62	28.98	−10.32	9.60	−11.78
China	0.00	0.00	0.00	0.00	−4.98	−14.65	−1.17	−8.59	0.19	0.00	−19.83	0.29	−14.26
Côte d'Ivoire	−28.37	−28.31	−28.33	−28.20	−58.13	0.00	−84.45	7.61	−50.24	−29.13	4.02	−15.52	−10.78
Egypt	4.21	4.11	4.23	3.99	0.00	−28.11	7.81	−25.00	17.61	0.00	−15.56	−1.39	−26.61
Ghana	10.81	10.82	10.81	10.81	−32.78	−27.95	81.66	−7.62	23.13	50.60	−27.78	−3.20	5.23
Israel	0.00	0.00	0.00	0.00	4.15	−8.02	−1.39	−18.86	−0.28	0.08	−25.35	−20.79	−3.12
Japan	0.17	0.17	0.17	0.17	0.73	−19.81	0.05	−22.28	1.09	0.93	−30.01	0.57	−17.20
Jordan	4.52	4.54	4.51	4.56	−16.04	−14.72	2.18	−10.14	−38.17	−8.90	−0.18	−17.47	−9.07
Malawi	0.00	0.00	0.00	0.00	2.52	2.28	0.00	0.00	0.00	−9.14	−30.39	−5.71	12.30
Malaysia	3.44	3.31	−0.44	3.30	−5.84	1.20	35.96	0.00	44.67	44.99	−13.86	6.93	5.26
Mozambique	24.99	24.26	36.65	23.61	21.10	−14.61	42.29	70.63	22.81	41.28	−16.59	−11.50	5.84
North Macedonia	0.00	0.00	0.00	0.00	−3.52	−15.69	1.61	5.80	0.00	−5.64	−9.21	−45.92	−23.43
Norway	0.81	0.80	0.82	0.80	−0.24	−9.98	−9.35	−24.52	0.83	−5.38	−0.80	−10.97	−15.30
Philippines	1.81	1.81	1.77	1.76	−0.57	−13.40	0.00	0.00	5.81	0.72	−13.02	0.08	−17.91
Russian Federation	2.87	2.86	2.78	2.85	−13.25	−42.12	−51.28	0.00	−9.83	36.72	0.00	0.00	−29.40
Saudi Arabia	52.73	52.73	52.77	52.73	42.67	−2.56	51.27	8.59	57.00	0.56	−1.64	8.90	0.00
Sri Lanka	−1.66	−1.75	−1.86	−1.87	0.00	0.00	−44.77	0.00	−50.09	−42.28	2.03	−9.73	−1.22

(continued)

Table 3.8 (continued)

DMU	Pm land	Pm LSU	Pm energy	PmGHG	Sr Land	Sr LSU	Sr Employment	Sr Energy	Sr Nitrogen	Sr Pesticides	Sr Production	Sr Dietary	Sr GHG
Switzerland	1.16	1.16	1.18	1.16	2.36	-19.32	0.42	-0.40	4.18	-0.09	-21.83	-1.49	-20.29
Thailand	-5.78	-5.84	-5.76	-5.83	0.00	0.00	4.13	0.00	10.73	-21.40	0.00	9.80	12.98
Turkey	-1.01	-1.02	-1.16	-1.03	-31.13	-12.93	-2.51	-21.52	-1.82	-6.96	-0.07	-1.71	-12.17
Ukraine	-20.35	-20.27	-20.04	-20.36	36.96	0.00	-9.21	0.00	10.06	-8.51	0.00	14.14	13.63
USA	0.05	0.06	0.04	0.05	-30.45	-27.68	-1.02	-26.70	-3.29	-0.30	-28.93	0.17	-30.04
Viet Nam	0.00	0.00	0.00	0.00	0.00	-14.70	0.00	0.00	0.00	0.00	-1.54	0.01	-8.23
Average	3.21	3.19	3.61	3.14	-5.03	-14.19	4.83	-4.97	5.78	4.32	-10.72	-4.27	-10.25

Table 3.9 The change in relative proportionate movement (Pm) and slack (Sr) values between 2005–2007 and 2016–2018 sub-periods in Cluster 2 (in pp.)

DMU	Pm land	Pm LSU	Pm energy	Pm GHG	Sr Land	Sr LSU	Sr Employment	Sr Energy	Sr Nitrogen	Sr Pesticides	Sr Production	Sr Dietary	Sr GHG
Australia	-5.23	-5.23	-5.50	-5.49	-31.39	-22.53	-0.91	-14.88	-1.91	-2.64	-13.53	0.45	-21.34
Azerbaijan	31.81	31.83	31.75	31.82	2.84	-14.38	48.85	-13.26	13.86	48.89	-24.10	-2.44	-10.39
Bangladesh	-5.49	-5.47	-5.49	-5.48	0.00	-45.63	-68.19	5.40	-54.68	-94.81	1.49	-12.53	-33.50
Belarus	-2.67	-2.71	-2.89	-2.69	-9.40	-24.97	-8.17	-5.85	-2.45	-8.18	0.00	-2.71	-1.25
Bolivia	0.00	0.00	0.00	0.00	-35.62	-24.97	-1.87	3.02	0.00	0.00	-16.90	0.85	-27.48
Brazil	0.00	0.00	0.00	0.00	-22.43	-23.98	0.71	-2.03	0.00	0.00	-16.71	-0.36	-25.63
Brunei Darussalam	0.00	0.00	0.00	0.00	0.22	-33.15	14.05	-35.45	-17.26	4.09	-34.93	-34.92	-30.40
Burkina Faso	64.18	63.99	61.95	64.18	0.00	-22.87	44.86	0.00	-45.21	33.56	-6.21	-0.41	-27.87
Cameroon	0.00	0.00	0.00	0.00	-2.05	-3.36	0.00	0.00	0.00	0.00	-21.18	0.23	-6.35
Colombia	30.43	30.36	30.28	30.28	-9.23	-6.69	-60.57	0.00	-10.21	-27.94	0.00	1.26	-10.53
Costa Rica	0.00	0.00	0.00	0.00	-12.48	-0.95	-1.08	5.48	0.00	0.00	-16.93	1.11	2.84
Ecuador	17.53	17.36	17.66	17.44	0.00	-1.03	55.35	63.40	32.18	36.54	0.00	-10.52	-10.44
El Salvador	-11.05	-11.20	-11.03	-11.13	4.69	-17.97	5.25	0.00	-14.77	90.08	0.00	0.89	2.99
Ethiopia	69.65	69.67	69.68	69.64	-9.11	0.16	81.44	0.00	1.73	85.95	-7.16	11.97	-13.92
Georgia	-14.44	-13.84	-14.82	-13.74	-16.51	0.00	-23.83	-34.33	38.84	63.57	0.00	9.72	0.05
Guyana	-19.78	-20.00	-20.04	-19.91	-7.95	0.00	-19.17	-9.83	-16.97	-8.17	0.00	1.90	9.32
Honduras	14.48	14.52	15.32	14.60	13.95	-11.10	44.03	0.00	12.11	43.17	0.00	-5.32	-10.48
Iceland	0.00	0.00	0.00	0.00	-28.66	-8.62	-0.87	-34.81	2.85	0.00	-7.09	1.14	-20.50
India	0.00	0.00	0.00	0.00	-25.74	-33.59	0.00	-30.61	0.00	0.00	-24.82	0.07	-31.49
Indonesia	0.00	0.00	0.00	0.00	-28.26	-8.72	1.97	-3.68	0.00	0.00	-27.04	0.66	-26.09
Iran	3.34	3.34	3.40	3.27	-22.50	-14.73	3.10	-24.92	3.03	3.10	-29.95	0.71	-6.74
Iraq	33.38	33.45	35.03	33.46	-23.17	4.60	3.23	18.46	52.72	20.64	0.00	8.99	0.00
Kazakhstan	3.71	3.71	3.70	4.21	-29.44	6.90	0.00	-20.00	4.72	8.87	-15.77	-7.92	-2.76
Kenya	24.85	24.64	25.17	24.62	-13.04	-3.99	32.48	-4.45	0.00	6.98	-11.15	15.82	-6.79

(continued)

Table 3.9 (continued)

DMU	Pm land	Pm LSU	Pm energy	Pm GHG	Sr Land	Sr LSU	Sr Employment	Sr Energy	Sr Nitrogen	Sr Pesticides	Sr Production	Sr Dietary	Sr GHG
Kyrgyzstan	0.00	0.00	0.00	0.00	-33.03	10.55	4.26	-24.98	0.00	-8.04	13.12	-6.07	22.34
Madagascar	21.28	21.28	22.19	20.98	0.99	0.00	72.86	0.00	69.79	61.02	-10.76	25.12	8.19
Mexico	42.69	42.69	42.77	42.72	1.97	-3.51	0.00	-8.61	23.82	81.96	0.00	-2.02	-0.86
Mongolia	27.26	27.22	28.48	26.93	-26.68	-21.92	7.75	0.00	3.00	-0.65	0.00	-26.11	-23.56
Morocco	-0.62	-0.65	-2.05	-0.19	-21.90	-2.06	5.61	1.20	-6.10	21.91	0.00	-0.39	8.08
Namibia	-0.29	-0.09	-0.39	-0.39	1.72	0.00	40.35	24.73	48.75	23.80	0.00	-5.28	4.64
Nepal	1.08	1.10	1.19	1.10	-5.25	-25.98	11.16	31.12	-3.36	14.31	-33.45	1.38	-28.57
New Zealand	0.00	0.00	0.00	0.00	4.28	-23.05	0.00	-14.93	-3.61	0.00	-20.21	-0.06	-24.84
Nicaragua	32.72	32.71	32.68	32.71	-1.00	11.07	0.00	0.00	67.51	90.48	-28.77	0.00	9.89
Niger	0.00	0.00	0.00	0.00	-32.18	-23.95	0.00	13.18	-2.82	-3.68	-19.39	0.34	-21.15
Pakistan	-8.61	-8.92	-8.98	-8.82	-2.26	-54.39	-9.42	6.57	-41.63	-20.73	6.15	-3.20	-37.82
Paraguay	31.35	30.99	31.19	30.67	28.44	-11.66	0.00	0.00	0.00	12.67	-2.45	3.48	-9.58
Peru	14.50	14.57	13.96	14.60	-18.70	-23.82	12.53	0.00	-2.92	8.00	0.00	-8.65	-10.70
Senegal	7.40	7.21	7.50	7.39	-34.16	0.00	37.38	0.00	-24.68	6.66	0.00	-7.04	-2.48
South Africa	22.49	22.49	22.48	22.39	-18.22	-18.29	-13.39	0.00	-1.03	17.71	0.00	5.33	6.98
Tajikistan	-15.02	-15.48	-14.69	-15.58	-33.83	4.50	-26.68	-63.33	-13.60	-52.12	3.11	-28.33	-1.34
Tunisia	0.00	0.00	0.00	0.00	-11.56	-3.58	0.00	9.78	0.00	0.00	-0.06	-3.94	2.13
Uruguay	-12.01	-12.22	-12.09	-12.19	-48.50	-41.18	0.13	-4.22	17.59	-4.82	0.00	-1.28	-37.37
Yemen	0.00	0.00	0.00	0.00	-14.55	-22.98	0.00	-45.28	0.00	0.00	-22.60	0.00	-13.54
Zambia	24.67	24.44	22.70	25.27	-10.10	0.00	-1.71	0.00	-30.80	23.86	0.00	-8.96	-31.55
Average	9.63	9.58	9.57	9.61	-13.18	-12.20	6.62	-4.84	2.24	13.09	-8.80	-1.98	-11.13

Table 3.10 The change in relative proportionate movement (Pm) slack (Sr) values between 2005–2007 and 2016–2018 sub-periods in Cluster 3 (EU) (in pp.)

DMU	Pm land	Pm LSU	Pm energy	Pm nutrient	Pm production	Pm GHG	Sr Land	Sr LSU	Sr Employment	Sr Energy	Sr Nitrogen	Sr Pesticides	Sr Production	Sr Dietary	Sr GHG
Austria	4.3	4.4	4.3	4.4	4.3	4.3	-0.1	-12.4	7.3	0.0	3.7	4.4	0.0	-3.0	-8.1
Belgium	-0.5	-0.5	-0.5	-0.5	-0.5	-0.5	2.8	-22.0	-44.2	9.8	-12.5	-7.4	0.0	-6.6	-17.8
Bulgaria	-10.5	-10.6	-10.4	-11.0	-10.1	-10.6	5.7	0.0	-15.1	0.0	22.7	21.2	0.0	7.5	19.8
Croatia	0.0	0.0	0.0	0.0	0.0	0.0	-31.6	-27.5	6.1	-13.6	-31.7	-63.2	0.0	-7.3	-32.0
Cyprus	-3.7	-3.8	-3.7	-4.7	-3.7	-3.9	34.3	0.0	0.0	11.9	-8.7	33.3	0.0	-4.0	-9.5
Czechia	-33.3	-31.2	-31.1	-33.5	-32.4	-31.4	9.0	-37.0	0.3	-45.7	9.6	2.4	4.8	-2.4	-14.9
Denmark	0.7	0.7	0.7	0.7	0.7	0.7	-8.8	-3.5	-16.5	-2.3	5.0	-46.0	0.0	-0.1	0.7
Estonia	-2.8	-2.8	-2.9	-2.7	-2.8	-2.8	3.9	0.0	-12.2	15.2	31.9	12.0	0.0	0.0	2.6
Finland	2.6	2.5	2.5	2.6	2.5	2.6	7.4	0.0	-1.1	4.6	-9.3	32.9	0.0	0.0	-1.2
France	-6.6	-6.5	-6.8	-6.3	-6.8	-6.5	-19.5	-22.7	0.2	-13.0	-22.6	2.4	0.0	0.1	-21.9
Germany	1.7	1.6	1.7	1.6	1.6	1.6	0.0	1.1	0.8	-2.8	-14.4	-9.0	0.0	1.5	-4.8
Greece	1.1	0.9	0.9	1.3	0.3	0.9	-29.9	-10.1	6.0	-20.4	8.6	-25.8	0.0	-8.5	-10.9
Hungary	5.0	5.0	5.1	4.8	5.1	5.0	-13.8	0.0	-6.3	31.2	24.0	15.2	0.0	0.0	14.5
Ireland	-0.2	-0.2	-0.2	-0.2	-0.2	-0.2	-7.8	-12.0	6.5	0.0	-3.8	0.0	0.0	2.4	-13.3
Italy	-8.8	-8.6	-8.8	-8.5	-8.8	-8.6	-4.0	-4.0	13.4	-8.9	6.2	-2.2	0.0	3.3	-3.0
Latvia	-2.1	-2.0	-2.1	-2.0	-1.9	-2.0	3.6	0.0	-32.2	10.0	22.3	10.7	0.0	0.0	6.7
Lithuania	-0.5	-0.1	-0.4	-0.4	0.2	-0.2	7.2	0.0	10.8	0.0	15.5	22.5	0.0	0.0	7.0
Luxembourg	0.0	0.0	0.0	0.0	0.0	0.0	-8.0	-7.3	7.0	0.6	-13.9	0.0	-1.2	-26.8	-11.1
Netherlands	-5.7	-5.6	-6.0	-5.7	-5.7	-5.6	0.0	-28.2	2.4	-29.6	-21.2	-6.7	0.0	0.1	-27.2
Poland	3.7	3.7	3.9	3.6	3.7	3.7	-2.9	0.0	1.0	-6.8	26.4	0.0	0.0	-3.9	7.2
Portugal	16.0	15.9	16.1	15.1	15.9	15.9	-16.3	-4.0	-7.7	-11.3	0.0	14.4	0.0	-0.4	-4.8
Romania	-15.5	-15.0	-14.1	-16.4	-16.3	-15.4	-17.0	-19.7	17.6	0.0	9.6	13.6	0.0	-7.0	-15.7
Slovakia	1.5	1.5	1.8	1.2	1.5	1.5	6.2	0.0	0.0	16.0	14.6	-2.5	0.0	3.8	9.8
Slovenia	-0.9	-0.9	-0.9	-0.9	-0.9	-0.9	15.8	0.0	-14.5	5.2	-5.7	-1.7	0.0	0.0	0.4

(continued)

Table 3.10 (continued)

DMU	Pm land	Pm LSU	Pm energy	Pm nutrient	Pm production	Pm GHG	Sr Land	Sr LSU	Sr Employment	Sr Energy	Sr Nitrogen	Sr Pesticides	Sr Production	Sr Dietary	Sr GHG
Spain	3.1	3.2	3.0	3.2	3.0	3.1	7.5	−2.5	1.5	−11.4	28.7	2.5	0.0	4.0	−11.0
Sweden	5.8	5.7	5.6	5.6	5.5	5.7	0.7	0.0	0.0	−3.3	3.0	−2.4	0.0	−1.0	1.9
United Kingdom	−8.5	−8.8	−9.2	−9.0	−8.7	−8.8	−25.2	−24.6	0.0	0.0	−29.4	−6.4	0.0	0.4	−26.7
Average	−2.0	−1.9	−1.9	−2.1	−2.0	−1.9	−3.0	−8.8	−2.5	−2.4	2.2	0.5	0.1	−1.8	−6.0

land and LSU, the average changes in the slacks were slightly negative (–3 and –8.8 pp.), but, in contrast to Clusters 1 and 2, changes regarding these variables differ a lot between countries. There were countries such as Croatia, Greece or the UK which reduced the slack on land by more than 20 pp. but for 13 countries this slack remained unchanged or even increased. Similarly, the slack change on LSU was positive or neutral for 12 Member States but it greatly decreased for other countries. In some cases it was more than 20 pp.

The average slack change regarding employment was –2.5 pp., which makes this cluster different from the others. However, the negative average is inflated by countries such as Belgium or Latvia where this slack was reduced by more than 30 pp. Similarly to other clusters, the average slack on nitrogen and pesticides increased. However, in both cases there are countries where a significant growth or decrease in slack value was recorded. For example, the slack on nitrogen was reduced by more than 20 pp. in Croatia, France, the Netherlands and in the UK, and increased by more than 20 pp. in other countries (Bulgaria, Hungary, Latvia and Poland). One needs to note that high increases in the slack values are often recorded in the so called 'new' EU Member States. This is a very problematic issue since these countries are trying to narrow the income and productivity gap with the old EU countries but this is not beneficial from the integrated efficiency point of view. This phenomenon confirms the old hyphothesis of Cochrane. The 'old' EU countries can be viewed as "early adopters" who were the first to introduce new technologies and they can benefit from lower costs and higher income. In contrast, farms in the 'new' EU members are not that innovative and they fall into a vicious cycle—they are investing but it does not translate into sufficient income growth. To overcome income problems, they invest even more but this lowers their efficiency.

To sum up this part, we may notice that, on the global scale, the adequacy of basic input use (such as land or LSU) is improving. Another positive trend that we may observe is that the slack on GHG emissions are decreasing, which gives a hope that in the future a significant reduction in emissions will be possible. A problematic issue is still the use of nitrogen and pesticides. There are many countries that managed to diminish slacks on these variables but, on the other hand, many countries still have large room for improvement in a more reasonable use of these inputs.

A comparison of the calculations above with the estimation using international dollars (see Tables A.5, A.6, and A.7 in Appendix) does not contribute a lot to this discussion. For Cluster 1 and 2, changes in slack are slightly in a more desired direction, i.e. if some slacks are decreasing, then these changes are usually bigger when using international dollars. When it comes to cluster 3 (the EU), however, the more desired changes are observed when constant dollars are used. Nevertheless, we need to remember that in this case the slacks are generally higher so it can be easier to decrease them.

3.3.5   Global Gains from the Integrated Efficiency Improvement

In this sub-chapter we aim to see to what extent improving integrated efficiency levels would contribute to reducing emissions and saving resources from the global perspective. On a world scale, agriculture is a sector which consumes lots of resources and is responsible for a significant amount of GHG emission. At present, progress in emission reduction in agriculture is not sufficient. As an example, we may refer to the data provided by the European Environment Agency. According to this organisation, in the 2005–2019 time period the agricultural emission of GHG in EU27 decreased by –1.8% (see Table A.8 in Appendix). Obviously, the situation differs between member states. For example, this reduction in Croatia, Malta and Greece was –17.7%, –16.6% and 11.9%, respectively. In contrast, in some member states the emission even increased. In Latvia it was 22.8%, Bulgaria 24.7% and in Estonia even 27%. Projections towards 2030 show that GHG emissions will be reduced on the EU level only by 2%, in comparison to the situation in 2005. If additional policy measures will be implemented, then this reduction could be –5%, which is still far from the ambitious goals of the EU.

In this book we claim that significant reductions in GHG emissions and savings in input use may be achieved by eliminating inefficiencies in the operations of the agricultural sector. In other words, elimination of slacks within existing technology could lead to lower emission and resources use on a global scale. In Table 3.11 we present calculations for our whole dataset (99 countries). In the first row of the Table we present the total value of inputs and outputs (2005–2018 averages), and in the second row we present the total value of slacks. Since we use the super-efficiency approach and slacks may be positive for inputs (and bad outputs) and negative for good outputs, in the fourth row we present also the total value of slacks when only negative slacks for bad output and inputs and positive slacks for good outputs are considered.

It turns out that if there were no inefficiencies in the agricultural sectors of the countries under analysis, GHG emissions would be 14.6% lower in the 2005–2018 period. Furthermore, a clear saving of inputs could also be achieved. The use of land could be lower by 22.9%, and livestock use could be lower by 12.7%. Less people could be engaged in agricultural activity (between 7.8 and 8.3%) and they could move to work in other sectors. Energy use could be lower by 11.7%, while nitrogen and pesticides use could decrease by 9.1 and 7.4%, respectively. At the same time, despite the lower use of inputs, it would be possible to achieve a higher level of dietary energy supply (1.8 up to even 6.1%).

Table 3.11 Potential saving of inputs and undesirable outputs in the integrated efficiency framework (2005–2018)

	Land (bln ha)	LSU (bln)	Employment (mln)	Energy (EJ)	Nitrogen (Mt)	Pesticides (Mt)	Production (bln USD)	Dietary*	GHG (Gt CO2e)
Initial values (physical units)	3.997	1.530	808.86	8.281	98.88	3.877	3451	123.8*	4.91
Total sum of slack	–0.914	–0.194	–63.12	–0.966	–8.99	–0.288	–239	2.178	–0.72
Percentage change	–22.9	–12.7	–7.8	–11.7	–9.1	–7.4	–6.9	1.8	–14.6
Only negative slacks for input and positive for good output considered	–0.914	–0.194	–67.17	–0.966	–9.08	–0.294	0.126	7.512	–0.72
Percentage change	–22.9	–12.7	–8.3	–11.7	–9.2	–7.6	~0.0	6.1	–14.6

*The average percentage values

3.3.6   The Evolution of Integrated Efficiency Change

In previous parts of this book we focused on slacks and their changes in our research period. Below we focus on changes of integrated efficiency as a comprehensive indicator of sustainability of the agricultural sector. We will study integrated efficiency change with the help of the Malmquist index (noted as MI in Tables 3.12, 3.13, and 3.14). Furthermore we aim to answer what is the role of efficiency change (the improvement in operation in agriculture of a given country) and technological progress.

Table 3.12   Integrated efficiency change and its components (2006–2018 average), Cluster 1

DMU	MI	EC	TC	OBTC	IBTC	MATC
Albania	1.037	0.998	1.040	0.971	1.008	1.062
Algeria	0.974	0.954	1.021	0.971	1.009	1.042
Argentina	1.036	0.989	1.047	0.953	1.003	1.095
Armenia	0.970	0.942	1.030	0.994	1.007	1.029
Canada	0.995	0.975	1.021	1.000	0.995	1.026
Chile	0.971	0.951	1.021	0.992	0.996	1.033
China	1.008	0.983	1.025	0.978	0.997	1.051
Côte d'Ivoire	1.084	1.069	1.014	0.997	0.998	1.019
Egypt	0.992	0.967	1.026	0.990	0.987	1.050
Ghana	0.948	0.931	1.018	0.989	1.015	1.013
Israel	0.979	0.955	1.025	0.993	0.982	1.051
Japan	0.992	0.974	1.019	0.992	0.990	1.037
Jordan	1.002	0.985	1.017	0.982	1.003	1.033
Malawi	0.998	0.974	1.025	0.967	1.009	1.051
Malaysia	0.967	0.946	1.023	0.998	0.993	1.033
Mozambique	0.982	0.959	1.024	0.998	0.986	1.040
North Macedonia	1.003	0.977	1.027	0.970	1.005	1.054
Norway	0.998	0.981	1.016	0.989	0.995	1.032
Philippines	1.003	0.989	1.014	0.987	1.000	1.028
Russian Federation	1.049	0.999	1.051	1.004	0.997	1.049
Saudi Arabia	0.960	0.944	1.017	0.994	0.999	1.024
Sri Lanka	1.046	1.023	1.023	0.999	1.001	1.022
Switzerland	0.999	0.982	1.017	0.993	0.991	1.033
Thailand	1.027	1.004	1.023	0.996	1.002	1.025
Turkey	1.012	0.997	1.015	0.980	1.007	1.029
Ukraine	1.053	1.029	1.023	1.000	1.000	1.023
USA	0.993	0.973	1.020	0.991	0.990	1.040
Viet Nam	1.032	1.001	1.031	0.968	1.003	1.063
Average	1.004	0.980	1.024	0.987	0.999	1.039

Note: MI, Malmquist index (sequential technology), *EC* efficiency change, *TC* technical change, *OBTC* output-based technological change, *IBTC* input-based technological change, *MATC* mixed technological change

Table 3.13 Integrated efficiency change and its components (2006–2018 average), Cluster 2

DMU	MI	EC	TC	OBTC	IBTC	MATC
Australia	1.001	0.980	1.021	0.988	0.993	1.041
Azerbaijan	0.969	0.947	1.022	0.991	1.001	1.031
Bangladesh	1.155	1.041	1.109	0.971	0.998	1.145
Belarus	1.022	1.001	1.021	1.000	0.999	1.022
Bolivia	1.044	0.996	1.048	0.958	1.001	1.093
Brazil	1.001	0.984	1.017	0.971	1.013	1.034
Brunei Darussalam	0.940	0.917	1.025	0.963	1.014	1.050
Burkina Faso	0.942	0.905	1.040	0.997	1.009	1.034
Cameroon	1.013	0.987	1.026	0.966	1.010	1.053
Colombia	1.057	1.003	1.054	1.002	0.994	1.058
Costa Rica	1.017	0.986	1.032	0.975	0.994	1.064
Ecuador	1.015	0.963	1.054	1.003	1.004	1.047
El Salvador	1.018	0.992	1.027	1.000	0.999	1.028
Ethiopia	0.937	0.900	1.041	0.964	1.014	1.065
Georgia	0.995	0.978	1.017	0.999	0.996	1.022
Guyana	1.060	1.052	1.007	1.002	0.999	1.006
Honduras	0.977	0.942	1.038	1.001	0.999	1.037
Iceland	1.016	0.989	1.027	0.969	1.006	1.053
India	1.024	0.985	1.040	0.958	1.004	1.081
Indonesia	1.013	0.976	1.038	0.958	1.006	1.077
Iran	0.997	0.971	1.027	0.973	1.001	1.055
Iraq	0.954	0.932	1.024	1.001	0.997	1.026
Kazakhstan	1.011	0.988	1.024	0.978	1.009	1.038
Kenya	1.005	0.975	1.031	0.985	0.995	1.052
Kyrgyzstan	1.063	1.002	1.061	0.984	1.003	1.075
Madagascar	0.965	0.935	1.032	0.996	0.994	1.041
Mexico	0.984	0.952	1.034	0.999	1.002	1.034
Mongolia	1.035	0.967	1.070	1.005	0.998	1.067
Morocco	1.085	0.975	1.113	1.011	0.994	1.107
Namibia	0.955	0.919	1.038	1.000	1.004	1.035
Nepal	0.987	0.963	1.024	0.980	1.003	1.042
New Zealand	1.002	0.982	1.021	0.984	0.995	1.043
Nicaragua	0.974	0.940	1.036	0.992	0.997	1.047
Niger	1.023	0.979	1.045	0.953	1.003	1.093
Pakistan	1.141	1.003	1.137	1.005	0.968	1.169
Paraguay	1.026	0.962	1.067	0.990	0.990	1.089
Peru	1.095	0.994	1.101	1.005	1.003	1.093
Senegal	1.054	1.009	1.044	1.003	0.992	1.050
South Africa	1.002	0.976	1.026	1.001	0.999	1.026
Tajikistan	1.045	1.018	1.026	0.998	0.999	1.030
Tunisia	1.017	0.998	1.019	0.979	1.003	1.038
Uruguay	1.029	0.978	1.052	0.988	0.996	1.070
Yemen	1.020	0.989	1.031	0.948	1.025	1.061
Zambia	1.032	0.982	1.051	1.002	0.998	1.051
Average	1.016	0.975	1.042	0.986	1.000	1.056

Table 3.14 Integrated efficiency change and its components (2006–2018 average), Cluster 3

DMU	MI	EC	TC	OBTC	IBTC	MATC
Austria	1.002	0.999	1.003	0.996	1.002	1.004
Belgium	1.004	0.982	1.022	0.982	0.998	1.043
Bulgaria	1.035	1.005	1.031	0.999	0.997	1.034
Croatia	1.000	0.980	1.021	0.986	0.994	1.041
Cyprus	1.012	0.999	1.014	1.000	1.003	1.011
Czechia	0.968	0.899	1.077	0.966	0.961	1.161
Denmark	0.992	0.972	1.020	0.988	0.995	1.037
Estonia	1.017	1.003	1.014	0.998	1.005	1.012
Finland	1.011	0.988	1.024	1.000	0.999	1.025
France	0.991	0.978	1.014	0.993	0.994	1.027
Germany	0.998	0.990	1.008	0.996	0.997	1.015
Greece	0.992	0.959	1.034	0.958	1.010	1.069
Hungary	1.007	0.989	1.018	1.000	1.002	1.016
Ireland	1.021	0.999	1.022	0.984	0.998	1.041
Italy	0.990	0.967	1.023	0.989	0.989	1.046
Latvia	1.018	1.007	1.012	1.003	1.000	1.009
Lithuania	1.035	0.995	1.040	1.000	1.002	1.038
Luxembourg	0.998	0.979	1.020	0.991	0.990	1.040
Netherlands	1.012	0.983	1.030	0.980	0.991	1.060
Poland	1.006	0.988	1.018	1.000	0.999	1.019
Portugal	0.984	0.961	1.024	1.002	1.001	1.022
Romania	0.980	0.955	1.026	0.991	0.988	1.048
Slovakia	1.046	1.003	1.044	1.001	0.997	1.046
Slovenia	1.011	1.003	1.008	1.000	1.001	1.007
Spain	1.025	1.006	1.019	1.002	0.992	1.025
Sweden	1.007	0.986	1.021	1.000	1.000	1.021
United Kingdom	0.997	0.972	1.026	0.989	0.986	1.052
Average	1.006	0.983	1.023	0.992	0.996	1.036

The general trend in MI evolution is similar across the Clusters (see Tables 3.12, 3.13, and 3.14 and Figs. 3.12, 3.13, and 3.14). The average MI change in all clusters is positive—in Cluster 1 and 3 it is rather negligible—0.4 and 0.6% p.a., while in Cluster 2 it is more apparent and equal to 1.6% p.a. Our results support the findings obtained by many authors (including Hoang and Coelli (2011) and Eberhardt and Teal (2012)) that progress in integrated efficiency in agriculture results mainly from the technological change, while efficiency change is often even negative. Our results have shown that average change in efficiency was –2%, –2.5%, and –1.7% p.a. for Clusters 1, 2 and 3, respectively. The higher pace of growth in integrated efficiency in Cluster 2 was the result of significant improvement in technical progress—the average TC growth was 4.2% p.a., while in Clusters 1 and 3 this growth was smaller—2.4 and 2.3%, respectively.

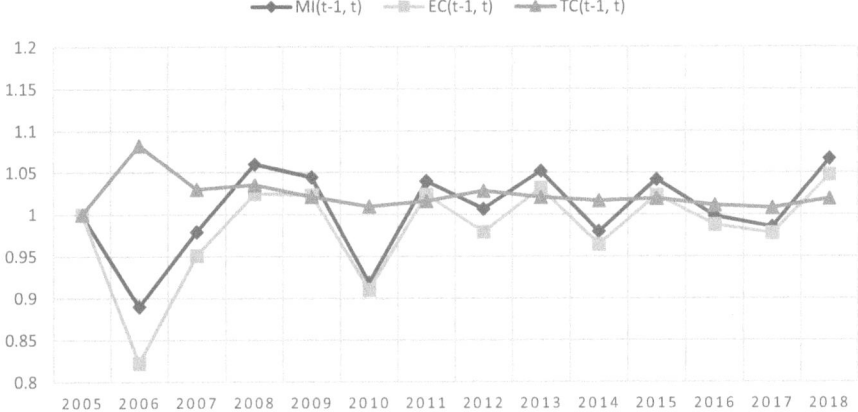

Fig. 3.12 Evolution of Malmquist Index and its component—cluster 1

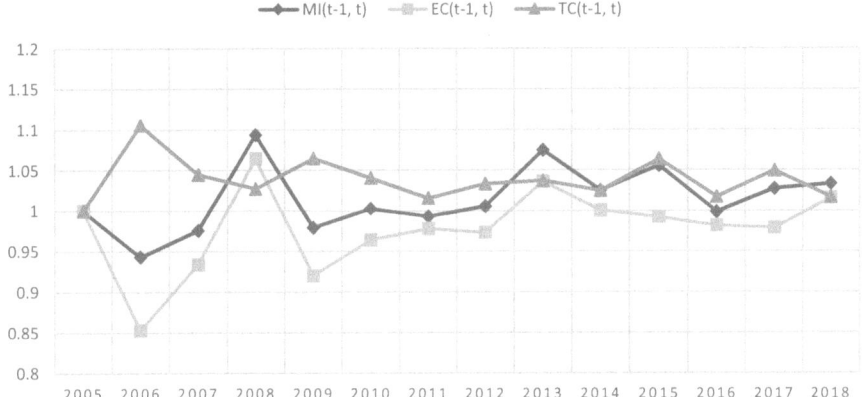

Fig. 3.13 Evolution of Malmquist Index and its component—cluster 2

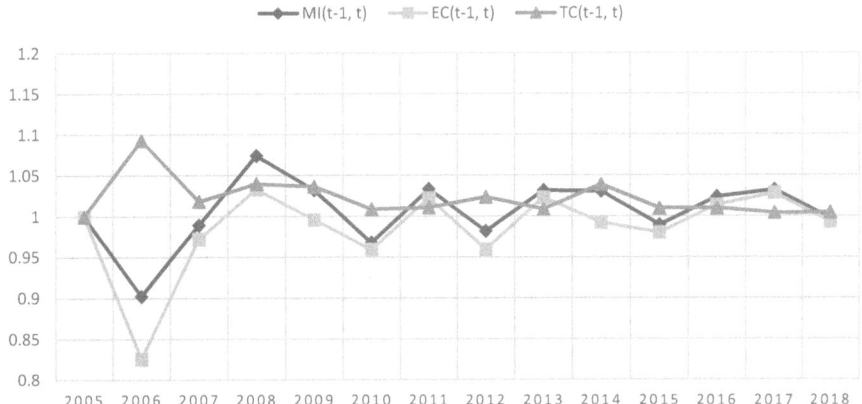

Fig. 3.14 Evolution of Malmquist Index and its component—cluster 3

Regarding evolution of the MI index in time, we may notice that the MI evolves in a similar way to the EC. In 2006, these indices decreased significantly in all the clusters. After this time, a sharp growth was observed in 2008. In 2009–2010, the MI index declined and fluctuated in subsequent periods around the value of 1. Interestingly, the TC index behaved differently. It grew significantly in 2006 and then declined in 2007. For the rest of the time period it was at a relatively stable level and clearly above 1.

Obviously, the situation regarding the evolution of integrated efficiency was different across countries in a given cluster. In Cluster 1, a particularly high pace of MI growth was observed in countries such as Côte d'Ivoire (8.4% p.a.), Sri Lanka (4.6% p.a.) and Ukraine (5.3% p.a.). The characteristic feature of these countries was the fact that a high level of MI growth resulted not only from positive TC change but also from significant improvements in efficiency, which contrasted with the overall tendency in the cluster. In contrast, in countries where the MI was decreasing (such as Ghana, 5.2% p.a.), a major cause of this was not a problem with the availability of technological progress but rather deteriorating efficiency.

Among countries that improved their integrated efficiency to the largest extent in cluster 2, one may find Bangladesh, Peru, Morocco, or Kyrgyzstan. In contrast to the leaders in Cluster 1, these countries were improving their integrated efficiency almost exclusively due to technical progress. Only in Bangladesh could one observe a significant EC growth—4.1% p.a., but the improvement of TC was 10.9%. In Kyrgyzstan the EC was 0.2% p.a., while in Morocco and Peru the efficiency change was negative— –2.5% and –0.6% p.a., respectively. Countries where the integrated efficiency change was particularly negative, such as Ethiopia, Burkina Faso or Brunei Darussalam were characterised by a significant negative change in efficiency, ca. 10% p.a. The TC in these countries were positive. However, we should recall that we use the sequential MI index, which means that the TC cannot be negative and all integrated efficiency drops are associated with the EC component.

In the EU countries changes in MI are generally smaller (MI is closer to 1). However, it can be observed that the highest pace of MI growth was recorded among the 'new' EU members, such as Bulgaria (3.5%) p.a., Lithuania (3.5%) or Slovakia (4.6%). Positive change in MI was mainly due to TC. The worst performing country was Czechia—the average MI change was –3.2% p.a. Interestingly, the TC change was highly positive (7.7% p.a.) but the EC was –10.1% p.a. Overall, it is a good sign that the average MI growth among 'new' EU members is higher than the EU average (1.1% p.a. vs 0.6% p.a.), but we need to remember that these countries still experience relatively large slacks and they need to improve further, in particular with regard to EC.

If we switched to calculations in which production is expressed in international dollars (see Tables A.9, A.10, and A.11 and Figs. A.1, A.2, and A.3 in Appendix), we can observe that for cluster 1 the average values of the MI, EC and TC changes for the whole cluster are identical to the constant dollars. However, there are some differences for particular countries and in a few cases (e.g. Canada, Japan, or Switzerland) the average change has the opposite direction. However, in these cases, the changes are usually around 0 with the exception of Malaysia for which the average MI change is 2.8% p.a. (international dollars) and −3.3% (constant dollars). When it comes to clusters 2 and 3, a slower pace of MI growth was observed when international dollars are used (1% p.a. for cluster 2 and 0.1% p.a. for cluster 3). This is mostly because the lower average values of the TC. These results may be caused by the fact that use of international dollars reduces the differences between countries to some extent.

Chapter 4
Approaching Agri-Environmental Policy Effectiveness

4.1 Impact of Agriculture on the Environment: In the Quest for Global Mitigation Strategy

Agriculture is a sector that contributes significantly to climate change, although it is difficult to accurately estimate the magnitude of this contribution and make precise projections. Globally, agriculture is thought to have been responsible for 9.3 trillion tonnes of carbon dioxide equivalent—9.3 Gt $COeq_2$.(including related Land-Use-Change), in 2018 (FAOSTAT, 2020). Total emissions from the sector have fallen, however, from 9.6 Gt in 2000. At the same time emissions from other sectors have increased, so it has translated into a decline in agriculture's share of global emissions from 24 to 17%. In the EU, this share is lower—the overall portion of agriculture in total GHG emissions in Europe is about 10.1% (EEA, 2021). Despite this decrease in emissions, agriculture remains one of the major emitters of GHG. Furthermore, it is important to note that while emissions from activities related to land use have decreased, emissions related to crop and livestock activities have increased by 14% between 2000 and 2018.

Agriculture, in particular, is a major source of methane (CH_4) and nitrous oxide (N_2O) emissions—49% and 66% of global emissions, respectively. In contrast, only 15% of carbon dioxide (CO_2) comes from anthropogenic management in agriculture. CH_4 emissions from enteric fermentation and N_2O emissions from agricultural soils stands for about 80% of total agricultural GHG emissions, of which CH_4 emissions from manure management has a share of about 10%. CH_4 and N_2O have very high conversion factors to CO_2 equivalent (21 and 310, respectively). Therefore, emission reductions can be achieved by improving the efficiency of resource management in agriculture, which would boil down to reducing energy intensity, fertilizers use and intensity of agricultural production.

Despite the fact that scientists from various disciplines are aware of the large share of agriculture in GHG emissions, the essence of this problem is not always

© The Author(s), under exclusive license to Springer Nature Switzerland AG 2022
B. Czyżewski, Łu. Kryszak, *Sustainable Agriculture Policies for Human Well-Being*,
Human Well-Being Research and Policy Making,
https://doi.org/10.1007/978-3-031-09796-6_4

properly understood. Taking into account the systematic growth of the world population and the accompanying increase of food needs, it should be assumed that emissions from agriculture have an increasing potential. Therefore, it will be difficult to reduce the already high share of GHG emissions from agriculture and forestry.

Many reports, including the last one of the Intergovernmental Panel on Climate Change (IPCC, 2021), state that active and extensive actions are necessary to stop global warming (i.e. global mean surface temperature) at the level up to 1.5 °C above preindustrial levels. To achieve this goal, several conditions must be met: (1) no more than 570 gigatones of carbon dioxide equivalent ($GtCO_2e$), (2) zero net global emissions of carbon dioxide by 2050, and (3) reductions in methane and nitrous oxide emissions that are difficult to quantify precisely, but which will be certainly significant (discussed further below). The agriculture plays a primary role in any scenario aimed at achieving the above assumptions, but the perspective of assessing the scale of the problem must concern the entire value chain of food production. It is not only about sustainable farming, but also about the issues of the consumption model, reduction of food waste and carbon sequestration in connection with forest management.

It should be emphasized that the challenges for agriculture in terms of GHG emission reduction are greater than for other sectors. Firstly, agriculture pursues a complex set of environmental and social objectives: apart from climate issues, there is biodiversity loss, broadly understood landscape protection, food safety and food security, the issue of agricultural income (relative deprivation and income gap) and the cultural heritage of the countryside. Secondly, the sectoral environment of agriculture has much greater potential for technological changes, which in agriculture itself is quite limited both in the technological sense and due to relatively slow diffusion of this progress to the level of farms, and especially "under the thatch" of individual farms, which are the dominant form of management in the world. Let us remember that agriculture engages ¼ of the world's population—which translates into a huge heterogeneity of food production processes and hinders technological progress.

Is there a general remedy, in the sense of universal guidelines for approximating climate goals in agriculture? It could be said that the first and key recommendation is not at all the search for breakthrough technologies, but simply to increase the efficiency of food production and improve farm management, especially in terms of resource management. The above-mentioned report estimates that the implementation of proven pro-efficiency solutions aimed at reducing the consumption of resources per unit of production could contribute to the achievement of the 20% required reduction in emissions in this sector. This postulate is even more important in view of the fact that the distances of individual countries from the technological frontier (i.e. production isoquants, including undesirable and desirable effects) are very different in the world, as shown in the analysis of the previous chapters, and there is a large room for improvement in the sense of Pareto-type progress referred to in the literature as "slacks". It can be assumed that a similar potential for optimization opportunities exists within individual farms in each country.

EEA (2021) reports that, "between 2005 and 2019, EU agricultural GHG emission levels hardly changed and this trend is projected to continue. By 2030, national projections from EU Member States show a modest 2% decline in agricultural GHG emissions, compared with 2005 levels", which is quite alarming information that may indicate the low effectiveness of agricultural and environmental policies in reducing GHG emissions from agriculture. However, let us remember that between 1990 and 2003, total GHG emissions from agriculture in the EU-25 decreased by 14% according to the ECCP Report (2016), mainly due to downward trends in animal populations, reduced fertilizer use and improved efficiency in manure management.

In analyzing the EEA data shown in Table A.8 (Appendix), we can see that in the period 2005–2019, the decrease of emissions for EU-27 amounted to only −1.8%, which in comparison with about 16% reduction of GHG emissions by the whole EU economy in the same period is a quite poor result (EC 2021a). Agricultural GHG emissions are addressed by the EU Effort Sharing Decision and Effort Sharing Regulation, which establishes annual targets for each Member State. The lack of harmonized targets means that the large emission reductions already achieved and projected to 2030 by some countries (e.g. Belgium, Croatia, Italy, Finland, Luxembourg, Malta, Spain, Sweden—see Table A.8) are offset by projected increases in the Czech Republic, Bulgaria, Estonia, Greece, Hungary, Latvia, Poland and Romania. The second group of countries is dominated by the so-called "New Member States" that accessed the EU in 2004.

The agriculture of these countries is currently characterized by lower intensity and lower productivity of production factors (mainly labor) than in Western European countries, and theoretically has great potential for sustainable development, especially in the environmental aspect. As can be seen, however, there are no shortcuts and it is difficult to take advantage of the backwardness by transforming it into a model of sustainable agriculture. As the projections in Table A.8 suggest, in order to get on the path of reducing emissions, it is first necessary to increase them. This contributes to the thesis that the implementation of environmental goals will always be ineffective when it contradicts the Pareto-optimal allocation of resources. Therefore, the targeting agricultural policy to improve efficiency under environmentally and socially adjusted production function can be seen as a second-best solution. Therefore, we are looking for a compromise between what is feasible and what is needed. The practice of environmental policy so far shows that their postulates, not only regarding sustainable agriculture, have been largely declarative. However, fully sustainable agriculture must also include a social aspect. While in developed countries, the most relevant element of the social aspect may be the income gap, hunger and nutrition remains key issues for most areas of the world. Data for 2019 indicate that approximately 690 million people are hungry, which translates to 8.9% of the world's population (FAO et al., 2020). Moreover, 2 billion people lack regular and safe access to quality food. After years of a fairly marked decline in the number of hungry people (from 825.6 million in 2005, to 628.9 million in 2014), a resurgence—of about 60 million people—has been observed since 2014. Additionally, the COVID-19 pandemic may add between 83 and 132 million hungry people, making the goal of Zero Hunger in 2030 unrealistic to achieve at this time.

4.2   How Effective Were Former Policies and How Effective Are Current Policies: The European Union Case

The previous point raises questions about the effectiveness of the agri-environmental policy to date. The choice of criteria for evaluating this effectiveness is still an open question. An inspiring field of discussion on the effectiveness of sustainable agriculture policy is the EU Common Agricultural Policy (CAP). In the case of the CAP, the problem of delivery of public goods by agriculture is a hotly debated issue. The conclusions of the available analyses of the hitherto achievements of environmental policy in agriculture are either so diverse that they make it impossible to formulate a consistent and unambiguous assessment (e.g. Bartolini et al., 2021; Batáry et al., 2015; Bonfiglio et al., 2017b), or the effectiveness of the pro-environmental components of agricultural policy is rated low.

In the architecture of the current CAP, there are three groups of environmental subsidies that are meant to play a double or even triple role: to increase the provision of positive external effects from agriculture, and to reduce negative environmental externalities, and at least not to negatively affect the historical goal of the CAP, which is social sustainability. The latter consists in reducing income inequalities in agriculture and assuring a fair standard of living for farmers. In particular, we refer to: (a) cross-compliance, (b) greening, (c) the agri-environment climate scheme (AECS), and (d) the previous agri-environmental scheme (AES) (see Chap. 6 for more detailed description of the CAP instruments).

Since the momentous reforms of the CAP, i.e. Agenda 2000 and the Luxembourg Agreement in 2003, successive evaluations (health checks) have outlined the need to reorient policy under the environmental component to a new more flexible delivery model (because the previous one-size policy did not fit all). In this model, measures would be more focused on environmental outcomes rather than being cost-oriented or value-oriented (Bartolini et al., 2021). However, this general recommendation did not change the fact that evaluations of the effectiveness of the tools used in the EU remained ambiguous.

A comprehensive analysis of the environmental effects of CAP at the regional level is periodically performed using the CAPRI partial equilibrium model (Gocht et al. 2017). In this model, a number of environmental sustainability indicators are included in the analysis, in particular, nutrient balances, greenhouse gas (GHG) and ammonia emissions, soil erosion, and biodiversity practices (as indexed by Paracchini & Britz, 2010). The results of this modeling, as well as other work in this area (van Zeijts et al., 2011; Chiron et al., 2013; Czekaj et al., 2014; Diotallevi et al., 2015; Mahy et al., 2015) indicate little impact of green CAP components on environmental improvement. These improvements are related to the issue of greenhouse gas emissions and biodiversity measured by the abundance of bird species in the biocenosis of rural landscapes (the case of France). In contrast, Cortignani and Dono (2015) highlight nitrogen consumption (in Italy).

In summary, although the effects of environmental policies under the CAP are positive on a per ha basis, providing an increase in farmland area, the sign at the impact coefficient of these policies may even turn negative (Gocht et al., 2017). In particular, the assessment of compulsory "greening measures" introduced into the CAP in 2013, accounting for 30% of the direct payments budget, is ambivalent. These included three instruments: crop diversification, maintenance of permanent grassland, ecological focus areas (EFA) (for detailed description, see EC, 2017b). As the cited European Comission's (EC) report states, "Overall the greening measures have led to only small changes in management practices, except in a few specific areas. As a result, their environmental and climate impacts have been limited, making a small contribution towards promoting more sustainable farming practices, although this effect is difficult to quantify and very locally specific."

Support for the development of permanent grassland undoubtedly had a positive effect on soil erosion, but on the other hand, it translated into an increase in the number of animals and a decrease in the number of forage crops, which worsened the balance of GHG emissions. In contrast, EFA had a positive impact on most environmental indicators except soil erosion. Nevertheless, it contributed to a negative trend in the petrification of set-aside areas in the EU, as farmers extensively used these lands under the EFA measure.

In the 1980s, growing concern about the environmental impact of industrial farming practices in Europe contributed to the introduction of AES (replaced by AECS in the 2014–2020 CAP programming period). European AES were initiated by the Agricultural Structures Regulation of 1985 (EU Regulation 797/85) as a mechanism to compensate farmers for loss of income caused by less intensive farming practices in environmentally sensitive areas. Since 1992, AES has been compulsory for all EU member states (EU Regulation 2078/92), but can vary significantly from country to country.

The AES objectives focus on the protection of biodiversity, the restoration of the natural rural landscape (to some extent, also the protection of the cultural heritage of the countryside), the reduction of nutrient and pesticide emissions and the prevention of rural depopulation. However, the problem is that the application of AES is greatest in areas of extensive agriculture, where biodiversity is relatively high anyway, and negligible in areas of high cropping intensity, where negative environmental effects are greatest.

The literature points out that it is methodologically difficult, if not impossible, to evaluate the effectiveness of AES. For example, Kleijn and Sutherland (2005) conducted a comprehensive review of studies testing the effectiveness of AES. They found that 76% of the studies were from the Netherlands and the UK, while only about 6% of the EU agri-environment budget had been spent by 2002 in these countries. In most studies, the methodology was inadequate to reliably assess the effectiveness of the programs. Indeed, about 30% of these did not include statistical analysis. Where an experimental approach was used, the designs tended to be weak and aimed at producing favorable results. Moreover, the most common experimental design was a comparison of biodiversity in agri-environmental programs and control areas. However, it was the program coordinators who chose the agri-environmental program sites, which meant that biodiversity could be higher (compared to control areas) even before the implementation of the program.

The main conclusion of the work of the cited authors, i.e. that the lack of reliable evaluation studies does not allow assessing the effectiveness of AES, may still be valid (Bartolini et al., 2021). In this case, biodiversity protection is the most problematic issue, as other environmental indicators can be easier subjected to evaluation.

In an evaluation conducted in parallel with the previously cited study, Primdahl et al. (2003) found that the most widespread effect of contractual obligations for farmers under AES was reductions in the use of nitrogen fertilizers and pesticides. The study was relatively large in scale, involving nine EU member states and Switzerland. Significant positive effects also included reduced livestock density and increased crop diversity. The authors noted, however, that there was a clear improvement in the environmental indicators mentioned, but the AES do not guarantee the sustainability of these results beyond the period covered by the contract with the farmer.

Another turning point in CAP environmental policy was in 2013, when the so-called "greening measures" were introduced. As a result of this reform, payments shifted towards regions dominated by grassland. Paradoxically, however, it is predicted that in the long run the intensity of agricultural land use in the EU may increase in the sense of higher average fertilizer application rates and that other negative effects, partly due to greening, such as decreasing topsoil organic carbon stocks and increasing greenhouse gas emissions, will occur (Kirchner et al., 2016).

Next strand of research links the effectiveness of AES to the scale of their use. Treatment effect analyses based on the propensity score estimator show that if the share of AES in farm income is high (e.g. greater than 5%, according to Arata & Sckokai, 2016), participation in AES programs is an effective mechanism for implementing sustainable practices. The previously cited study by Bartolini et al. (2021) also shows that the effectiveness of AECS varies according to payment levels.

A comprehensive assessment of the effectiveness of AES in the field of biodi-versity is presented in the EC (2017b) Report "Science for Environment Policy". Summarizing its conclusions, we can say that the effectiveness of environmental policy in the area of biodiversity largely depends on how diverse and tailored the agri-environmental programs are for the individual habitat needs of different species.

There are trade-offs and negative feedbacks in the long term between programmes supporting biodiversity and other environmental, climate and social objectives. Examples are the mentioned permanent grassland or set-aside, which on the one hand, develops habitats for soil invertebrates and generates a number of positive effects in terms of combating climate change. On the other hand, set-aside may reduce the productive value of land if maintained for too long, and in extreme cases, it may promote secondary plant succession (e.g. spruce), soil depletion and ultimately a significant decrease in species diversity.

The role of human capital in successful AES implementation cannot be overstated either, as farmers with more experience in agri-environmental programs and environmental knowledge create habitats that are more wildlife-friendly and have greater sustainability, according to the cited report. Similarly, farmers with knowledge of environmental policy are more likely to create wetland habitat. Thus, farmers' training in biodiversity conservation management, which imparts not only knowledge and skills, but also builds self-efficacy and establishes a value system, plays a significant role. With the development of human capital in agriculture, the creation of so-called "result-based payments" also makes more sense (EC, 2017b).

Evaluation of the AECS is proving to be as complex as its predecessor, the AES. A number of case-specific factors prevent generalizations of the results of evaluation studies. These include: location, dominant agricultural systems, weather, political mechanisms, and path dependency. Bartolini et al. (2021) conclude that the AECS has significant weaknesses consisting of poor tailoring, low average participation and individual farmer involvement, frequent opportunistic behavior and asymmetry of information in the farmer-program coordinator relationship. This conclusion is also supported by other authors who had researched both AEC and AECS (Finn et al., 2009; Uthes & Matzdorf, 2013; Batáry et al., 2015; Vergamini et al., 2020).

Interestingly, in studies where attempts have been made to create composite indices of sustainable practices in agriculture (e.g., Biodiversity Friendly Agricultural Practices Index, BFP—Paracchini & Britz, 2010; Index of Extensification IE— Bartolini et al., 2012 and 2021; Sustainable Value SV, and Modified Sustainable Value MSV- Grzelak et al. 2019, and Czyzewski et al., 2020 and 2021), AES or AECS had unambiguously positive effects on these indices, although the authors pointed out the issue of poor targeting of green CAP's measures, as well as the problem of moral hazard, referring to agency theory (Raggi et al., 2015). These latter issues may contribute to the low sustainability of environmental policy effects. The works previously cited also note that farmers' use of a combination of several schemes produces better results than single contracts.

There is also a consensus in the literature that CAP subsidies to production were inefficient, costly and had negative environmental effects (Bureau & Mahé, 2008; Pain & Pienkowski, 1997; Rizov et al., 2013). As a result, several CAP reforms were carried out, culminating in the decoupling of payments (in the Luxembourg Agreement in 2003) and the introduction of the so-called "decoupled subsidized", i.e. the Basic/Single Payment Scheme (BPS/SPS) and Single Area Payment Scheme (SAPS), and the expansion of the environmental and rural development oriented second pillar. As the previous considerations have shown, the evaluation of CAP in terms of environmental impact after this key reform is ambiguous, but certainly more positive (these are the programming periods 2004–2006, 2007–2013 and 2014–2020). Similarly, from the farmers' side, the evaluation of the effectiveness of the environmental measures in the second pillar of the CAP varies. Farmers mainly consider the sanctions related to non-compliance with the CAP rules, combined with some ambiguities regarding the interpretation of the rules, as its weaknesses.

Researchers generally analyze separately the policies related to different types of public goods in rural areas. It is assumed that their specificity requires the application of different types of instruments. Actually, four main groups of public goods (PG) connected with agriculture and rural areas can be indicated: (1) climate-related goods, including GHG emission issues and nutrient surpluses; (2) health-related PG, including dietary and hunger aspects (i.e. food security), food safety, soil, air and water pollution; (3) natural landscape and biodiversity; (4) rural culture.

It is puzzling that despite the obvious interactions between these groups (synergies and trade-offs), studies of policy effectiveness tend to look only at individual components, and the net effect is hardly taken into account. Meanwhile, it is quite likely that the success of environmental policies in the area of biodiversity or any of the other fields mentioned above may be offset by negative externalities in other areas. Therefore, it might be worthwhile to look for holistic measures to assess the effectiveness of environmental policies, as this is also the essence of the definition of sustainable agriculture.

In this context, the question of creating result-based measures is very problematic, as it is not really clear how to approach the results. For example, Hasund (2013) and Birge et al. (2017) focused on the landscape and biodiversity component (4), arguing that active management is necessary to maintain landscape public goods because they are positive external effects of agriculture, and without such management, this type of public goods will disappear.

According to the cited authors, landscape and biodiversity component differs from the other types of public goods (i.e. climate-, health- and culture-related), which are caused by some activity that creates negative externalities. On the one hand, imposing restrictions and limitations on landscape would be neither efficient nor consistent with a common conception of justice. Farmers cannot be forced to maintain unprofitable land management without adequate compensation. On the other hand, new production technologies and the market-driven orientation have contributed to a dramatic regression in the provision of environmental public goods, particularly those related to landscape, biodiversity and cultural heritage (see more in Chap. 1). It could even be perceived as a secular process. Indeed, EU CAP reforms introducing decoupled subsides at the turn of the 20th and 21st centuries (with cross-compliance in the first pillar and AES in the second pillar), have largely slowed down or stopped most environmentally damaging processes.

Currently, AECS schemes are the leading tool in the EU for developing environmentally sustainable agriculture, and their role cannot be overestimated (Batáry et al., 2015). In particular, Nature Management Grassland as a scheme promoting both biodiversity and water protection receives positive feedback. This instrument is considered to play a leading role in enhancing biodiversity in European agriculture (Birge et al., 2017). Nevertheless, also in this case, an accusation can be made about lack of precision: the scheme does not distinguish between old grassland and rotational grassland—since 2015 the support is equal for all plots. Moreover, for much of the program's operation, inspectors considered natural vegetation as "weeds", which may result in withdrawing the payment. Although natural vegetation is now allowed, the requirement for mandatory mowing of weeds still remained unclear, due to the lack of a precise definition of "weeds".

Thus, the weaknesses of the AECS include moral hazard and lack of incentives to achieve real results, as environmental payments are cost-oriented, insufficiently targeted and of low flexibility in adapting measures to the diverse conditions that exist in farming. In addition, lack of clear objectives and long-term strategy is evident (European Court of Auditors, 2011). The system is also criticized for low cost-effectiveness from the perspective of particular indicators.

There are ongoing discussions on the introduction of result-oriented measures, especially in the field of biodiversity, cultural heritage and landscape protection. EC experts generally agree that AECS should be more result-oriented (European Network for Rural Development, 2010). Similarly, according to the cited recommendation of the European Court of Auditors (2011), the agri-environmental component of the CAP should be more precisely targeted and clearer objectives of individual schemes should be formulated, which would be adapted to local conditions and give the possibility of precise measurement of effect.

According to Allen et al. (2014), result-oriented agri-environmental payments are already applied in some Member States, in addition to CAP subsidies (e.g. Germany, France and the Netherlands) and concern biodiversity management of agricultural landscape ecosystems. The undertaken measures introduce effect indicators, as well as a list of prohibited activities. The supporters of this solution point out as its main advantages the increase of farmers' interest in achieving real objectives, the building of broadly understood social capital and the stimulation of local innovation (we will return to the thread of result-based measures in the next section).

To conclude, it is questionable which perspective on the effectiveness of the CAP is appropriate: (1) aggregated, environmentally oriented—including synergy effects of many policy measures, (2) holistic, focusing on all three dimensions of sustainable development, or (3) the specific indicator-based approach. The latter (3) is most often taken up in research and generally gives a rather critical evaluation of the selected schemes. The former two (1 and 2) are analyzed much less frequently, but bring generally positive assessment.

4.3 Why Analyses of Policy Effectiveness Fail to Produce Clear-Cut Results: Premises for Adopting the Pareto-Inefficiencies Criterion

4.3.1 Cost-Effectiveness and Cost-Benefits Analysis

There is an ongoing debate in the EU on how to measure the eco-effectiveness of agricultural policies for sustainable agriculture. So far, mainly cost-based measures have been applied, the aim of which is to compensate the farmer for the opportunity and transaction costs of shifting to more sustainable farming practices. Such costs are generally associated with a reduction in the degree of intensification of agricultural production and a decline in profitability. For years, the question has been posed whether such instruments used in the CAP have been effective. At this point, we come back to the core of the problem of first two chapters, how to measure sustainability in agriculture so that the measurement criteria are as coherent as possible with the definition of sustainable development.

Cost-effectiveness analysis (CEA) is a concept well established in the literature. CEA, as a cost-effectiveness analysis of economic policies, aims at choosing the best way to allocate public resources while minimizing expenditure in order to achieve the desired result. *Ex ante* CEA applies when the objective of public policy has been defined and the cheapest (most cost-effective) path to achieve those objectives is sought. *Ex-post* CEA answers two questions: to what extent the objective has been achieved and at what cost of public policy? In each case, it is necessary to calculate the marginal or average effect of the public policy, which usually has a physical dimension, e.g. the number of animal or plant species present, the scale of reduction of emissions of a given type of pollutant, or the increase in caloricity of food produced. Thus, policy effects may relate to the reduction of environmental pressures, the avoidance of the effects of negative externalities, or the provision of a specific positive external effect (i.e., public goods).

The different options for achieving a given effect are then compared in order to select the cheapest one per unit of effect. It is worth noting that CEA does not provide an answer as to whether a given policy is justified in terms of social costs and benefits and welfare effects. Thus, the CEA method can be called a second-best solution in situations where it is not possible to conduct a full analysis of social costs and benefits due to their immeasurability in monetary (or other) units that could be referred to public policy expenditures (Görlach et al. 2005). In CEA, effects are often limited to being one-dimensional. Therefore, in order to assess a multidimensional effect, as in the case of CAP, a system of weights must be defined and then an aggregate indicator must be constructed. The Integrated Efficiency (IE) score we used is an example of such an indicator. In the frontier-based approach, the weights are set according to so-called "shadow prices". This is considered one of the biggest advantages of this approach.

Some authors equate cost-benefits analysis (CBA) with the CEA approach (Kronbak & Vestergaard, 2013). However, we think that these are two different methods adjusted to different decision-making situations. CBA is also performed to compare the effectiveness of alternative public policy actions. *Ex ante* analysis is undertaken prior to the implementation of public policy in order to evaluate alternatives and choose the best solution. An *ex post* analysis is conducted after the policy has been implemented and after it has been in place for a period of time and is expected to yield results.

According to Whitehead and Blomquist (2006) and Kronbak and Vestergaard (2013), a typical environmental policy CBA consists of a number of steps that include:

1. Identification of goals, criteria for welfare change and populations covered by the policy, alternatives, and implementation time.
2. Identification of potential environmental effects in physical units.
3. Valorization of effects (in monetary units).
4. Discounting of cost flows and social benefits and environmental effects over the assumed impact period.
5. Project selection based on Net Present Value criterion.
6. Sensitivity analysis, which assesses how robust the assumed effects are to changes in the values of the model parameters.

In this approach, the discounted benefits of a given activity (as the present value of cash flows from the analyzed period) are set against the also discounted, broadly defined costs. Discounting, of course, requires the use of monetary units linked to a specific discount rate (interest rate) for both benefits and costs. Thus, the greatest difficulty in CBA is the monetary valuation of environmental or social benefits and the determination of an appropriate discount rate.

Many public goods—environmental and social (e.g. biodiversity, clean air, climate stability, cultural heritage and food safety) are difficult to value in monetary terms because they are generally not tradable in the market or quasi-market (i.e. as tradable permits). In this situation, the market value/price can be estimated using stated preferences methods—e.g. contingent valuation which estimates WTP, or revealed preferences methods—e.g. hedonic pricing or factor/cost-based approaches. CBA assesses whether a public scheme is worth pursuing, i.e. whether its benefits outweigh its costs. Cost-benefit analysis can be performed as division or subtraction.

The idea of CBA in the case of environmental policy is captured by Formula 4.1. This formula is borrowed in part from Kronbak & Vestergaard (2013), and explains how environmental CBA is calculated:

$$Bn = \sum_{t=0}^{T} \frac{\Delta Bb_t - \Delta Cb_t}{(1+r)^t} + \sum_{t=0}^{T} \frac{\Delta EBn_t}{(1+r)^t} \tag{4.1}$$

where: Bn stands for benefits net; ΔEBn_t stands for changes in environmental benefits net; ΔBb_t indicates the present value of changes in benefits gross; and ΔCb_t indicates the present value of changes in costs gross.

The necessary condition for recommending a chosen policy option claims that with net benefits, $\Delta Bn \geq 0$, or that the present value (PV) is the difference ΔCb-$\Delta Bb \leq$ PV of ΔEBn. Thus, when viewed from the perspective of environmental objectives, the environmental benefits must be at least equal to the net costs of the program in question.

As can be seen, apart from the issue of expressing environmental benefits in monetary units, CBA is rather limited to being a one-dimensional type of effect. This can also be problematic. The cited authors transformed the above condition by dividing both sides by the change in the assumed policy effects (ΔE), and by concluding that the marginal value of the additional unit of effect is constant and corresponds to WTP per unit of effect. This resulted in $\Delta EBn = WTP \cdot \Delta E$, and after rearranging: $(\Delta Cb\text{-}\Delta Bb)/\Delta E \leq WTP$. Hence, a cost-effectiveness ratio that reffers to the CEA approach is generated. The cited authors called it "environmental cost-effectiveness analysis" (ECEA).

The similarity to CEA is apparent, however a value estimate of ΔBb and ΔEBn is still necessary, whereas in a typical CEA environmental effects can remain in non-monetary units.

4.3.2 Complexity Effect

However, evaluating the effectiveness of environmental policies is always complex because of the interactions between:

- optimization of public expenditures and agricultural policy criteria,
- levels of payment (willingness to accept vs. willingness to pay) and their calculation method (e.g. cost-based, result-based, value-based compensation),
- farmer behavior,
- expected improvements in environmental quality.

Hence, Bartolini et al. (2015, 2021) argues that the relationship of assumed effects and incentives (negative or positive) should be considered separately from factors that reduce policy effectiveness, among which he lists:

- farmers' attitudes and motivations toward environmental goods,
- adverse selection, moral hazard, including agency problem, asymmetric information,
- inappropriate design of priorities, payment mechanisms, and level of commitment (WTP vs. WTA relationships—see Fig. 4.1).

We think that there is one more aspect missing from this list, which we have called the "complexity effect" (or "fallacy of composition"). This effect occurs on three levels: First, each measure of environmental policy is generally designed to address one type of environmental pressure e.g. biodiversity loss, nutrient use, greenhouse gas (GHG) emission etc. Then, the measure of effectiveness is evaluated by looking at the achievement of this main objective, as we have shown in the works cited above. However, it often turns out, as in the case of EFA or grassland measures, that there are trade-offs or synergies between the realizations of different environmental objectives that were not taken into account in the policy design. These trade-offs or synergies of environmental effects constitute the aforementioned complexity effect that can cause the fallacy.

The second layer of the issue concerns the true willingness to accept feedback from the farmer for particular environmental policy measures. An environmental policy will only be accepted by a potential provider of public goods if it satisfies a weak or strong Pareto optimizing criterion (Johansson, 1993). A farm is not a black box—each farm is internally differentiated, and although it converts the same set of resources into outputs, in each case the production process may proceed differently. In the process of managing available resources, the farmer more or less consciously applies multidimensional optimizing and will prefer solutions that make him/her better off in all management aspects (strong Pareto criterion) or at least in one aspect while the others remain as good (weak Pareto improvement).

This is the search for progress in the Pareto sense that we wrote about earlier in the context of removing inefficiencies (i.e. slacks). Such a process of optimizing is per se multi-criteria, while environmental policy design is generally single-criteria, or bi-criteria (we write about this further referring to the social and environmental constraints of CAP (Mouysset, 2014).

Usually, policymakers formulate a specific environmental goal (or possibly a concomitant social goal, such as reducing income inequality) and then implement negative incentives (e.g. taxes or tradable permit requirements) or positive incentives (e.g. subsidies) to induce farmers to achieve it. In some cases, attempts are made to gain the formulated objectives at minimal public cost (as in the case of the US Conservation Reserve Program, see Sect. 4.4). Policy design, however, does not take into account the multidimensional system of intra-farm inefficiencies and how they are affected by specific measures.

Shifts in the multidimensional system of inefficiencies (i.e. in slacks) are generally an overlooked side effect of environmental policy that contributes to the fallacy of composition. Even if, for example, subsidies for public goods increase the supply of those goods to some degree, the effect of specific measures may be to increase inefficiencies in the management of other production factors and either compel trade-offs in environmental indicators, or generate opportunity costs in environmental or economic benefits. As a result, a farmer using the Pareto efficiency criterion may withdraw from providing public goods or succumb to moral hazard and only create the appearance of sustainable practices. Therefore, it is important to answer the question of how environmental policy affects the level of inefficiency and potential development opportunities in accordance with Pareto improvement criterion.

We take the position that this criterion is central to a farmer's production decisions, but because of its multidimensionality (improvement in one aspect, while the others are at least not deteriorated), we recognize that it is difficult to take into account at the environmental policy design stage.

At the same time, it is worth asking whether environmental policy itself increases inefficiency (suboptimality) in the sense of distorting the market mechanism. This brings us to the third dimension of the complexity effect. It is clear that microeconomic goals are not aligned with macroeconomic goals, especially those at the global level. This is what the classically understood fallacy of composition comes down to, according to Grzelak (2015). Most commonly, farmers at farm level are assumed to follow a neoclassical Constant Elasticity of Substitution CES type production function. This family of models includes the popular Cobb-Douglas production function and its more recent trans-log version, as stated by Kryszak & Herzfeld (2021).

If production technology is defined using the CES function with constant returns to scale and factor-specific technology parameters (Dudu & Krsitkova, 2017; Czyzewski et al., 2019b), we can formulate the following factor cost minimization problem:

$$PROD_{c,t} = \left[\alpha_D (A_F F_{c,t})^{\frac{\sigma-1}{\sigma}} + \alpha_K (A_K K_{c,t})^{\frac{\sigma-1}{\sigma}} + \alpha_L (A_L L_{c,t})^{\frac{\sigma-1}{\sigma}} \right]^{\frac{\sigma}{\sigma-1}} \tag{4.2}$$

$$\min C_{c,t} = PF_{C,T} F_{c,t} + PK_{c,t} K_{c,t} + PL_{c,t} L_{c,t}.$$

where: C stands for total costs of agricultural production, and $PROD$ stands for farm production, F is land, K is capital stock, and L is labor; while PF, PK and PL stand for the prices of land, capital and labor, respectively. Furthermore, αD, αK and αL mean the distribution parameters; A_F, A_K and A_L are factor-augmenting technology parameters for land, capital and labor, respectively; σ is the elasticity of substitution between F, K, L.

Such a specification is usually used in CGE models. Thus, farmers focus on optimizing the proportion of resources employed in agricultural production at given market prices and technologies. If environmental policy introduces into this model additional outputs, i.e. undesirable and desirable ones, and additional constraints associated with them, it is obvious that the conditions of microeconomic equilibrium are disturbed.

While the introduced additional variables are justified from the perspective of collective rather than individual utility, they lead to inefficiency. Therefore, a simultaneous corrective action on the right side of the model is necessary by introducing policy measures as factor-augmenting technology. However, turning off the autopilot in the form of the market mechanism and switching to manual control is not easy and does not necessarily remove the inefficiencies caused by the change in optimisation perspective from individual to collective.

Thus, the assessment of inefficiencies under the conditions of the above-mentioned change in the model is critically important. We believe the assessment of the impact of agricultural policy on inefficiencies under the environmentally adjusted production function should be an important criterion of environmental policy evaluation. As it was said before, despite the adoption of collective utility criteria by policymakers, farmers in practice can still optimize individual production functions and choose Pareto-optimal solutions from the microeconomic point of view. If the proposed policy measures do not offset those inefficiencies, then public goods will not be delivered or moral hazard and adverse selection will occur, which will draw only the appearance of sustainable practices.

In the context of the above conclusion, we cannot fully agree with the thesis posed by Bartolini et al. (2015) that when evaluating the effectiveness of agri-environmental policies, it is necessary to separate the optimal policy from the various sources of policy failure that determine the suboptimal level.

It is not sufficient to simply match the expected environmental quality of an area (e.g., the number of bird species of a rural landscape, as an indicator of biodiversity) with the level of the policy measure from that area, because the complexity effects may occur simultaneously. As a result, halting biodiversity loss in rural landscape may be accompanied by higher inefficiencies (slacks) in other areas, which will translate either into perishability of the positive environmental effect or its appearance due to moral hazard, or into trade-offs of biodiversity and other environmental indicators. Therefore, a necessary criterion for environmental policy evaluation is its impact on the level of inefficiencies of agricultural production function and its particular components. A decrease in this efficiency will diminish and an increase will boost the effectiveness of public policy.

An important factor in the above discussion is the mindset and attitude of farmers undertaking agri-environmental practices. According to various studies (Sutherland et al., 2016; Dessart et al., 2019), so-called "active" beneficiaries of AES or AECS may be so strongly motivated by their beliefs and concern for the environment that they will fulfill their commitments under the AECS agricultural policy scheme even if they are not compensated for doing so, i.e. after the AECS contract ends. The so-called "passive" beneficiaries, on the other hand, carry out environmental care commitments only when it is profitable for them.

However, it is not clear from the research cited above how the level of inefficiency in the multidimensional production function changed for "active" beneficiaries due to implementing a particular policy. This thread has not been investigated. So, it is possible that active farmers' attitudes resulted partly from the fact that the level of inefficiency was decreasing. In the case of "passive" farmers the level of inefficiency in connection with the implemented scheme may have increased and that could be the reason for the observed behaviors. Thus, we think that changes in inefficiencies resulting from public policy implementation require further research.

Similarly, in the case of farmers reluctant to change in terms of adopting sustainable practices, the problem may lie in perceiving the change as a source of multidimensional inefficiencies in the realized production function (even if policy measures, e.g. subsidies, are included), or as an effect of not spotting the chance of Pareto improvement. **We can hypothesize that identifying pareto-optimal changes in farming system related to particular policy measure is a necessary condition for implementation of sustainable practices in agriculture.**

It can be concluded that the multidimensional (holistic) perspective is closer to farmers, and although problems of adverse selection and moral hazard occur in the case of individual measures, the effectiveness of environmental policy in agriculture should be assessed from a broader perspective. This would coincide with the earlier conclusion that farmers optimize a multidimensional production function, and therefore a multidimensional inefficiency criterion (slacks) is reasonable for policy evaluation as it helps to avoid the fallacy of composition.

If we talk about the holistic perspective, then transaction costs gain in importance. In this case, the whole is also not the sum of the parts, so there is a complexity effect as well. Bartolini et al. (2021) states that it is desirable for public institutions to incur higher transaction costs for matching and targeting particular agri-environmental schemes. However, one wonders whether it would not make more sense to allocate additional resources to a precise evaluation of the entire support system for sustainable agriculture.

If the evaluation of elasticity of substitution between available resources affected by the implementation of certain policy scheme seems too difficult or impossible for farmers, this means that the transaction costs of such a policy are so high that they make it impossible to execute a transaction for the provision of public goods. Therefore, the assessment of the impact of the policy on the multidimensional inefficiencies should be the responsibility of public institutions designing environmental policies.

Thus, a general postulate should be considered—let us design environmental policy in agriculture in such a way that it reduces the level of inefficiencies in the realization of the production function modified by the consideration of collective utilities. In this context, we should consider whether it is worthwhile to focus on optimization of environmental subsidies in the context of single ecological indicators, or whether only holistic evaluations using synthetic measures of sustainability including economic and social aspects, make sense. An example of such a measure could be the integrated efficiency score and the slack-based analysis that we developed in this book.

Institutional perspectives on agricultural policy also demonstrate the complexity of the question of public policies effectiveness. Alesina and Giuliano (2015) analyzed how different types of institutions interact with particular aspects of culture defined broadly as: "those customary beliefs and values that ethnic, religious and social groups transmit fairly unchanged from generation to generation" (Guiso et al., 2006). Empirically measurable cultural traits include: family ties, generalized moral individualism, trust and work vs. luck (Alesina & Giuliano, 2015; Hofstede, 2001; the World Value Survey, 2021). Studies of the relationship between culture and formal institutions (which include agricultural policy) point to the bidirectional interaction and endogeneity of this relationship.

Generalized morality, individualism and trust go hand in hand, although the correlation is not always strong. At the same time, there is a positive bidirectional interaction between these cultural traits and the propensity for cooperative behavior (collective action) that results in the provision of public goods. On the other hand, generalized morality is negatively correlated with family ties (Alesina & Giuliano, 2015) and the related concept of limited morality (which means there are different moral norms towards family members than towards the rest of society).

In general, morality, defined as "ethical norms in society about the inappropriateness of behaviors that can cause harm to others" (James, 2015) fosters prosocial behavior—this was a conclusion drawn from several empirical studies (e.g., Capraro & Rand, 2018; Tappin & Capraro, 2018). According to Platteau (1993), in a developed democratic society, norms of honest and altruistic behavior are applied in every social situation, extending beyond narrow social groups such as family and friends, so that limited morality has little role. Developed generalized morality makes people willing to take into account the opinions of others when making decisions, while being guided by trust and faith in the morality of other people. On the other hand, deviating from the ethical code and breaking the norms evokes feelings of guilt, while free riding meets with social disapproval and stigmatization of such behavior.

Generalized morality and the other cultural traits positively correlated with it have a positive impact on economic development in the long term, which has been empirically confirmed by many publications (e.g. Platteau, 2000; James, 2015; Alesina & Giuliano, 2015). From the perspective of sustainable agriculture, however, the most important issue is the relationship of morality and related traits with the provision of public goods and the effectiveness of agricultural policies in this regard.

Environmental goals for the agricultural sector are set at the national, international or global level. It is worth considering to what extent their achievement is the result of collective action of individual farmers, and to what extent the sum of individual actions. The answer to this question lies somewhere in the middle— farmers to some extent maximize their own utility functions or are guided by individual altruism, but the effect of their actions is also burdened by spatial dependence, in practice, by the attitudes of farms in the neighborhood. Neighborhood effects can induce positive or negative spillovers in environmental indicators, particularly, groundwater pollution, GHG emissions, and biodiversity. Interaction and similar attitudes of farmers towards sustainable farming principles can thus enhance positive environmental effects, but opposing attitudes can weaken these effects. The typical collective action dilemma comes down to how to mobilize individual participants in a given farming system to actively participate in the provision of public goods and overcome free riding effects.

In this situation, the prisoner's dilemma may apply (Buchanan, 2009; Ostrom, 1990; Olson's 1965). Collective action theory, however, assumes that people are fully rational, with little socialization, and that decisions are made under conditions of atomization of individuals. Both the complete rationality of decisions and their atomization is a rather extreme assumption (Durkheim, 2020). In the case of individual farmer behavior, the most realistic seems to be the Embeddedness theory approach (Granovetter, 1985), which emphasizes the importance of the social context (social network) of decision-making, or in the language of economics, maximizing individual utility. The economic equivalent of this perspective is, to some extent, bounded rationality (Williamson, 1985). This brings us to the crux of the issue—does the social context with regard to sustainable farming consist of generalized morality, or limited morality (i.e. locally-specific form of morality depending on family/neighborhood ties)?

The World Value Survey cited above gives some insight into the global distribution of cultural traits and regions where generalized morality, individualism and trust dominate, and where family ties are important. We think that these traits may endogenously influence the effectiveness of agricultural policies and therefore explain to some extent empirical differences between clusters that we describe in Chaps. 3 and 5.

Society in the European Union has long recognized the need to protect the Earth's natural resources (EC, 2011), although the problem of biodiversity loss affects the whole world. Nevertheless, conservation activities should be effectively managed, which is difficult if insufficient financial resources are allocated. For example, between €550 and €1150 million per year was spent from the EU budget between 2007 and 2013 to maintain the Natura 2000 network—the largest transnational habitats conservation and protection area in Europe covering around 18% of land of the EU countries. It met only 9 to 19% of its over all financial needs, according to Hermoso et al. (2019), and in an earlier similar assessment by Kettunen (2011).

Following the discussion on the complexity effect of environmental policy, one may come to the conclusion that the low cost-effectiveness of particular schemes is due to the lack of integration of the overall environmental strategy of the EU and member states with sectoral strategies, including the environmental component of the CAP.

The current Habitats and Birds Directives cover too few species—only about 16% of all endangered bird species. In addition, the potential of existing Natura 2000 protected areas is not realized due to more endangered species than declared in the objectives of the EU Biodiversity Strategy (EC, 2011) Funds from other sources should be allocated to conservation activities in this area (European Court of Auditors, 2017). Furthermore, the Natura 2000 network is underinvested—as only about 19% of its funding needs are covered (Kettunen, 2011).

The literature indicates that the EU Biodiversity Strategy is inconsistent with environmental policies under the CAP (Hermoso et al., 2019). Other studies show that AECS are more effective at the local scale than programs designed and coordinated by the national agencies of environmental protection also supported with EU funds (Czyzewski et al. 2020, 2020; Czyzewski et al., 2022). On the other hand, however, the share of reserves, national parks, Natura 2000 bird and habitat areas, or ecological uses in the rural area is a factor that positively influences the eco-efficiency of individual farms in agriculture (Matuszczak, 2021), as it protects agricultural landscape species and habitats potentially threatened by intensified agricultural production and creates conditions that increase farmers' willingness to participate in agri-environmental programs.

Thus, the effectiveness of the environmental policy should be assessed together with its sectoral components, because even if the Natura 2000 program *per se* is not cost-effective, it increases the effectiveness of other environmental policy expenditures. For example, within the agricultural areas located within the Natura 2000 area, more sustainable agricultural management is carried out, which does not adversely affect the natural values of the area and at the same time promotes the preservation and improvement of biodiversity.

On the other hand, decoupled payments have a negative impact on the environmental sustainable value (ESV) of farms derived from a ratio of agricultural output and polluting activities such as use of mineral fertilizers, pesticides, increasing stock density and deforestation (Czyzewski & Guth, 2021). This observation was confirmed to some extent by the report of the European Court of Auditors, which indicates that "greening" payments (constituting about 1/3 of direct payments) do not have a positive impact on the environment (European Court of Auditors, 2017). The audit was conducted in five EU countries: Greece, Spain, France, the Netherlands and Poland, and showed that greening payments contribute poorly to improving indicators of environmental impact and climate change. Hence, the recommendation appeared that decoupled payments should be more strongly linked to the fulfillment of environmental standards by farms and be better specified than in the current cross-compliance rule (Matuszczak, 2021).

Thus, it may be concluded that when evaluating the effectiveness of agricultural policy, the aggregated effect of various environmental programs (sectoral and nationwide) should be considered, taking into account the interactions between them. In other words, it should be remembered that the environmental policy pursues a broader set of objectives beyond single-criteria programs, such as the biodiversity-oriented Natura 2000 network.

From the perspective of economic theory, the problem of low effectiveness of agri-environmental policy boils down to the fact that its impact on the production function realized at the microeconomic level is poorly recognized. Environmental problems begin at this level and are caused by individual entrepreneurial decisions, including that of farmers and consumers. It is therefore a question of weak integration into an economic model of equilibrium in which objectives set at the global and transnational level are not incorporated at all.

The top-bottom approach has just such a weakness. Therefore, policies referred to as bottom-up are characterized by greater cost-effectiveness. Here we mean, for example, cost-based measures under CAP, which attempt to adjust the microeconomic production function to collective utility criteria by incorporating the input costs of providing production goods into the production account. From the point of view of integrating social and environmental objectives into the input-output calculus at the microeconomic level, cost-based or value-based measures seem to be the best option after all. In contrast, result-based measures would only be viable if the outcomes were quasi-market tradable (e.g. as tradable permits). We discuss marked-based measures in the next section.

4.3.3   Result Based Measures: A Remedy for the Low Effectiveness of Agri-Environmental Policies?

Payments-by-results are considered by many authors to be an effective solution to the low cost-effectiveness of particular agri-environment-climate programs, so we decided to take a closer look at this issue. For instance, Birge et al. (2017) provides a list of indicator plant species as a criterion for identifying sites with high natural value in terms of vascular plants as an example of a measurable effect on which environmental subsidies can be based. The farmers under cited study was mostly positive about this approach. The perception of the local community was similarly positive. In contrast, concerns have been raised about the transaction costs of incorporation into the current institutional framework of the CAP, particularly verification of biodiversity outcomes. Those working for the national control agencies were most critical especially when it came to monitoring the results of the program entrusted to the farmers themselves (farmers were to be responsible for self-monitoring).

Interestingly, farmers pointed out two premises that could determine the success of the program (for more details, see Birge et al., 2017): (1) provision of access to practical training and information support (online materials, smartphone application for species recognition, obligation of continuing education), (2) determination of the amount of the result-based bonus to be paid based on the costs incurred, including the costs of foregone income (e.g. average income from field crops or subsidies for buffer zones, which is currently about 400 €/ha). Having in mind the last point, one may have the impression that in this case, the result-based measure still remains a kind of cost-based measure, only with a more precise definition of the program objective and a quite transparent system of evaluation of its achievement. Moreover, the assessment of the management declared by the farmer in the environmental contract has been replaced by the assessment of the achieved biodiversity indicator.

The local acceptance of this measure confirms our thesis that it is very important for farmers that a given environmental policy instrument does not reduce the allocative efficiency of the realized production function. The attitude of farmers in the described case testifies that it is possible to combine agronomic and environmental objectives in terms of biodiversity (what is technically called "strong disposability"), or even to generate synergy effect in this regard.

Therefore, we can conclude that sustainable agriculture policy should place greater emphasis on building social capital and spreading the idea of Pareto improvement as a costless progress toward environmental goals. Of course, each aspect of environmental policy from the previously mentioned (climate-, health-, landscape- and culture-related) would require a separate set of indicators to be developed in a result-oriented approach. Perhaps, it would be easier from a practical point of view to develop aggregated measures for each of the areas (e.g. measurement of GHG emissions at the farm level) and use them in monitoring and evaluating environmental programs.

While taking these measures into account in the realization of the environmentally adjusted production function, one should consider to what extent each of them is disposable (i.e. whether or not generates trade-offs) and what relative weight it has. It should be noted that, from a theoretical point of view, agri-environmental schemes based on outcome indicators may provide a socially optimal supply of public goods, but their main drawback is a potential large variation regarding the amount of payment at the farm level, which is not in line with CAP regulations (Hasund, 2013).

The cited author points out three conditions that must be met in order to design socially effective policy measures for environmental services: (1) the consumption of public goods takes place under conditions of non-rivalry and non-excludability, (2) environmental goods are considered as positive external effects generated by agriculture—which makes it impossible to introduce top-down requirements and restrictions on agricultural production, because a farm cannot be forced into unprofitable land use and management, (3) the value of environmental public goods set up by a public institution differs due to the heterogeneity of the rural landscape (Hasund, 2013).

To this list we would add what we wrote about earlier: (4) that the measures implemented must not deteriorate allocative efficiency in the environmentally adjusted system of farm inputs and outputs; in other words, they must be consistent with the idea of the improvement in Pareto sense.

The second point raised by the cited author is quite debatable. According to the analysis of problems with the implementation of the greening scheme in the EU, trade-offs may occur between particular public goods supplied and environmental indicators (see Chap. 5, Fig. 5.5). When we talk about trade-offs, we mean a situation in which the provision of one kind of good negatively influences other environmental indicators, i.e. other public goods, e.g. permanent grasslands support biodiversity, but may contribute to higher GHG emissions due to an increase in livestock production. **Hence, a condition for the design of socially efficient policy measures for environmental services should be a balance above zero of positive and negative external effects in the provision of public goods in agriculture.** Similarly, in (3), we should point out that the value of public goods estimated by public institutions will be adjusted to the balance of negative and positive externalities.

We agree that this is the most difficult condition to implement due to the issue of measuring externalities, as well as the politically difficult to implement differentiation of payments to farmers of different countries and regions. It would be worth considering whether maintaining this condition is necessary.

One could say that differentiating subsidies according to effects, i.e. the amount of net public goods, is a condition for strong environmental policy effectiveness. On the other hand, if we were to operate at "weak effectiveness," one could accept that some farmers receive too much subsidy in relation to the actual net effect, while others are somewhat undervalued in this regard.

We should also ask the question as to what extent moral hazard (resulting in free riding in the form of apparent sustainable farming practices) and altruism of producers compensate each other. To some extent they certainly do, and there is no reason to believe that free riding and moral hazard prevail in the EU. From the point of view of an optimal provision of public good (Fig. 4.1), the key is the aggregated WTP/WTA (for the moment, let us assume WTP and WTA are equal) and not the individual levels of the willingness. If we take maximizing social welfare as the guiding policy criterion, then at the point of optimal supply of public goods, aggregate marginal willingness to pay (not individual) is equal to the marginal cost of providing public goods. Therefore, increase in social welfare depends on two terms: marginal increase in the utility of household and marginal utility of income of household (the social welfare maximizing problem is described in detail by Johansson, 1993, pp. 16–17 and developed in Sect. 4.4 herein).

There are indications that in most EU countries, due to high levels of generalized morality (World Value Survey, 2021), altruism will be the third force after the state and the market in the provision of public goods and will largely offset moral hazard. There is, however, a lack of direct research on this issue. The main problem does not lie in the mismatch between the institutional valuation of public goods and their net utility (i.e. after including negative externalities) at the level of individual households, since the optimal point is determined by the aggregate marginal willingness to pay. A bigger problem is the differences in marginal utility of household income between richer and poorer countries.

These differences are significant in the EU if we compare the so-called Old and New Member States. Wages in the new member states, even after taking into account differences in prices, are only 20–40% of the EU average, and GDP is 50–70% of the EU average. In determining WTP, the marginal utility of public goods is weighted by the marginal utility of income, which is relatively higher in poorer countries. This means that in poorer countries aggregate WTP is also lower. It is necessary to reckon with a relatively higher cost of providing public goods due to a higher level of transaction costs, which generally characterizes less mature market economies correlated with a lower level of generalized morality and trust (see World Value Survey, 2021). Thus, with a uniform system of support for the provision of public goods by the CAP, one should expect their lower provision in poorer EU countries. Therefore, it would be more appropriate to focus on differentiating environmental policies according to the marginal utility of household income across countries than on matching them to the heterogeneity of net public goods at the farm level.

From the above two premises (i.e. aggregated WTP and differences in marginal income utility), it then follows that result-oriented measures would make more sense if outcomes were measured in net terms (as the balance of positive and negative external effects) and the valuation of this outcome was matched to differences in marginal income utility of households between countries in the European community. Of course, the challenge would be to accurately estimate marginal cost and aggregated willingness to pay for public goods in a "gross" or "net" approach (without taking into account interactions between external effects or with consideration of these interactions). Therefore, it will never be an exact calculus. Hence, the assumption that a policy will be effective if it reduces slacks (i.e. creates conditions for Pareto improvement) and increases the multidimensional efficiency of the adjusted production function seems more feasible, albeit, oversimplified to some extent. However, such an indirect approach may be a second best solution.

Olson's (Olson, 1995) traditional approach to evaluation distinguishes the following stages, which are easily adapted to slack-based evaluation:

1. The objectives of the evaluation are established, e.g. whether the applied policy measures increase the aggregate efficiency of adjusted production function and whether they reduce particular slacks;
2. Alternative policy measures to be evaluated are identified;
3. Operational evaluation criteria are developed—in slack-based evaluation, there are two criteria: aggregated efficiency and particular slacks level;
4. A criterion weighting system is introduced—the advantage of the frontier-based approach is that no arbitrary choice of weights is necessary;

5. Effects are assessed in relation to the adopted criteria—the assessment is purely quantitative;
6. Comparative analysis is performed for different measures.

The main argument in favor of application of result-based measures is that setting a tariff for specific public utilities means that authorities do not have to estimate the marginal costs which are the basis for setting the value of an environmental subsidy. On the other hand, farmers would be rewarded for the public goods provided ex post without any guarantee that land sustainable management will continue in subsequent years. The weakness of environmental indicator-based payments is thus, the assumption of 'ceteris paribus' when setting the tariff for designated environmental utilities (e.g. the permanent grassland indicator, the forest edge indicator, the biorich trees indicator, the historical relics indicator) (see Hasund, 2013). However, it is uncertain whether the listed indicators do not interact with other environmental issues, e.g. the issue of carbon sequestration, GHG emissions by higher livestock density, or the decline in biodiversity due to secondary succession of specific tree species. Hence, the tariffs inherently take a piecemeal approach to environmental quality.

Moreover, there remains the issue of slacks in the realization of farm production functions. Calculation of fees on the basis of selected environmental indicators (valued on the basis of revealed or stated valuation methods) may result in not taking into account inefficiencies that occur as a side effect in managing other farm resources. In such a situation, the introduction of a public goods tariff would not have the expected effect because farmers will simply not use it. In other words, if subsidies are estimated on the basis of the environmental effect as perceived by consumers or authorities, rather than the cost to farmers, this may result in increased inefficiency in the production and management process. The introduction of result-based measures would have to be based on the assumption that WTP is equal to WTA.

An additional complication is that result-based measures are to some extent incompatible with CAP Regulation 1974/06, Article 27, according to which payment subsidies may constitute no more than compensation of costs up to 20% of the transaction costs. It is also questionable whether linking payments to environmental indicators does not contradict the WTO Green Box regulations, which allow only for compensation of costs (including opportunity costs).

Even if a system was developed that was feasible (assuming WTP = WTA), satisfactory in terms of environmental effects, and, taking into account potential inefficiencies in the production function (including transaction costs on the part of the farmer and the public institution), socially acceptable, fair, and in compliance with CAP or WTO rules, the question arises as to how internationally replicable it would be? Can the valuation of individual public goods be standardized or is it context-specific? With respect to the types of indicators proposed by Hasund (2013) and Birge et al. (2017), we have some doubts about this.

The mentioned forest edge indicator, bio-rich trees indicator, historical relics indicator, but also indicators of occurrence of certain plant and animal species can only function in a specific geographical and historical context, which gives them the characteristics and value of public goods. A common payment tariff per indicator unit of public goods for all EU countries, therefore, seems to be an unfeasible undertaking (although other authors express the opposite opinion—e.g. Randall, 2002). An additional complication could be political resistance related to the difficult to predict distributional effects of result-based measures between regions and countries of the European Community.

4.4   Why Is an Optimal Supply of Goods Impossible to Achieve?

In summary, in determining the optimal level of public goods, the researcher faces a number of theoretical and practical problems (implementation issues) which means that without active management, public goods from agriculture may not be provided at all. These problems are listed below. Even if active management is implemented, it is unlikely that the optimum level of public goods can be established. The following paragraphs refer to Fig. 4.1; the list is divided into theoretical issues and implementation issues.

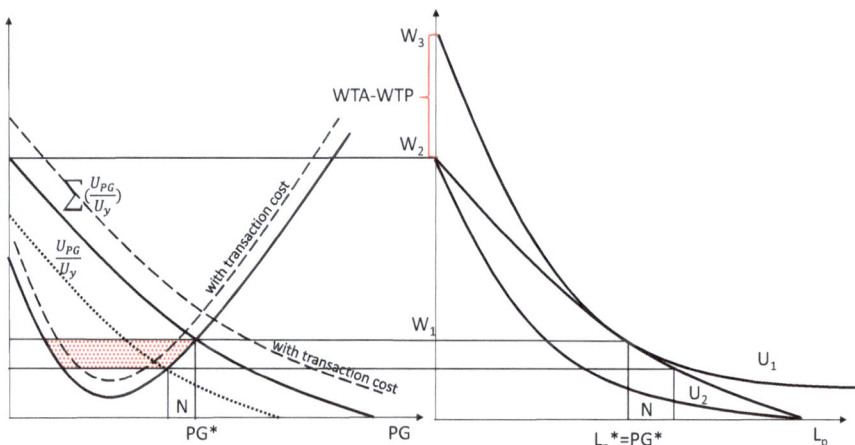

Fig. 4.1 Determinants of optimal public goods provision. Source: own elaboration, c.f. Johansson (1993), Miceli and Minkler (1995), Hasund (2013). Note: $\frac{U_{PG}}{U_y}$ is a ratio of marginal utility of public good to marginal utility of income of household, i.e. individual WTP; $\sum \frac{U_{PG}}{U_y}$ equals marginal social benefits, i.e. aggregated WTP; Dashed lines indicate respectively marginal gross social benefits with transaction costs of the public authorities and farmer's costs including transaction costs; PG*— optimal public goods supply; L*$_P$—bundle of exclusively private rights associated with ownership of land that ensures optimal PGs provision (private entitlements L are the reverse of public entitlements PG); U_1, U_2—indifference curves; W_1, W_2, W_3—prices, i.e. compensation values for private entitlements to land taken in favor of public rights; N—shortage of PG with regard to its optimal provision

4.4.1   Theoretical Issues

1. Marginal social benefits (i.e. aggregated willingness to pay WTP) is not a simple sum of individual public goods utilities because there is a complexity issue—i.e. synergies and trade-offs between individual environmental, economic and social objectives.
2. The individual WTP/WTA curve runs below the collective WTP/WTA curve and therefore the social and environmental production function may face inefficiencies (slacks); as a result, calibrating the production function with factor-augmented technology, including policy variables, may be difficult.
3. Farmer transaction costs may be non-additive with the other cost categories and take a qualitative form (e.g. behavioral lock-ins), in which case they should be classified as uncontrollable variables.
4. Marginal rates of technical substitution (MRTS) of inputs vary across farms, and average rates of technical substitution vary across agricultural sectors within countries; the marginal rate of substitution tells how much one type of input must be increased when another type of input decreases by a unit for revenues to remain unchanged. Moreover, the elasticity of substitution E_{ml} is thus different for various levels of inputs and so the CES assumption is not satisfied. In the line with this assumption $E_{ml} = 0.90$ implies that a 1% decrease in the ratio of fixed assets to employment shall be offseted by an increase in the marginal technical rate of substitution of 0.90%, in which case, net revenues will remain constant.
5. The reason for the issue from the previous point is the attempt by agricultural policy to adjust the production function with environmental and social outputs, which results in a change of optimization criteria from individual to collective. Moreover, altruism comes into play, as common goods are categories that refer to generalized morality. As a result, slacks are expected, and the optimal level of desirable outputs is not achieved by itself, hence, the determination of a single curve of the marginal cost of agricultural production is not possible.
6. There may also be endogenous relations between the inputs; in particular, they may concern human and social capital, which in agriculture may constitute a barrier to the implementation of technical and organizational progress.
7. In light of the above, assessing agricultural policy effectiveness in isolation from a multidimensional production function may be subject to omitted variable bias; hence, an aggregate and holistic perspective is less risky.
8. There are no pure public goods, only common goods; thus, the social benefits line may not be additive with private benefits. The condition for the provision of public goods is that the efficiency of the use of other resources is maintained at least at its current level (Pareto improvement condition).
9. Endowment effects in agriculture mean that WTP may be less than WTA—which may deter farmers from providing public goods in a sustainable agriculture model. This is illustrated in Fig. 4.1, which is expanded herein based on the analysis by Knetsch and Borcherding (1979) and Miceli and Minkler (1995). On the right-hand side of Fig. 4.1, the X-axis shows the extent of private

entitlements for the land resource that could potentially provide public goods. Obviously, the greater the extent of private entitlements, the less public (common) goods are provided—the opposite of the X-axis on the left side of the figure. The Y-axis represents the compensation value for private entitlements taken. The indifference curve U_1 depicts the combination of wealth and private entitlements to resources indifference for the landowner. Hence, the owner is indifferent at point W_3 when he is completely deprived of private entitlements (only public entitlements remain, i.e. the entire land resource is a common good) in exchange for a price W_3 that equals his WTA. The slope of the budget line represents either the market price or the institutional price of the farmer's public land entitlement, or simply WTP. At the point of optimal PG provision, the compensation equals fair WTP value W_1 as based on the market or institutional valuation. If the entire private rights to land were taken, the owner would receive W_2-W_1, whereas he values this expropriation at W_3-W_1. This means that market/institutional valuation under compensates the owner by W_3-W_2 with regard to his WTA, and forces the farmer to move to a lower indifference curve U_2. The farmer may be reluctant to do so and then public utilities are not provided at all. Hence, W_3-W_2 reflects the difference between WTP and WTA.

10. WTP and WTA are equal to the ratio of the marginal utility of public goods and the marginal utility of money (Johansson, 1993). The latter varies significantly depending on the economic development and wealth of the country. In poorer countries, the marginal utility of money is higher, which may translate into a decrease in WTP/WTA for public goods. Not only does the CAP not propose uniform hectare subsidy rates for all EU countries, it offers much lower rates to the Central and Eastern European countries that joined the European Community in 2004, by weighting subsidies by purchasing power parity (as wage rates and prices are still lower in these countries).

11. This policy results in farmers in the Central and Eastern European countries benefitting relatively little from agri-environmental programs. It is particularly the case of small farms that have high environmental potential, but are accompanied by high marginal utility of money. In relation to consumers, the marginal utility of money in poorer countries is also relatively high, which lowers the WTP and the possibility of market-based valuation of PGs (e.g. poor development of market for organic products).

4.4.2 Implementation Issues

1. The transaction costs of implementing specific programs (including political frictions) lift the marginal WTP curve and can be difficult to estimate.
2. A dissonance emerges between the holistic nature of the sustainable development goals and the single-criteria evaluation of the effectiveness of individual policy instruments.

3. There is a similar dissonance between environmental and social goals at the global and local levels; however, some hierarchy of goals must be established, and in our view, a holistic and global perspective should take the lead.

4. In practice, the supply of public goods may turn out to be less elastic than assumed in the theoretical model of their provision. Referring to Fig. 4.1, the PG* set at a given aggregated WTP may not be able to increase even if the WTP line moves to the right. The tendency for such a WTP movement will increase as the market demand for eco-utilities from agriculture increases. Over time, a market for public goods may develop and an increasing proportion of the eco-utilities will be capitalized at market prices, thus cushioning agricultural policy to some extent. However, it is difficult to assume that agriculture will be able to increase the supply of eco-utilities proportionally, given that natural resources are limited and largely exhausted already, especially in the highly developed countries where the above trend of the increasing demand is drawn. Therefore, if the WTP line shifts to the right and the supply of PGs cannot be increased accordingly, the suppliers of public goods will realize a producer's rent resulting from the price increase on eco-utilities (supposing a constant supply of them). However, this mechanism also works the other way, which is very often overlooked in research. In less developed countries (we are talking in this context about the so-called EU-12, i.e. the post-communist countries of Central and Eastern Europe that joined the European Community after 2004), there are a large number of small family farms. In the EU, they are defined by standard output (SO) classes (average production value in 5 years in a given region)—small farms have up to 15,000€ SO per year. Such entities provide a relatively large amount of public goods mainly because they are small and *per se* extensive. They provide public goods even if they do not receive any payment for it—therefore, this group of farms can be called "forced providers of public goods". The individual WTP line in Fig. 4.1 illustrates a situation in which the individual consumer is not willing to pay for public goods (which does not mean that he does not use them as a free rider). In such a situation, public goods should be provided at the PG_0 level instead of PG*. Given the problem of factor immobility, however, the farmer still supplies an amount of PG*, incurring a loss whose value is determined by the shaded box in Fig. 4.1. This loss is, at the same time, a kind of consumer surplus, which can be called a "social surplus", because the consumer has the opportunity to benefit from certain environmental utilities even if he does not make an individual demand for them (from a social perspective, however, it is desirable that these utilities be supplied—a more extensive discussion can be found in Czyzewski et al., 2021).

The case of small-scale farming is important globally (as we advocated in Chaps. 1 and 2), so we will zoom in on the essence of "forced provision" of public goods based on the case of small-scale field crops farming in Poland.

4.4.3   The Case of Small-Scale Field Crop Farms in Poland

Because there is no single official definition for a small-scale farm, various criteria may be used to determine the number of small farms in Poland. According to the official national statistics agency Statistics Poland, small farms represent almost 54% of the 1.3 million existing Polish farms, and used approximately 13% of the country's registered agricultural land area. They have a considerable potential with regard to participating in agri-environmental schemes (AES) or converting to organic farming. We define a small-scale farm by its economic size using the Farm Accountancy Data Network (FADN) typology of very small and small farms (nES9 Type 1 and 2), which is equivalent to an economic size of 4000–15,000 EUR SO.

To draw a clear picture of the agri-environmental potential of such a farm, let us set the additional criterion that at least 50% of the houshold's family work unit is engaged in farm activities. Such farms, where the majority of household labor is devoted to agriculture, are fewer in number, about 160,000 in Poland, but in Romania (as the second largest country in Eastern Europe with regard to the agricultural sector), about 1.19 million. Based on the above criteria, we can describe an average small farm that focuses mainly on agricultural activity in Poland as one that has, on average, 14 ha of utilized agricultural area, where 1.6 Family Work Units (FWU) of labor input are engaged in the farm activity of raising field crops—38% of all small farms, or of mixed production—33% (according to FADN type of farming typology), with 53,000 PLN (€11,600) of yearly agricultural products value. Operating capital endowment: i.e. mineral fertilizers, pesticides, energy expenditure accounts for 25% of the output value (which may bring about a relatively good eco-efficiency score). There are, on average, 6.27 livestock units. In addition, the farm manager had received a general vocational school education.

Moreover, the farm has potential to provide environmental public goods through sustainable land use and protecting biodiversity, as permanent meadows represents about 20% of UAA, winter land coverage accounts for 42% of arable land and soil organic matter index is positive and equals 1.48 (so there is a room for improvement).

Such a farm receives, on average, 35% of agricultural income from CAP subsidies, of which, only 6% can be classified as AES (it could definitely be a higher share). The share of agricultural income in total household income equals, on average, 82%.

Considering the value chain, we see the predominant type of distribution channel consists in this case of selling to a processing plant or other intermediary (82% of all small farms in Poland). Regarding contractual integration, ad hoc sales without any contract dominate (74%) and the farm also has different suppliers without any long-term contracts (75%). Thus, there is a vast room for improvement in terms of value chain from a farmer's perspective. Small farms can relatively easily convert into organic farming or implement AES, increasing their incomes at the same time (EC, 2019). Nevertheless, organic food production has been stagnating or regressing in

Central and Eastern European countries, as described in Chap. 2, and the problem mainly touches small-scale farming.

There is no easy answer as to why a small-scale farmer accepts the losses shown in the Figure 4.1, in the case of a lower WTP than required for PG*. If this were the case in Western European countries with an established capitalist system, we would say that it is the result of altruistic attitudes related to the need to care for the common good. In the case of Eastern Europe, however, the answer that was once given by Czajanow (1966) is to some extent still valid.

In Europe, but also in many parts of the world, the mentality of more peasant than farmer prevails. Small subsistence family farms are not profit-oriented, but motivate their activities by the level of satisfaction of family needs. Therefore, they do not go bankrupt in the traditional sense, but become self-sufficient in various ways under adverse market conditions. The factors of production in smallholder agriculture, especially labor, are not very mobile. Peasants are tied to their land and their farms, which makes them resistant to low living standards and income decline. On the other hand, their chances of finding a job in the city are low due to lack of education and money. This is particularly evident in the post-transition countries of Central and Eastern Europe, where State Farms dominated until the 1990s. Moreover, it is also a side effect of agricultural policy and social subsidies that help small, unprofitable farms to survive.

4.5 Dilemmas of Effective Policy Designing: Synergies and Trade-Offs

4.5.1 Direct GHG Emissions

Climate change is a major environmental challenge for national and international communities. Since 1990, the European Union has pursued a strategy of reducing GHG emissions by 20% by 2020. The data presented in paragraph 4.1 show that this has been achieved with some improvement (i.e. reduction by about 21%), of which the reduction in agriculture alone amounts to about 16%, of which 14% falls in the initial period 1990–2004. Let us remember that most of the so-called "New Member States of the EU-12" from Central and Eastern Europe joined the Community in 2004, and since then the process of reduction of emissions from the agricultural sector has significantly slowed down or even stopped (see Table A.8).

As we have mentioned, the agricultural sector makes a significant contribution to GHG emissions, but also holds a large technical potential for its mitigation, which can be exploited in the short term at low opportunity cost by removing existing inefficiencies in production and management. Many authors take a similar view (Freibauer et al., 2004; Smith et al., 2008; Moran et al., 2008; Bakam et al., 2012).

Market-based instruments used to reduce GHG emissions can be more effective than command-and-control (CAC) instruments as long as the transaction costs of public policy are well calculated. Market-based policy measures are supposed to encourage the implementation of specific behaviors through signals about changes in the profitability of production. However, they face the barrier of transaction costs due to uncertainty, lack of knowledge and skills, or high resource specificity in farm ownership. In general, transaction costs are underestimated both at the farm level and at the level of the public institution (Bakam et al., 2012) and may reduce the profitability of agri-environmental programs (Fig. 4.1), but they are also responsible for reluctance to adopt sustainable farming practices.

The theoretical importance transaction cost (in this, we also mean contracts for the provision of a public good) has been well recognized since the precursor work of O. Williamson (1985). In the context of the sector, this issue has been raised by many authors (e.g. Stavins, 1995; Falconer & Whitby, 2000; Falconer & Saunders, 2002; McCann et al., 2005; Bakam et al., 2012). However, in empirical analyses of CEA and CBA, the issue of transaction costs is rarely taken into account, and thus an underestimation of the administrative costs of individual policies comes about (Falconer & Whitby, 2000; Falconer & Saunders, 2002).

Interest in the use of market incentives in environmental policy design is high both in academic discussions and on the policymakers' side. Policy measures to reduce pollution in general can be categorized into several groups of instruments: (1) pollution charges, (2) tradable permits, (3) market friction reductions and (4) government subsidy enhancing the reductions or public goods provision (Povellato et al., 2007; Bakam et al., 2012).

If the polluter pays a fee for the unit's emissions, it is in its own economic interest to reduce emissions (e.g. CO_2 emission tax or N fertilizer tax). The tradable permits system, in turn, sets an allowable overall level of pollutant emissions and distributes them to polluters as permits for a given level of emissions. Production units that have emission levels below the allowed level can sell the surplus permits to other entities that pollute more. The least recognized are the mechanisms of friction reduction to which we could include the slack-based approach used in this work. As a rule, market friction generates transaction costs. Hence, its reduction improves the operation of the market mechanism, enables progress in the Pareto sense and increases production efficiency.

The literature mentions the following public policy instruments aimed specifically at reducing market friction: (1) market development in the sense of a valorization mechanism (i.e. price setting) for inputs and outputs related to environmental quality. Bakam et al. (2012) provided as an example measures enabling the voluntary exchange of water rights; (2) public institutions can promote corporate responsibility, which encourages companies to assess the potential environmental damage caused in the production process and show the market benefits of reducing it. Such benefits can, for example, be achieved by labeling products for positive externalities (e.g. environmental qualities of food, a friendly way of production, etc.) or for energy efficiency.

However, the benefits of including the agriculture and forestry sectors in national tradable permits systems for CO_2 emissions are largely theoretical and not empirically substantiated. In practice, however, the potential efficiency gains may be undermined by transaction costs that are difficult to estimate ex ante, including monitoring costs that may be so high as to offset the assumed improvements in system effectiveness.

In the EU's agricultural sector and that in other highly developed countries, subsidies are the most commonly dispersed policy measures. These, in a theoretical sense, mirror pollution charges and thus generate incentives for sustainable practices. Obtaining subsidies or their level depends on: (1) the application of specific good practices and meeting environmental standards in typical farming systems (e.g. the cross-compliance principle in the CAP covering public health, animal and plant health, animal welfare, environmental protection and good agricultural culture), (2) abandoning production (e.g. fallowing) or changing to a more ecological farming system (e.g. AES, AESC, including "greening" and ecological farming) or maintaining a less profitable but more extensive farming system (e.g. Less Favoured Areas (LFA) payment).

Interestingly, the CAP lacks measures specifically targeted at GHG reduction, as the reduced emissions from agriculture are assumed to be a positive external effect of a combination of other sustainable practices. Perhaps this assumption should be considered flawed, and an attempt should be made at designing result-oriented policy instruments particularly targeted at the farm generated GHG emission regardless of what types of management practices are applied. The comparative CEA for potential emission taxes and tradable permits schemes conducted by Bakam et al. (2012) shows that a 'caps-and-trade scheme', where farmers could freely trade permits among themselves and meet standards set by public authorities, would theoretically be the most effective. In the case of tradable permits, the market interaction of supply and demand sets the permit price as the Nash equilibrium value and mainly for this reason this solution is considered a cost-effective way to reduce emissions to the target ceiling.

A systemic literature review published by Povellato et al. (2007) shows that three different perspectives prevail in assessing the effectiveness of environmental policies in reducing agricultural-sourced GHG emissions: (1) a technical one focusing on the effectiveness of inputs; (2) sectoral partial equilibrium models or general equilibrium models used to examine the welfare effects of policies, (3) bottom-up linear programming models and more complex agent-based models that theoretically allow for up-scaling of results.

The technical approach (1) focuses on assumptions about trade-offs (or synergies) between inputs and their relation to the physical output of the activity, neglecting the economic efficiency aspect (where the output is a value). The second and third strand of literature mentioned above focus on the cost dimension at the sectoral level (in which case, costs are endogenous in the equilibrium model), or on the marginal abatement cost at farm level. They aim at answering the question as to what instruments enable a given target to be achieved at the lowest monetary cost (e.g., a level of carbon sequestration or a specific reduction in CH_4 emissions).

However, the final conclusion of the cited authors was that it is necessary to make methodological progress in the field of holistic treatment of spatial and temporal variability of phenomena, as the contextual conditions at the farm level make it impossible to generalize the results of the abovementioned approaches. This conclusion harmonizes with our proposed cognitive perspective that assumes a comprehensive approach to allocative efficiency of the multidimensional system of inputs and outputs, including desirable and undesirable external effects. In turn, the measurement of the latter should strive to use aggregate measures to avoid trade-offs between ecological indicators. GHG emission may be an example of such a complex outcome measure, which is the result of different stages of farming practices.

In the next decade, climate change will determine global environmental policy. Agriculture, together with forestry, plays a special role in this case, as both sectors are, on the one hand, strongly dependent on climatic phenomena, and, on the other, significantly affect the climate as GHG sources, sinks and sequestration systems.

As mentioned previously, agriculture has the potential to reduce emissions, particularly of methane, through increased efficiency of production, while forest management can improve carbon sequestration. The latter may be supported, to some extent, by agricultural activities in terms of afforestation, set-aside of agricultural land, permanent grassland management and bioenergy harvesting, although it must be remembered that the disposability of these activities is weak and there may be trade-offs due to competition with other land uses and social objectives.

CO_2 sequestration in cropland, for example, is very effective under conditions of zero tillage or bioenergy crops, and can provide a biological potential for CO_2 storage of 90–120 Mt. per year. Nevertheless, socioeconomic and other constraints reduce this potential to only 20% of the full capability (Povellato et al., 2007). Furthermore, referring to the study by Brink et al. (2005), it is worth noting that activities supporting CO_2 sequestration or reducing CH_4 emissions can simultaneously increase N_2O emissions. This is another voice for the necessity of holistically capturing the effects of environmental policies in agriculture. Indeed, on the example of the EU, decoupling agricultural subsidies from production might, therefore prove to be a beneficial solution mainly due to the decrease of N_2O emission, and to a lesser extent, CH_4.

The cited author is skeptical about the possibility of comprehensively estimating and improving the effectiveness of environmental policies in agriculture, as integrating models to account for both social welfare and spatial heterogeneity seems to be a challenge that must be addressed by the next generation of model makers.

Methane emissions from livestock farming slightly exceed N_2O emissions from soils when converted to CO_2 equivalents, with cattle farming being the main source. Accurate prediction of N_2O emissions from soils in agricultural cultivation is difficult, however, because of the complex bionic processes involved—chemical fertilization on the one hand and microbial nitrogen fixation by soil microorganisms on the other. Therefore, projections of GHG emissions from agriculture are subject to large errors of up to several tens of percent (EEA, 2021).

The consensus in the literature is that CH4 reduction is associated with improved management efficiency in the livestock sector. Still, the largest relative decrease in GHG emissions can be achieved by reducing the use of chemical fertilizers. Emissions in terms of N_2O are closely related to mineral fertilizer management, which is the main source of surplus nitrogen and phosphorus on arable land. A side effect of the dual development of agriculture stimulated by the model of the EU CAP (on the one hand, extensive use of land, on the other hand, very intensive farming systems) brings an increase of phosphorus and nitrogen consumption on arable land (phosphorus from processed crop residues) and a reduction in the amount of these nutrients that are returned to arable land from livestock manure due to increasing specialization in crop production and consequently less sequestration. It should be noted that in other parts of the world, rice cultivation is a major contributor to N_2O emissions. As for methane, livestock farming is mainly responsible for this, while conversion of forests into cropland (deforestation) increases carbon dioxide emissions.

The CAP is gradually being adapted to the Climate Action coordinated by United Nations Framework Convention on Climate Change (UNFCCC) that focuses on mitigation and adaptation policies towards global climate change (UNFCCC, 2021). The UNFCCC identifies three leading strategies for adaptation of the agricultural sector to global climate policy goals: (1) direct reduction of GHG emissions, (2) carbon sequestration—carbon captured from the atmosphere can be sequestered in biomass, soils and agricultural raw materials from certain crops, and (3) substitution of fossil fuels by bioenergy and energy conservation—burning of fossil fuels enables the CO_2 that was stored in fossil deposits to be released into the atmosphere. Moreover, the implementation of these strategies should be guided by a CEA analysis aimed complementarily at maximizing social utility and minimizing inputs (OECD, 1989; Povellato et al., 2007).

Still, the cost-effectiveness of particular policy schemes does not always imply optimal allocation (in the Pareto sense). As argued by Tietenberg (2004), efficient policies usually are cost-effective, but cost-effective policy is not always efficient.

The distributional effects of environmental policy, macroeconomic issues and administrative transaction costs, as well as those at the level of the policy recipients must also be taken into account. The difficulty, however, is to integrate environmental goals with economic and social goals, which are also part of sustainable development. Unfortunately, policymakers have trouble demonstrating the social benefits associated with environmental policy, with the result that such integration may become politically unfeasible because it will not receive broad public support. Thus, environmental goals are often merely declarative and their implementation is ostensible.

An *ex-ante* analysis of the effectiveness of the environmental policy in terms of GHG reduction is extremely difficult to carry out, as it has a complex impact and it is hardly assessable in the time range and on the territory that the effects of the policy should be measured. It seems relatively easier to perform an *ex-post* analysis.

An interesting thread is the assessment of the potential consequences of climate change on agricultural productivity in Europe (Olesen & Bindi, 2002; Ewert et al., 2005), which may interact with projected reductions of GHG emissions. Projected long-term climate changes in temperature, precipitation and CO_2 concentrations should, on average, increase resource efficiency in European agriculture due to average yield increases (increased crop productivity of 25–163% (Ewert et al.,

2005)) and technological developments. However, climate change will mainly be beneficial in agricultural terms in the northern EU member states, while the southern ones will lose out.

In summary, there is no consensus in the literature on the mitigation potential expressed by marginal abatement cost curves. Generalizing, one can assume that a tax on CO_2 equivalent emissions that is equal to the carbon price threshold, could be profitable. However, studies conducted both in Europe and in the USA confirm that reducing CH_4 emissions is cheaper than doing so with N_2O. Moreover, there is no doubt that technical and economic efficiency is crucial in determining the marginal costs of emission reductions and for the effectiveness of different policy options for reducing GHG emissions from agriculture.

Equally important is the previously mentioned theme of "positive" and "negative" synergies induced by environmental policy between different emission sources—e.g., there is a positive synergy in efforts to reduce CH4 emissions, as ammonia emissions are also reduced. On the other hand, trade-offs can occur between reducing ammonia and N2O emissions (e.g. Brink et al., 2005).

The effectiveness of policies to reduce emissions of key GHGs from agriculture is thus a very complex issue. Considering this issue from the perspective of single emission sources may lead to contradictory conclusions. There is no doubt, however, that the potential for GHG reductions from agriculture is large and so far little utilized. However, the programs of the agri-environmental policy should not act against the economic optimization of production processes in agriculture, as the increase of allocative efficiency itself belongs to the main factors of GHG reduction.

4.5.2   Carbon Sequestration

Forested land, agricultural land covered with trees and bushes and afforestation on agricultural land play important roles for carbon sequestration. Therefore, it is obvious that there will be some trade-offs between the economic and environmental dimensions of sustainability due to competing land use and accompanying opportunity costs. Thus, maximizing benefits should not be taken as a single optimization criterion. For example, from an environmental point of view, the most fertile fields should be devoted to the production of biomass for carbon sequestration. Such management obviously raises high opportunity costs, after which sequestration on pasture and uncultivated land becomes a more obvious option (Povellato et al., 2007). However, an alternative to this option is to promote afforestation on private farmland (or counter deforestation) and forest maintenance. It should be noted that a forested square with a side of only 100 meters, produces about 700 kilograms of oxygen per day. Indeed, forests in the territory of the EU absorb the equivalent of almost 10% of total EU GHG emissions yearly.

The so-called "land-use, land-use change and forestry sector" (LULUCF) was included into greenhouse gas accounts by the Kyoto Protocol adopted in 1997. Up to 2020, individual EU Member States were obliged under the Kyoto Protocol to ensure that greenhouse gas emissions from land use were compensated by an equivalent removal of CO_2. The LULUCF Regulation (EU) 2018/841 sets a "no debit" rule that is a binding commitment for each Member State to ensure the above compensation through active management (EC, 2021b). Forests on agricultural land have a multifunctional character and, due to their recreational and landscape values, enrich the agritourism offer. Thus, they represent a way of additional capitalization (in the market prices of agritourism services) of environmental public goods, building coherence between economic and environmental perspectives.

At the same time, if the agri-environmental policy also supports the reduction of the tree rotation age and the development of wood production for professional purposes (e.g. in construction, furniture industry, etc.), additional economic value is created for the farmer (or forest owner) while maximizing the benefits from carbon sequestration (Garcia-Quijano et al., 2005; Tassone et al., 2004). The Rural Development Programme (RDP), within the second pillar of the CAP, provides EU Member States with a framework to pay silvicultural premiums for forests on private land ("Support for investments increasing the resilience of forest ecosystems and their environmental value") and thus to incentivize more climate-friendly land use.

On the example of Poland, it should be noted that although the idea seems cost-effective in theory, the transaction costs associated with obtaining such subsidies from RDP are so high that farms owning relatively small afforested plots (1–5 ha) may not be interested in this solution. The main barrier is the fact that RDP funds are disbursed after the investment has been carried out, and after meeting a number of detailed conditions and after the on-the-spot check has been carried out successfully. Although forest utilities may to some extent capitalize in the price of the harvested wood, this perspective may be too distant and uncertain for the farmer.

The other side of the coin is the impact of the change of land use in a land registry and in a local development plan on its market value. For example, in Poland, agricultural land that is afforested spontaneously or intentionally by the owner is subject to reclassification even without the owner's consent during periodic inventories conducted by county authorities. Due to the special protection of forests, this process is, in practice, very difficult to reverse from a legal point of view. Agricultural land, once reclassified as forest, remains forest no matter what happens in reality to the trees growing on it, even if they wither due to disease, pests, drought, etc. The owner should then plant new trees. Trees also cannot be cut down without the permission of the administrative authority under the threat of very high financial penalties.

From the legal and administrative point of view, deforestation in the EU (similar legislation is also in force in other countries of the Community) is therefore very difficult, and in many cases impossible. The land designated as forest in the local land management plan is subject to numerous restrictions—in particular, it is hardly possible to apply for the right to develop it in any way. As a result, it is almost completely excluded from any investments made by the farm, except for forest management. This is reflected in a significant drop in a market price in relation to land designated in the local development plan as agricultural or building land— from 50% to 75% (not counting the potential value of timber harvestable in the future). Afforestation is therefore associated with a significant financial risk for the value of an agricultural holding in Poland, including the risk of the permanent loss of value.

These conditions, to a large extent, offset the above-mentioned benefits of capitalizing public goods in the prices of agro-environmental services or forest management products. The provision of a public good in the form of a forest and its environmental services is thus associated with a significant reduction in the owner's ownership rights, which obviously reduces his/her willingness to accept the incentives offered by the agri-environmental policy towards forest management on farmland. It is therefore necessary to look in the future for a more reasonable compromise between the protection of the forest ecosystem as a public good on private land and the property rights of private owners. As it stands, the protection of these public goods is overly expropriatory, and may significantly slow down the development of carbon sequestration potential, which might be a very effective pathway.

Carbon sequestration efforts are theoretically tied to the amount of a carbon tax. Once the tax threshold is exceeded, carbon storage in forests becomes economically effective and allows, in theory, for a lower tax on CO_2 emissions. The two mechanisms—carbon sequestration policy and carbon tax are thus theoretically coupled.

One of the outcomes of the Kyoto conference was the world's first international emissions trading system "the EU ETS" set up in 2005. Under this system, power plants and industry must hold a permit for every tone of CO_2 produced. The price of these permits is determined in a quasi-market system depending on the level of supply and demand. After the global financial crisis in 2008–2009, the demand for and price of permits fell steadily (between 2013 and 2017, they oscillated around €5, but by the end of 2021 they reached a record €75). The low price of CO2 certificates discourages companies from investing in green technologies, but the high price compels companies to relocate production outside the EU (the so-called "carbon leakage"), or to pass the costs on to the final customer (rising energy prices).

The EU's efforts to reduce its carbon footprint as part of the European Green Deal and move towards climate neutrality by 2050 are, therefore, undermined by less ambitious countries. To address this problem, the EU intends to reform the ETS by introducing a carbon tax to be levied on products from countries with less ambitious regulations than the EU, making the imported product no cheaper than a similar EU product—i.e. the Carbon Border Adjustment Mechanism (CBAM). The project is aimed solely at achieving global climate targets and promoting a level playing field in the world and should not be seen as protectionism. This mechanism will become part of the new EU industrial strategy. According to the European Parliament, the CBAM should by 2023 cover energy and energy-intensive industries, which account for 94% of EU industrial emissions and still receive substantial free allocations (EP, 2021a).

In terms of carbon sequestration, there are also some simple solutions that could be implemented without significant changes to the current CAP subsidy system. In areas of compulsory set-aside which determine the first pillar of the CAP, an integrated farming system consisting of short rotation coppice, biofuel or biomass oriented crops, or at least biofuel strips separating fields, would suffice.

There is a general consensus in the literature on a certain hierarchy of carbon sequestration strategies in terms of their cost-effectiveness: conversion of agricultural land to forests is relatively more expensive than other strategies; biomass energy harvesting as a long-term option is more effective than afforestation because of the areal production efficiency and greater ability to reduce emissions. Nevertheless, CO_2 sequestration is only a complementary strategy to direct agricultural mitigation (Povellato et al., 2007). Carbon storage in forests or agricultural soils is more costly than direct action in the livestock sector and fertilizer management (e.g. reducing the stocking rate and limiting nitrogen fertilizer), so the potential for emission reductions in these areas should be exploited first. In turn, the introduction of a tradable permits system (see Sect. 4.4) can reduce the potential costs of direct emission reductions.

4.5.3 Bioenergy

The first European directive on renewable energy sources (EP and Council Directive 2009/28/EC), established that 20% of the EU's total energy consumption must come from renewable sources by 2020. Moreover, in the transport sector, Member States committed to achieving a 10% share of fuels from renewable sources. Subsequently, in December 2018, the revised Renewable Energy Directive (EP Directive 2018/2001) came into force as part of the "Clean energy for all Europeans" package, which raised the binding target to 32% by 2030 and also increased the target in the transport sector to a 14% share of energy from renewable sources (i.e. biofuels). In July 2021, in a package called the European Green Deal, the EC proposed to amend the Renewables Directive by raising the binding target to 40% by 2030 (EP, 2021b).

Currently, the cost-effectiveness of the strategy of replacing fossil fuels with biofuels depends on the prices of the latter. So far, it has been too costly relative to other solutions for reducing CO_2 emissions, but with the price relationship between agricultural commodities and non-renewable energy being as it is at the end of 2021, it is becoming increasingly cost-effective.

The competitiveness of bioenergy production in the EU has always depended on the level of its subsidization because bioenergy prices have been higher than their fossil fuel energy equivalent. In addition, there is generally a negative net energy balance when substituting petrol for bioethanol (this is not the case with sugar beet-based bioethanol). The biggest drawback of the strategy of limiting GHG emission through the production of biofuels is the competitiveness of this production for agricultural raw materials in the food chain (land use competition). This process brings about not only an increase in food prices, but also more intensive and input intensive management on land, which remains in typical agricultural use (i.e. crop production), causing negative environmental effects. Therefore, other ways of using biomass energy can be more effective. Here we mean, for example, biogas plants, which can provide energy for individual farms or even entire villages. However, due to the relatively low uptake of this way of using biomass in agriculture itself, there is a lack of representative research on its environmental and economic costs and benefits. For a biogas plant to be economically viable, individual farms must take collective action to provide sufficient biomass feedstock, which must otherwise be procured. This significantly increases the transaction costs of the whole process.

The biogas market is growing in many EU countries due to the favorable financial support system in the form of subsidies for biogas energy production. The leading positions in agricultural biogas production are held by Germany, Italy and the United Kingdom, where financial support has been above €270/MWh for several years. Currently, similar to Poland, some countries are switching from a system of certificates and guaranteed prices, to an auction system. For example, in France it has been assumed that this system will generate investments worth €170 million, which will have a production potential of about 480 MWh of electricity per year (Gostomczyk, 2017). According to cost-effectiveness estimates by Börjesson and Ahlgren (2012), however, only a small part of the technical biogas potential can be effectively utilized without subsidies. Although low subsidies can give rapid increases in biogas utilization, biogas installation with full technical potential requires high subsidies that reduce significantly its cost-effectiveness (Börjesson & Ahlgren, 2012). It can be said, therefore, that the EU policy to promote biomass energy (and especially biofuels) seems to be currently not fully justified from the point of view of economic and environmental accounts regarding the opportunity costs of other strategies to reduce GHG emissions in agriculture.

To conclude, the cost-effectiveness of a policy depends on the type of instruments used. The most cost-effective are market-based instruments such as taxes, subsidies or tradable permits as opposed to CAC instruments such as emission standards and emission quotas. Market-based instruments have the virtue of equalizing the marginal costs of abatement for different sources of pollution, i.e. in practice, they reduce harmful emissions first where it is most cost-effective for farmers to do so, although not necessarily where the environmental need is greatest. Consideration of the transaction costs of public institutions, however, may change the above assessment and result in the selection of CAC instruments that are easier and faster to implement. This theme is particularly relevant to the design of strategies to support biodiversity and deliver landscape public goods.

4.5.4 Biodiversity and Landscape Public Goods

The long-term EU policy on biodiversity protection is largely based on the global guidelines set out in the documents of the Convention on Biological Diversity (CBD) and the findings of the tenth Conference of the Parties to the CBD in Nagoya 2010, which resulted in the EU Biodiversity Strategy to 2020 adopted in 2011. Although the evaluation of the latter strategy is ongoing, previous evaluations of EU biodiversity conservation policies have not fared well, having failed to halt the progressive biodiversity loss beyond the natural rate of species replacement (Braat & ten Brink, 2008; Klassert & Möckel, 2013).

According to the Braat and ten Brink report cited above, the cumulative loss of ecosystem services worldwide is valued at €14,000 billion by 2050, equivalent to about 7% of projected global GDP. The EU Biodiversity Strategy to 2020 (2011–2020) was aimed at halting biodiversity loss and at restoring as many ecosystem services as possible. In particular, the Strategy implemented such trajectories as: protecting species and habitats, maintaining and restoring ecosystems, achieving more sustainable agriculture and forestry, making commercial fishing more sustainable and combating invasive alien species (EC, 2020c).

On 20 May 2020, the European Commission adopted the draft version of the EU Biodiversity Strategy 2030, which assumes, among others, three key points: (1) establishment of protected areas covering at least 30% of the EU land and sea areas (i.e. much wider than the current Natura 2000 areas), (2) restoration of degraded ecosystems, in particular, through 50% reduction in the use of pesticides and by planting trees (three million trees), (3) allocation of EUR 20 billion a year for the protection and promotion of biodiversity from EU and national funds (including private funds) (EC, 2020c). As can be seen, these assumptions are much more precise compared to the objectives of the previous strategy. Interestingly, they emphasize the issue of reducing pesticide use, which is very often mentioned in the literature as a Pareto-optimal direction for implementing sustainable practices in agriculture (Chèze et al., 2020).

Regarding the specific tools for implementing a biodiversity strategy, many authors point out that market-based instruments (MBI) provide an opportunity to increase its cost-effectiveness and achieve environmental objectives beyond the level made possible by CAC regulations (Ring & Schröter-Schlaack, 2011; Klassert & Möckel, 2013). As we mentioned earlier, market-based instruments can be classified as price-based instruments, such as taxes and subsidies, and quantity-based instruments, such as permit trading. The latter are assumed to be particularly desirable when environmental degradation is more costly to society than distorting the market mechanism (Weitzman, 1974).

The effectiveness of biodiversity policy depends on several specific elements that do not apply to other ecological indicators. Therefore, effective implementation of this policy is perhaps the most difficult of all aspects of environmental sustainability. To be effective, habitat and ecosystem conservation policies must not only provide a certain level of ongoing protection, but also inter-period and spatial coherence. Intertemporal relationships involve so-called "threshold effects", which can cause ecosystems to collapse when they are stressed (Ring et al., 2010; Dickie & Tucker, 2010). Market instruments, e.g. tradable permits, but also subsidies, lead to a dual allocation of environmental activities (in some agricultural sectors there is an accumulation of protective activities, in others, there is an accumulation of pressure on the environment), as a result of which the mentioned thresholds may be irreversibly exceeded.

Spatial coherence, on the other hand, concerns the maintenance of habitat networks necessary for populations with different dispersal patterns to thrive (Reid,

2011; Wissel & Wätzold, 2010; Hartig & Drechsler, 2009), and here, too, market-based instruments may fail. Problems with inter-period and spatial consistency affect CAC-type measures to a lesser extent.

4.5.4.1 Tradable Permits

Klassert and Möckel (2013) considered two new directions of biodiversity protection measures in the EU, which could improve the effectiveness of the support system, but which may also be promising solutions in other parts of the world: (1) Site selection under a tradable permit scheme, and (2) Habitat banking. Let us try to refer to the possibility of their implementation from a slightly different perspective than the cited author, who considered mainly the legal aspects.

In the first case—i.e. site selection, this type of transferable entitlement system in relation to the current European Natura 2000 protected area would allow for agricultural management in this area under the condition of buying permits from another landowner who has land outside the current protected area, but with high environmental values and potential for their development under protected conditions. This would remove the contradiction between economic objectives in agriculture and the provision of biodiversity public goods. Protected areas could be designated more precisely and, most importantly, would be given a quasi-market value. In our opinion, the system could be complementary to current protected areas and concern the buffer belt around the unique habitats of the Natura 2000 area. It must be noted, however, that full freedom of land exclusion from the protected area through site-certificate purchase would cause loss of those unique ecosystems that should be strictly protected. We mean here, for example, river valleys inhabited by rare species of fauna and flora.

Another solution could be limiting the types of investments and land management that a farmer could carry out in the areas excluded from the protection system. We mean here, for example, investments in renewable energy sources, e.g. photovoltaic farms, whose construction in Natura2000 areas encounters numerous administrative and legal barriers. This solution seems worth considering in the context of the EU Biodiversity Strategy 2030, according to which protected areas are to be extended to cover at least 30% (!) of EU land and sea areas. The criteria for designation of protected areas in individual member states are not entirely clear in the cited strategy, and a system of transferable site-certificates would allow for less arbitrary expansion of existing protected areas. This would significantly increase the cost-effectiveness of environmental policy, as well as the efficiency of resource allocation in agriculture (including public goods).

To sum up, a Natura2000 area should consist of a hard core of unique ecosystem elements (areas with the highest level of protection) to which CAC-type instruments would be applied, and a so-called "buffer belt" with a lower level of protection, where it would be possible to conditionally purchase a site-certificate only for a specific area of land. In this way, the protective belt area would be more flexible and more adjusted to the diverse needs of local communities.

Some approximation of the magnitude of cost savings achievable by distinguishing between sellers and buyers of bio-certificates can be found in the work of Ando et al. (1998), who estimated that marketization of public utilities could reduce the cost of a given component of a conservation system by as much as 25% given the achievement of a target indicator.

The system of transferable site-certificates could be further developed in the direction of a hybrid model (combining CAC and MBI tools) by introducing the obligation to possess a specified number of certificates, as in the case of CO_2 emission rights, which would constitute an additional element of the cross-compliance rule. In such a variant, firstly, the allocative effectiveness of resources in agriculture would increase; secondly, a greater part of resources considered valuable in terms of nature (theoretically the whole) would be subject to quasi-market valuation. Small farms located on environmentally valuable land outside Natura2000 areas would particularly benefit from this solution.

We should remember that natural market processes have already caused that small-scale and semi-subsistence farms are generally located in naturally valuable areas which are at the same time less-favored areas from the point of view of agricultural productivity. This is a normal course of things, because on fertile lands without natural production limitations the process of land concentration begins quite quickly and economically strong entities oriented on a large scale of agricultural production are formed. In other words, if we consider environmentally adjusted production function, allocative inefficiencies will be mainly concentrated in areas with low agricultural potential and high environmental values. This is because small and economically weak farms in these areas cannot in any way capitalize in market prices and in output value the environmental resources (i.e. public utilities) they own. At the same time, the support system in the form of AES and AECS for various reasons (e.g. transaction costs, risk perception, not profit-oriented system of value, cognitive lock-ins, low self-efficacy) does not reach the owners of such entities, as shown by the data from Poland and Romania (Chap. 2). Public goods in these areas are therefore undervalued or not valued at all. There is therefore little risk that the "threshold effects" that could threaten ecosystems will occur in these areas, as there are no natural conditions for large-scale profitable agricultural production. If site-certificates are introduced, it is rather to be expected that the disposers of environmentally valuable resources in agriculture will sell them to highly productive farms. The latter usually operate in areas with relatively low natural values, so the risk of degradation of the natural resources of unique value would be limited.

The second solution, habitat banking, is defined by EC as: "A market where credits from actions with beneficial biodiversity outcomes can be purchased to offset the debit from environmental damage. Credits can be produced in advance of, and without ex-ante links to, the debits they compensate for, and stored over time. The term 'habitat banking' can refer to both species and habitats" (IEEP, 2010, p. 14).

This instrument is based on the idea of compensation, which is allowed by Article 6(4) of the Habitats Directive (Directive 92/43/EEC). Although this solution has great potential to increase the cost-effectiveness of agri-environmental policy

(Klassert & Möckel, 2013), the Habitats Directive limits its application to developers as potential buyers of permits.

Habitat banking in relation to wetlands operates in the USA. Opinions on its cost-effectiveness are often positive (Santos et al., 2011, 2015)—for instance, the US Wetland Mitigation Banking enabled creation of thousands of hectares of wetland that would not have existed otherwise. However, some authors point out negative environmental side effects of habitat banking systems (McGillivray, 2012) or their low ecological relevance, e.g. the BushBroker and BioBanking schemes with regard to the systems operating, respectively, in Victoria and New South Wales in Australia (Madsen et al., 2011). Indeed, Santos et al. (2015) came to the conclusion that "drawing a definitive and general conclusion on the environmental effectiveness of habitat banking or tradable development rights in terms of their contribution to achieving fixed conservation objectives (for example 'no net loss') is not possible".

Typical habitat banks should involve three types of main stakeholders: the damager (debits provider who apply for credits), the credit provider (seller) and the regulator agency that carries out certification of credits and assessment of debits and makes decisions as to whether or not to accept debits. Offset regulation and legal obligation that force potential damagers (e.g. farmers, developers) to buy credits on the market are crucial for the effectiveness of this type of scheme (Carroll et al., 2008).

Many EU countries also have an institutional framework and legal regulations enabling compensation pools: i.e. Germany, France, Sweden, Hungary, Bulgaria, the UK and the Czech Republic (Dickie & Tucker, 2010), but the schemes in operation are local and do not provide systemic solutions for the agricultural sector. In our view, a systemic solution like habitat banking could contribute to the effectiveness of the EU CAP (we will return to this issue in the last chapter of the book). However, offset schemes will not work properly without an adequate method for measuring and assessing the biodiversity values gained and lost due to credits trading. One method to accomplish this measurement could be slacks in terms of desirable outputs that reflect biodiversity level (e.g. soil organic matter slacks). Such a measurement would show how much the trading scheme has reduced inefficiencies in the supply of environmental goods in relation to the possible Pareto-optimal allocation determined by the frontier for the selected impact area.

Another debatable issue is the geographical scope for the selected habitat bank or tradable permits scheme potentially implemented in agriculture—the question arises whether it should be local, regional, national or international. Transnational solutions seem politically unacceptable, although technically the wider the market (more transactions), the better. However, it would be difficult to explain to voters that the offsetting of their national natural resources degraded by more intensive farming takes place in another country. This could simultaneously lead to environmental exploitation in less developed countries where the return on investment is higher and environmental awareness is lower. Thus, if we assume that the potential agricultural productivity of land is negatively correlated with its natural values, the national level seems most appropriate. Regions with more extensive and less profitable agriculture (but with greater natural wealth) would provide credits to the system that would be purchased by regions with large-scale intensive agriculture, offsetting the environmental pressure generated.

4.5.4.2 Contractual Nature Conservation and Conditionality of Direct Area Payments

At the moment, however, the dominant agri-environmental solution in EU CAP is contractual Nature conservation under AES, AECS (area payments in the second pillar of the CAP), RDP (support mainly for foresters also in the second pillar), and the principle of cross-compliance maintained in the new programming period 2023–2027 as the "strengthened cross-compliance for direct payments" that replaces greening and first pillar eco-Schemes (EC, 2020b).

In December 2019, EC announced the objectives of The European Green Deal, following which the EC in May 2021 adopted the Roadmap for the Elimination of Water, Air and Soil Pollution, which presents an integrated vision of the world by 2050, aiming to reduce environmental pollution to levels that do not threaten human health and natural ecosystems (for more detailed discussion see Chap. 6). The plan brings together all relevant EU policies in this area and provides for a review of EU legislation for better implementation. The European Green Deal touches upon various policy areas, of which the key one for agriculture is the Farm to Fork (F2F) strategy, which forces a change in the hitherto architecture of the CAP, directing it towards the targets of reducing: (1) the use of plant protection products by 50%, (2) nutrient losses by at least 50%, while ensuring that there is no deterioration in soil fertility, (3) the use of fertilizers by at least 20%, (4) sales of antimicrobial agents to farmed animals by 50%. The strategy also calls for at least 25% of agricultural land to be dedicated to organic farming by 2030, which may prove to be its biggest challenge. As a result, between 2021 and 2027, around 40% of the total CAP budget will be directed towards environment and climate action.

The changes in the architecture of the CAP will consist primarily in the introduction of strengthened cross-compliance with a number of new and reinforced "Good agricultural and environmental conditions" GAECs and "eco-schemes" to which at least 25% of the budget for direct payments (EC, 2020a) will be allocated. The new and strengthened GAECs, for example, relate to: (1) management of plowing or other appropriate cultivation techniques to reduce the risk of soil degradation, (2) prohibition of conversion and plowing of designated wetlands and peatlands, (3) minimum soil cover in the most sensitive periods and areas, (4) crop rotations and diversification, (5) a minimum share of 5% of arable land to be used for non-productive facilities or areas or catch crops or nitrogen-fixing crops grown without plant protection products, (6) the retention of landscape features, (7) a ban on pruning hedges and trees during the bird breeding and rearing season.

If a GAECs violation is found, sanctions are generally 3% (inadvertent) and 20% (intentional) of the total payment amount, but can be increased up to 100% in special cases.

The list of 'eco schemes' is long and includes, for example, issues such as: creation of areas with melliferous plants, extensive grazing on TUZ with stocking rates, green stubble, winter catch crops, intercropping, developing and following a fertilizer plan, improving crop structure, practicing sustainable management on all farmland on the farm, plowing manure on arable land within four hours of application, simplified cultivation systems, pesticide- and fertilizer-free strips, management of crop and post-crop residues, mid-field afforestation and agro-forestry systems, water retention on permanent grassland.

The voluntary contractual Nature conservation scheme, together with conditional direct payments are among the MBIs with the greatest potential for effective habitat conservation under the CAP. Their cost-effectiveness could theoretically be improved by linking the payment directly to the desired environmental outcome, rather than basing the payment on the average cost of achieving that outcome or on the average cost of undertaking the recommended action, although in practice, this solution raises a number of difficulties (this has been discussed previously in Sect. 4.3.3).

A payment oriented towards average costs of achieving an outcome, in principle, attracts those land users who can provide the desired outcome at the lowest cost (Klassert & Möckel, 2013), and thus introduces enhanced ecological protection in areas that least require it. Besides, an average estimate of the costs of introducing sustainable practices usually leads to higher compensation than individual WTAs except for marginal participants in the system. Properly constructed result-based measures may differentiate incentives between low-cost and high-cost providers of habitat protection, supposing that they precisely define the public goods required and do not focus only on the potential of an area/activity to provide them. Such practices encounter, however, several limitations (discussed previously in Sect. 4.3.3).

Another example of contractual Nature conservation is the voluntary U.-S. Conservation Reserve Program (CRP) under which farmers are paid to take environmentally sensitive land out of active production and plant it with grasses or trees for a contract period of 10–15 years (Stubbs, 2013). The CRP covered between 30 and 37 million acres (12.14–14.97 million hectares) from 1990 to 2011, but by 2015, the area had dropped to about 24 million acres (9.71 million hectares) (Miao et al., 2016). This is because the willingness to participate in this program depends on the opportunity cost of cultivating agricultural land, which increases when agricultural commodities are more prosperous. Eligibility for subsidies under the CRP is made on the basis of a synthetic environmental benefits index (EBI) that includes issues such as impacts on wildlife (biodiversity), water quality, air quality, erosion susceptibility, carbon sequestration potential, and opportunity costs in the form of potentially lost rent. If the latter are high, they may reduce the EBI so much that an application to take the land out of agricultural production is denied. In this way, the potential of agricultural land for food production is protected, on the assumption that its excessive reduction could result in social welfare losses. It has been assessed that the CRP has so far been an effective tool for protecting and delivering environmental public goods in terms of soil, groundwater, biodiversity and other landscape utilities.

4.5.4.3 Offsetting Policies

Interestingly, as Miao et al., (2016) demonstrate, the CRP is coupled with the Federal Crop Insurance Program (FCIP). The latter covers more than 280 million acres (113 million ha) of land with an annual average insured value of about $95 billion (Risk Management Agency 2016). Support under the FCIP consists of farm insurance premium subsidies averaging 60% share (see Shields 2015 for more details). The linkage between CRP and FCIP is that the withdrawal of land from agricultural use under CRP mostly results in public savings on insurance premium subsidies under FCIP. Consequently, if insurance subsidies were similar to subsidies under CRP, land could be taken out of production and provide public goods at very little budgetary cost to the public institution. Considering that the insurance subsidy saved by participating in the CRP offsets public expenditures, the system for allocating subsidies under the CRP could be streamlined by subtracting the insurance subsidy from the rental rate, which is taken into account when calculating the EIB (Miao et al., 2016). In this way, the land covered by the insurance support program would be rightly favored for exclusion from production as the one that generates budget savings.

The aforementioned is an example of how consistency across different agri-environmental policy instruments can increase cost-effectiveness. It also follows that evaluation of the effectiveness of environmental policy should be done comprehensively, rather than for individual programs separately, as there may be other, unintended couplings involving other agricultural support programs (e.g. the Counter Cyclical Payment and Agriculture Risk Coverage Payment in US).

Similar linkages between different agricultural support programs may also exist in EU agricultural policy. There is a lack of research in this area. For example, in Poland (which is one of the leading producers of agricultural raw materials and food in the EU), according to the Act of 7 July 2005 on insurance of agricultural crops and livestock (Journal of Laws of 2019, item 477), farmers are entitled to a subsidy from the state budget to offset insurance premiums of up to 65% of the premium. At the same time, farmers who received direct payments to agricultural land are obliged to conclude an insurance agreement on at least 50% of the area of agricultural crops (i.e. crops of cereals, corn, rapeseed, canola, hops, tobacco, ground vegetables, fruit trees and bushes, strawberries, potatoes, sugar beets or leguminous plants, from sowing or planting to their harvest, against the risk of damage caused by flood, hail, drought, winterkill and spring frost). A farmer who fails to do so will pay a fee for not fulfilling the insurance obligation. It is therefore worth noting that in this case there is an offset mechanism analogous to that described in the US.

In the EU, all farms over 1 ha receive a single area payment for each hectare of land that meets the GAECs. Moreover, in specified areas there are LFA payments and/or farmer-specific payments such as "young farmers" or "structural pensions". In addition to these CAP payments, farmers may receive supplementary payments for specific crops or animals, or alternatively subsidies for the provision of public goods under AES, AECS, set-aside premiums, afforestation premiums, "greening", "eco-schemes" (from 2023) or some RDP programs. It follows that the mentioned offset between support programs concerns both supplementary payments and subsidies for public goods, but also the latter and subsidies for insurance premiums. In other words, a farmer who applies for subsidies for public goods generates at the same time savings of public funds due to subsidies for crop insurance (because in most cases, farming systems providing public goods are not subject to the obligation of insurance, except for ecological crops) and savings due to supplementary payments for a specific type of farming, which was replaced by provision of public goods.

The offset effects of agri-environmental policy are complex and difficult to quantify in a cost-effectiveness calculus. As a rule, however, each exclusion of agricultural land from typical agricultural production and its allocation for public goods delivery causes offset savings and is partially self-financing. Thus, subsidizing public goods delivery appears to be a cost-effective direction of agri-environmental policy. Moreover, as discussed in Chap. 2, replacing typical agricultural production with public goods delivery releases the farm from the market treadmill and may increase the farmer's share of the value chain if the public utilities are capitalized in prices by the market mechanism.

4.5.5   Dual Development in Agriculture

However, it is important to consider the broader market implications (including international and global) of a dual model of agricultural development in which some farms specialize in extensive production and provision of public goods, while others focus on large-scale intensive production of agricultural commodities. Such a polarization process can be triggered by subsidizing the production of public goods related to the broadly defined rural landscape. From the point of view of the cost-effectiveness of public policies and the allocative efficiency of resources in agriculture, this process seems to be beneficial from a theoretical point of view, but its environmental effects and interaction with global food markets require in-depth research. In the context of global demographic growth, some production constraints in the EU and other developed countries will need to be compensated elsewhere. Potential environmental risks in two-track agricultural development can be reduced through minimum cross-compliance requirements on all farms. With respect to global agricultural commodity and food prices, the potential upward pressure on prices may be mitigated by concentrating agricultural production in the dual model on the most fertile land.

The dualism of agricultural policies and the consequent dual development of this sector was also pointed out by other authors e.g. (Chamberlain et al., 2000; Donald et al., 2001; Stoate et al., 2009; Strijker, 2005 and Mouysset, 2014). The last author cited wrote that public policies on agriculture are generally directed towards two different farming systems; the first concerns crops that have high relative profitability but generate negative impacts on biodiversity. The second concerns grasslands, which are associated with lower average agricultural incomes but have positive impacts on biodiversity. The cited author noted that the above division coincides with socio-economic polarization because the poorest regions are specialized in grasslands and the richest in field crops.

Generalizing this conclusion, we can say that poorer regions are at the same time rich in public goods that could potentially be provided to society. In order to maintain social equilibrium, it is necessary to focus subsidies for the provision of public goods on grasslands (contractual nature conservation), because it is relatively easy to achieve high environmental results there—an orientation towards maximizing the environmental effect is therefore postulated. At the same time, the introduction of conditional subsidies (i.e. cross-compliance) in the area of field crops is subject to few social constraints and is associated with little welfare loss (Mouysset, 2014), but faces some structural constraints as it is impossible to achieve high environmental outcomes—hence an orientation towards just meeting basic threshold requirements in terms of environmental pressure is postulated.

Returning to the thread of evaluating the effectiveness of agri-environmental policy, the issue of offsets between different programs makes such an evaluation much more difficult and points to the validity of a holistic approach rather than a single-scheme perspective.

The modification of the optimization problem by adding environmental and social criteria has been raised many times in the literature (as we showed in Sect. 4.2). From a social perspective, the goal is to protect the poorest farmers, while from an environmental perspective, environmental capacity should not be overpassed. Particularly close to our view is the approach of the cited Mouysset (2014) (and also Mouysset et al., 2013), who constructs a multi-scale model for optimizing the farmer's utility function with social and environmental criteria treated as constraints by the decision-maker (the so-called "bioeconomic model").

In addition, there is a fresh stream of literature being developed in this direction devoted to the so-called "green efficiency" (called also Agricultural Green Total Factor Productivity (AGTFP)), which differs from the former concept of eco-efficiency in that it modifies the classical production function to a greater extent by introducing desirable and undesirable public goods AGTFP is oriented towards both outputs and inputs optimization (Ge et al., 2018; Han et al., 2018; Shen et al., 2018; Xie et al., 2018; Liu & Feng, 2019;).

To sum up, in the holistic approach, the key criterion is raising the allocative efficiency of agricultural resources, including environmental and social utility, within the framework of the environmentally and socially adjusted production function. The multidimensionality of such a function gives opportunity to capture the offset effects occurring between particular resources participating in the creation of widely understood outputs within the agricultural and forestry sectors. We assume that the increase in the integrated efficiency that includes both economic and social criteria brings us significantly closer to meeting global environmental targets. The proposed approach does not replace the single-scheme effectiveness assessment, but complements it, drawing attention to the fact that if the toolbox of agricultural and environmental policies induces allocative inefficiencies, it will always be ineffective to some extent.

Chapter 5
Cost-Effectiveness Assessment of Environmental Expenditures in Different Regions of the World: Slack-Based Approach

5.1 In Search of the Holistic Approach to Policy Assessment

As can be seen from the previous considerations, there is no unified and comprehensive approach to assessing the effectiveness of agri-environmental policy. Research generally focuses on the ecological effectiveness of such a policy, while national and international strategies take into account both ecological and socioeconomic criteria, as exemplified by the two pillars of the European Union's Common Agricultural Policy (CAP). Hence, there may be dissonance in assessing the effectiveness of public policies.

The literature review has shown that assessment of the cost-effectiveness of a particular policy option may be biased because there may always be new categories of transaction costs, different cognitive lock-ins, multiplier effects, spatial dependence, policy synergies or trade-offs. Models cannot keep up with the reality. Maybe we have to accept the fact that it is impossible to tailor policies precisely and that it is impossible to create tools that are fully effective and precisely targeted at a given type of externalities? At the micro level each instrument will perform better or worse in a given context. The free-rider effect will be compensated by the irrational behavior of some farmers who maximize a different objective function than the market approach.

It is clear that the neoclassical production function does not correspond to social (collective) utility criteria. However, it is difficult to expect spontaneous altruistic changes in the individual utility functions usually followed by farmers. A change in the WTP/WTA must therefore be created by the appropriate public policy. However, a measure of the effectiveness of changing the production function to incorporate broader social criteria is to maintain the efficiency of the production as expressed in

The research in this Chapter was partly funded by the National Science Centre, Poland under the project No. UMO-2021/41/B/HS4/02433.

B. Czyżewski, Łu. Kryszak, *Sustainable Agriculture Policies for Human Well-Being*,
Human Well-Being Research and Policy Making,
https://doi.org/10.1007/978-3-031-09796-6_5

the Pareto improvement embodied by slacks. In sum, public policy, while modifying the production function, should affect the mechanisms and the system as a whole, but not the individual outcome. We can pay, for example, for mid-field woodlots, and they may be created, but no new, lasting mechanism will develop if we do so in isolation from resource relations in agricultural production and their efficiency. However, if we take the introduction of Pareto improvement as a criterion, the effect may be long term. Policy should therefore improve the efficiency of the market allocation mechanism rather than harm it.

Another issue is the convergence of the adjusted production function adopted in particular public policies with the three-dimensional definition of sustainable development. Ecological and social criteria are now integrated with the historical economic objectives of the CAP. More precisely, they are part of a structure referred to as the "second pillar", with the first pillar being historical production and income support. The instruments of the second pillar, while appearing socially acceptable and adequate, are criticized for their environmental ineffectiveness, as we have shown in earlier discussion. In criticizing, it is often forgotten that the declared objectives of the CAP go beyond a narrow ecological perspective. Similarly, the definition of sustainable agriculture as a component of sustainable development *sensu largo* also goes beyond the criteria of environmental sustainability. At the outset of a policy cost-effectiveness analysis, it is therefore necessary to identify the broader context of the policy as part of an overarching sustainability strategy, bearing in mind that greening and sustainability do not mean the same thing. In addition, broadening the perspectives of the assessment to include an international aspect can also show that environmental constraints causing a decline in welfare on the national scale can contribute to an increase in welfare internationally and globally.

The historical goal and genesis of the CAP was to provide farmers with a minimum level of income. Thus began the subsidization of production that made this objective possible. However, such a strategy, as we wrote earlier, encouraged intensification, leading to industrial agriculture and social polarization. In regions with potential for intensification, the CAP stimulated the development of the scale of agricultural production, land concentration and increased yields through mechanization and fertilization, generating high agricultural incomes. The side effects of this development have been landscape homogenization, loss of semi-natural elements, water pollution (Carpenter et al., 2012; Volk et al., 2009), and loss of biodiversity (e.g. Foley et al. 2005; Tscharntke et al., 2005; Carpenter et al., 2012), particularly concerning mammals, plants and birds due to degradation of their habitats (Mouysset, 2014). On the other hand, in regions where there were no natural and structural conditions for the intensification process, agricultural incomes did not increase and a process of land fallowing and land abandonment began (Mottet et al., 2006), widening the gap between rich and poor farmers and the relative deprivation of the agricultural sector. Similar processes to those described above are also taking place in developing countries, as we believe that avoiding a kind of path dependency is impossible and the realization of the backwardness rent seems theoretically possible only in the academic debate. In practice, however, less developed countries largely repeat the agricultural development path of highly developed countries. Therefore, an exclusively ecological orientation of agri-environmental policy cannot be sufficient. The role of greening increases with the economic and social development of a country and a given continent. Nevertheless, the definition of sustainable development will always have in it a trade-off between environmental, economic and social objectives. Therefore, the evaluation of the effectiveness of public policies in this area must also strive for the broadest possible perspective and the most uniform methodology.

Chinese agriculture is currently experiencing similar problems, only on a slightly larger scale, facing declining fertility and the undesirable side effect of rapidly spreading soil and water pollution, caused in particular by phosphorus consumption, nitrogen consumption, chemical oxygen demand, and GHG emissions (Du, 2014; Pang et al., 2016; Wang & Zhang, 2018; Yang et al., 2019). Conservative estimates only put the direct economic losses associated with natural resource degradation in the agricultural sector at 0.5–1% of China's GDP, without taking into account indirect multiplier effects (Xie et al., 2018). Therefore, many focusing on this issue in the context of Chinese agriculture attempt to adapt the traditional production function to the deepening constraints on the use of the agricultural land resource by introducing undesirable outputs into the analysis to complete the input-output calculus (Du et al., 2016; Ge et al., 2018; Han et al., 2018; Shen et al., 2018; Xie et al., 2018; Liu & Feng, 2019). This develops the mentioned concept of agricultural 'green efficiency', similar to the one applied in this book, which is used as an indicator of sustainable development in Chinese agriculture. Full convergence with the definition of sustainable agriculture, however, requires also taking into account social balance, which can be expressed by assessing the relative deprivation of farmers (income gap) or more widely and probably more universally by food security indicators with regard to different levels of economic development and a different hierarchy of social needs throughout the world. The authors quoted above usually use as a macroeconomic measure of desired effects the gross production value of agriculture, or possibly agriculture, forestry and fishery together. Due to differences in agricultural prices and in the purchasing power of agricultural incomes, but also due to the increasing specialization of production, the increase in the value of production does not always go hand in hand with the satisfaction of food needs. Therefore, in our research we applied an extended version of efficiency measurement taking into account the issue of food security—the Integrated Efficiency (IE) score.

However, our approach is two-step, as we also consider the risk of exceeding environmental thresholds by the estimated target values (see Chap. 7). Using the language of mathematics, we distinguish between necessary and sufficient conditions. Building sustainable agriculture means first of all removing inefficiencies. We put forward a thesis that a necessary condition is a reduction of inefficiencies defined as slacks. They do not require a fundamental technological change. However, this change, although it results in getting closer to the frontier, in some cases may not be sufficient. In the second stage, it is necessary to assess whether the requirements concerning the desired levels of emissions, which do not exceed the environmental capacity, are met.

In this chapter we address the call made in the previous section for a holistic cost-effectiveness assessment. Here we mean the evaluation of how the spending of environmental and agri-environmental policy influences the multidimensional system of variables that are taken into account in the agricultural production, including public goods and undesirable goods. We refer to the thesis that only such a policy can be effective which positively influences the efficiency of the production process by creating conditions for progress in the Pareto sense (improvement of efficiency in the use of a given type of resource, including public goods, without losses in the efficiency of the use of other resources). If this condition is not met, public policy may not bring the expected results or these results will be short-lived. This is a necessary, though perhaps not sufficient, condition for an effective policy. However, our analyses have shown that the increase in resource efficiency in agriculture alone brings the world economy significantly closer to achieving environmental targets (see Chap. 7), without exceeding its critical environmental capacity, e.g. in terms of fertiliser pressure per ha, or LSU per ha (see also Chap. 7). We therefore take the position that the economic rationality of individuals cannot be fooled in principle and any action by institutions against it will be ineffective.

Therefore, we will further attempt to assess the impact of environmental and agri-environmental expenditures on the efficiency of a multi-scale system of agricultural output/input data (including public and undesirable goods) across the previously defined clusters (see Chap. 3 or Appendix). We will particularly focus on the EU CAP schemes, due to the availability of precise and long data series in this area. We have called the procedure used—'holistic cost-effectiveness' analysis (Fig. 5.1) and it can be reproducible for different economic sectors and types of public policies.

As we wrote in previous chapters, statistical evaluation of the impact of uncontrollable factors such as public policy on measures of relative efficiency is problematic because of the clash between deterministic and stochastic approaches. Data envelopment analysis as a tool for measuring relative efficiency has come under criticism because the input/output variables are used in a deterministic way whereas they may be contaminated with stochastic noise (Simar & Wilson, 1998, 2007; Dharmapala, 2018). Therefore, each DEA score is a result of the deterministic model, which makes statistical inference difficult to apply. The same problem touches the cost-effectiveness analysis (CEA) or cost-benefit analysis (CBA) (see Chap. 4) using discounting techniques in a deterministic approach. Thus, the influence of stochastic noise in policy effectiveness analysis cannot be excluded, but one can attempt to reduce it by the following means:

- Bias-correction of efficiency score and standard errors by bootstrapping, respectively a one-stage or two-stage approach (Kneip et al., 2008; Simar & Wilson, 1998, 1999, 2000, 2007; Dharmapala, 2018)
- Applying Stochastic DEA including Stochastic Frontier Analysis (SFA) (Olesen & Petersen, 2016);
- Using long-term panel data increasing the number of observations;

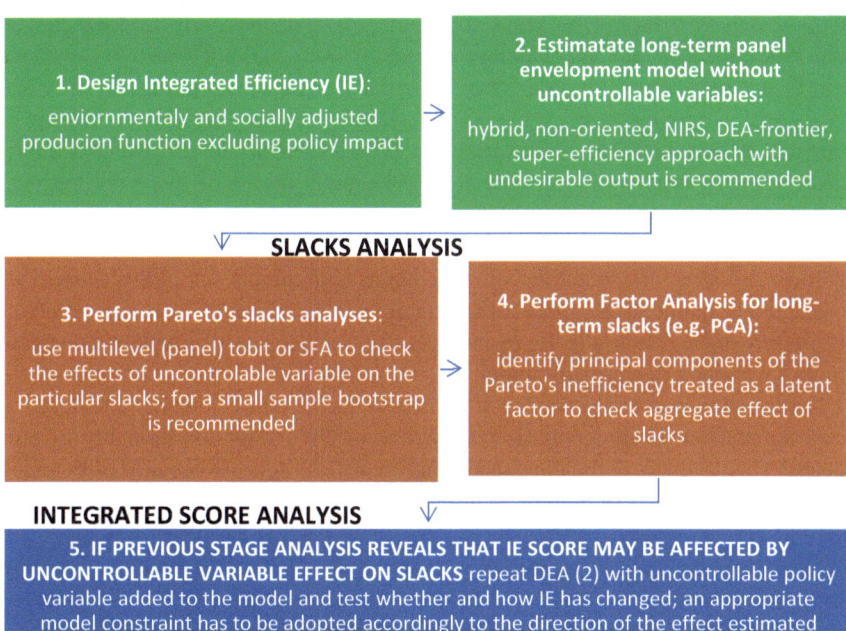

Fig. 5.1 Strategy of assessing a holistic cost-effectiveness of agri-environmental policy as an uncontrollable variable

- Performing Pareto's slack ('real slack', see Chap. 3 for definition) regression instead of regressing efficiency scores.

Of course, each method has its limitations and side effects, which in some cases may outweigh the benefits. For example, there are no bootstrapping methods for the non-radial models, which excludes e.g. hybrid approaches and significantly limits the use of slack-based approaches. In other words, bias-correction by bootstrapping implies the need to make assumptions about the existence of a "proportionate movement" of inputs or outputs, which significantly reduces the feasibility of model assumptions with regard to the implementation of technical progress.

In contrast, SFA requires the adoption of a specific type of production function (e.g., Cobb-Douglas or trans-log), which is debatable when including specific public goods and undesirable outputs in the analysis because the course of such a function is not clear.

As for lengthening the time series and expanding the cross-section of data, this is obviously the simplest solution, which undoubtedly reduces the potential random bias to some extent, but it is not clear how much.

The 'real slacks' regression, on the other hand, although seemingly biased as the efficiency score regression, may have additional advantages if one considers the theoretical determinants of the occurrence of slacks. First of all, there is no simple link between a 'real slack' value and efficiency score. In the case of radial input/

output variables, reducing the 'real slack' has no effect on the score. From a technical point of view, it is a movement along the frontier, but it does not change the distance to it. In this sense the score in the case of radial variables is an imperfect and quite inaccurate measure, because even a significant decrease of environmental pressure or emission of undesirable variables may not be reflected in an increase of the efficiency score if these variables were defined as radial (i.e. implying a proportional change of other radial variables). As we wrote earlier, 'real slack' should be understood as a potential improvement in a particular production input or output that does not affect the quantities of other inputs and outputs and in this sense does not require a change in the production technology. Therefore, we treat it as a Pareto-like improvement. Thus, it is, for example, a costless increase in output, or a decrease in input without a decrease in output.

The question then arises why such a change (slack) is only potential and does not occur spontaneously in economic reality, due to the economic rationality of producers? This leads us to the conclusion that 'real slack' occurs because of bounded rationality. The causes of bounded rationality can vary, but most often they involve cognitive burdens and are systemic in nature. In Eastern European countries, for example, we have institutional hysteresis (Setterfield, 1997) after the transition from the socialist to market economy in 1989, which translates, among other things, into low levels of trust among farmers and the reluctance to change. Such systemic constraints on rationality are very durable and, even when there is a random change in the quantity of inputs or outputs, they persist. This means that the structure of real slacks in the study sample may be theoretically robust to stochastic noise, and certainly more robust than the efficiency score alone. For example, if in the time series under study agricultural production were lower in some years due to drought, the efficiency score would certainly change after data correction. However, the structure of the 'real slacks' will not necessarily change after adjusting the input data. This is some theoretical assumption that has not yet been tested empirically in detail. The undoubted advantage of slack-based regression, however, is the decomposition of the efficiency score so that the risk of the regression being affected by statistical noise is spread—in the case of stochastic bias within one type of inputs/ outputs, the regression for the remaining inputs/outputs remain unaffected by this bias.

Returning to the diagram in Fig. 5.1, we will now describe its subsequent stages. The first stage is the estimation of the underlying model, which was described from the point of view of our study in Chap. 3. The key issue is to design the adjusted production function so that it includes the public utilities whose realization is desirable from the social welfare perspective. In the case of sustainable agriculture, this function should integrate the three aspects (economic, social and environmental) and take into account the trade-offs between them, referring to the definition of Chap. 1; hence we proposed IE approach as the relation of desirable and undesirable outputs to classical agricultural production factors. Public policy in this case creates a production environment over which the producer has no direct influence (an uncontrollable variable), but in which the institutional framework may create better or worse conditions for the efficient transformation of inputs into outputs

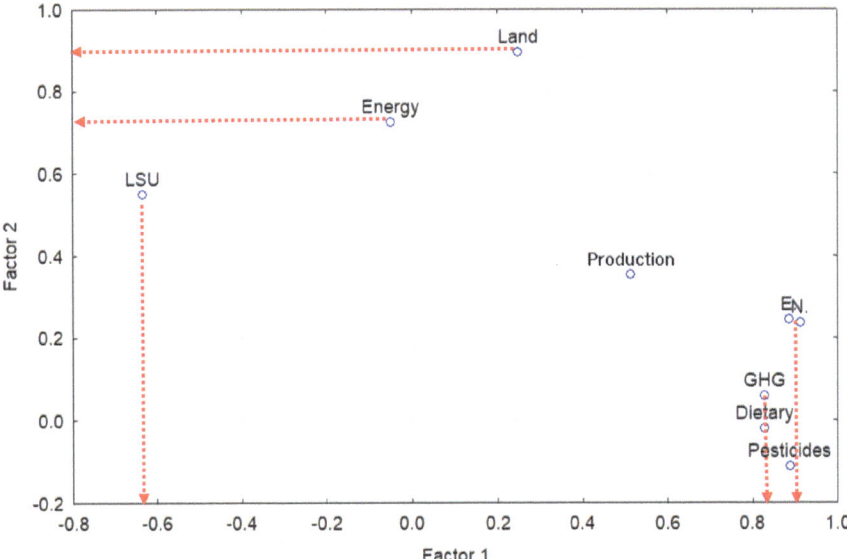

Fig. 5.2 Cluster 1: Factor analysis (PCA) for Pareto's slacks. Notes: E employment, N nitrogen; Factors' explained variance: F1 = 0.50, F2 = 0.21

(especially bearing in mind public utility effects). In the case of the baseline model, we ignore this issue for now, focusing on the pure efficiency of the adjusted production function.

To reduce the stochastic bias of input/output data we suggest using panel data with the longest possible time series. In doing so, one should keep in mind the use of constant prices in the case of economic output, which, as we argued in Chaps. 2 and 3, is relatively best suited to the concept of production efficiency and the microeconomic optimization perspective in agriculture. The assumptions on radial and non-radial variables are crucial. We propose a hybrid approach because it is obvious that some kind of inputs and outputs cannot change completely in isolation from other variables and to some extent there is a proportionate movement. Here we mean, for example, animal production, energy consumption, GHG emissions, land resources—even under conditions of technical progress it is difficult to reduce any of the above-mentioned inputs without affecting the others. In the case of active agricultural policy (e.g. the CAP), the interactions between factors are even more far-reaching. This is confirmed by Figs. 5.2, 5.3 and 5.4 with the results of factor analysis (PCA) for the mean slacks of the long period. Some authors propose a technical approach in this regard and select non-radial or radial variables based on correlation analysis (Wang et al., 2019). We, however, take the position that expert selection of assumptions based on literature reviews, knowledge of agricultural economics and local conditions is more appropriate because of the potential stochastic noise that we wrote about above.

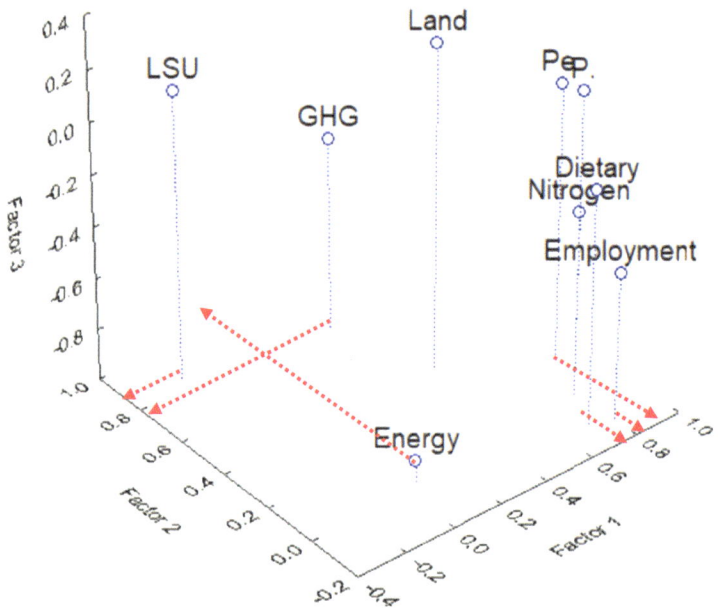

Fig. 5.3 Cluster 2: Factor analysis (PCA) for Pareto's slacks. Notes: *Pe* pesticides, *P* production; Factors' explained variance: F1 = 0.42, F2 = 0.18, F3 = 0.15

At the same time, we suggest to adopt the assumption of a non-increasing return to scale with reference to the argumentation in Chap. 3 (the issue of a self-accelerating degradation of environment), and to adopt the super-efficiency approach to avoid the hard-to-interpret large number of efficient DMUs with score = 1 (in practice it may be 25–50%) and to find outlier observations.

The suggested lack of specific orientation towards inputs or effects stems from the belief that it is difficult to arbitrarily assess which orientation model dominates in the surveyed sample, especially when we conduct sectoral research in different countries. This is also a reference to the directional distance function, which is preferred in the subject literature. Such a solution is thus a compromise between the postulate of minimizing environmental pressure and maximizing the value of production and public goods provision. From a technical point of view, it should be remembered that in the case of the non-oriented model, the proportionate movement will be divided in half (in %) between radial outputs and inputs.

The next step is to perform a detailed 'real slacks' analysis, which is the biggest added value of the proposed approach. It consists of a regression analysis of public policy spending on slacks and factor analysis for slacks averaged over a long period.

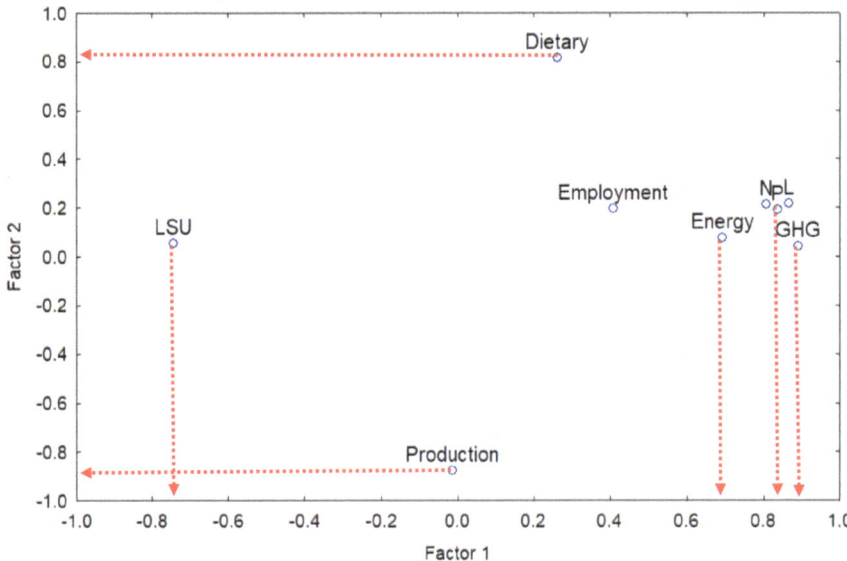

Fig. 5.4 Cluster 3: Factor analysis (PCA) for Pareto's slacks. Notes: N nitrogen, P pesticides, L land; Factors' explained variance: F1 = 0.46, F2 = 0.18

The results for the clusters we analyzed are shown in Tables 5.2 and 5.3 and Figs. 5.2, 5.3 and 5.4 when it comes to factor analysis. It is worth emphasizing again that the purpose of the regression analysis is to determine how public policy expenditures affect systemic inefficiencies and whether they create conditions for progress in the Pareto sense (improved management of one type of resource without trade-offs). From a modeling strategy perspective, we suggest an approach that censors fully efficient DMUs lying at and above the frontier and those cases where slacks do not occur at all (i.e. panel or multilevel tobit regression), following the study of Yang & Pollitt (2009). We are aware that theoretically one could assume that public policy should stimulate efficiency gains also for those fully efficient DMUs (with score $\geq$ 1) as the super-efficiency approach allows us to do it, but this would probably be an overly excessive and debatable and even dangerous assumption in some cases. For example, should the positive slack for energy use, which means that a country has a margin to increase energy use, be as large as possible? The proposed regression analysis tend to assess how public policies affected the different types of slacks (energy slack, nutrient slack, GHG slack, etc.) and to identify potential trade-offs between policy schemes that can be taken into account when designing more efficient solutions.

Factor analysis is complementary while testing the aggregate effect of slacks and relationships between slacks. We performed our analysis using principal component analysis (PCA), which involves the orthogonal transformation and reduction of a set of variables under study into a smaller number of latent variables, which are linear combinations of observable variables (i.e. slacks). Pareto-type inefficiencies (i.e. slacks) can just be seen as peculiar latent variables because of the systemic and hard-to-measure causes that give rise to them. These causes, e.g. the mentioned lack of trust and unwillingness to change, may manifest themselves in the suboptimal allocation of different types of production resources. Therefore, the PCA approach seems to fit well here. The principal components are set to explain as much of the latent variable as possible, and the variances of subsequent components are made smaller and smaller. The variables forming the principal components are highly correlated with each other and are identified according to the value of their factor loadings determining their contribution to explaining the latent variable. Factor loadings are standardized coefficients of regressions of a latent variable on observable variables (the minimum accepted loading level is generally defined as 0.6). The determined principal components are orthogonal to each other and weakly correlated. The sum of variances of components is equal to the sum of variances of primary variables. The starting point of PCA is usually, as in our case, the covariance matrix formed from the initial set. The results of PCA are presented in Figs. 5.2, 5.3 and 5.4 (the axes represent the identified principal components, and the scales represent the values and directions of the factor loadings). The slacks selected on the basis of factor loadings as indicators of particular latent variables (inefficiencies), depending on the sign, either have a strong positive (synergies) or negative (trade-off) interaction with each other.

The last step of the analysis, the integrated score analysis, makes sense depending on the result of the slack analysis. In DEA analysis with uncontrollable variables one has to assume in advance what type of constraint we introduce into the model. The slack analysis should provide such knowledge by checking the statistical significance and direction of the influence of the uncontrollable variable on the efficiency score. This influence can occur directly through individual non-radial variable slacks, or indirectly through radial variable slacks (aggregated effect of slacks), if they form a principal component with non-radial data slacks. Theoretically, this effect can also appear through the proportionate movement of input/output data, but if this happens, our necessary condition for public policy effectiveness is not fulfilled (we search for the Pareto improvement, not proportionate changes in input/output data). Hence, if the slack analysis step does not reveal any potential channels of public policy impact on the Pareto slacks, then the increase in IE, even if statistically significant, may not be long-lasting.

Introducing an uncontrollable variable into an efficiency analysis (in this case, an input variable) means that the efficiency ratio can change as a result of the assumed positive effect of that variable, with slack and proportionate movements for uncontrollable variables being zero by definition. This is because these variables are beyond the control of the DMU and therefore the inefficiencies associated with them cannot be reduced. Technically, it is just a matter of introducing an additional condition which limits the role of uncontrollable input to ensuring the rationality of the reference, i.e. "the production environment of the projection point is not better than the evaluated DMU", assuming that we introduce an uncontrollable variable coded like a typical input (Cheng, 2014, pp. 208–209). As a result, there is no direct causal influence of the uncontrollable input on the efficiency value, but indirectly the variables slack and proportionate movements of other variables can change.

The third stage is the most generalized (holistic) approach to assess the cost-effectiveness of public policy, but therefore also the most simplified and at risk of statistical bias. The impact of an uncontrollable variable on efficiency has already been studied in this way in the literature (e.g. Yang & Pollitt, 2009; Mamoon, 2012).

However, to avoid spurious results one must have a theoretical rationale that policy influence will be reflected in changes in the IE score and what direction it will take. In our approach, the channels of this influence are also important, as stated above. Therefore, we recommend this general approach when there is a significant impact of public policy on at least one non-radial variable slack from the second stage (it is worth recalling that changes in radial variable slacks are not reflected in the efficiency score). It is possible, however, that a statistically significant radial variable slack forms the principal component of Pareto's inefficiency together with non-radial slacks. In this case, we can also assume that the policy effectiveness condition is satisfied to some extent and that the analysis of the IE score changes makes sense. Therefore, the third stage should be treated as a complement to the second, provided that in the second stage a significant effect of public policy on slacks is demonstrated and the assumed direction of its impact is confirmed.

5.2 Environmental Policy Impact on Inefficiency Slacks

5.2.1 Global Perspective

We performed the analysis of slacks in several cross-sections starting with the most general approach, in which we estimated three level tobit models. These models explain the effect of environmental expenditure (from the FAO database) on particular slacks while addressing random individual effects to three successive levels: (1) clusters (described in Chap. 4), (2) countries—DMUs, (3) years (Table 5.1). Environmental expenditures in the FAO database are internationally comparable but are a highly aggregated measure. In many cases typical agri-environmental expenditures are included in environmental expenditures, or even if not, it can be assumed that they are strongly correlated with them (e.g. EU Natura 2000 and AES/AECS).

Thus, in the global approach we used one additional level of panel regression with random effects. The main feature of the random effects model (RE) is that the unobserved component (residuals) are assumed to be uncorrelated with the explanatory variables. Hence, RE can estimate an unobservable (randomly distributed among countries) effect of time-invariant variables (Wooldridge, 2002, pp. 251–252). The above assumption fits the conditions under which inefficiency slacks arise, assuming that they are caused by hard-to-measure systemic, cognitive (bounded rationality) or behavioural factors. These determinants are unlikely to be correlated with current public policy and are largely time-invariant by definition. Besides, environmental and agri-environemntal expenditures are also likely to be

Table 5.1 Global three-level (i.e. cluster, DMU, period) tobit regression model for environmental policy effect on slacks (2005–2018, 3 clusters, 900–985 obs)

Environmental expenditures in const. Million (central gov.) effect on slacks:	Coef.	Std. Err.	z	P > z	[95% confidence interval]	
GHG (kilotonnes):	2.399171	0.7479844	3.21	0.001	0.9331483	3.865193
_cons	−12212.2	3415.539	−3.58	0.000	−18906.53	−5517.865
Land (thousands of ha):	3.544335	1.205528	2.94	0.003	1.181544	5.907127
_cons	−11551.98	4092.666	−2.82	0.005	−19573.46	−3530.506
Employment(thousands of employees):	0.5021051	0.2410134	2.08	0.037	0.0297275	0.9744826
_cons	1814.18	533.0623	3.4	0.001	769.3967	2858.962

Notes: For the other slacks under study a significant influence of environmental expenditures were not reported; Stata syntax: [metobit 'slack var.' 'expenditures' || Cluster: || DMU: || period:, ul(0)]

time-invariant to some extent due to budget programming periods of several years. The significance of random effects was confirmed by econometric tests (Twisk, 2006). In Table 5.1 we show only those models for which also the fixed effect (regression coefficient) turned out to be significant. For the global approach these are GHG, land and employment slacks. The estimated regression directions (sign '+') suggest that the necessary condition for the effectiveness of public policy expenditures was met only in the above mentioned types of inefficiencies. In other words, an increase in environmental expenditures globally reduced inefficiencies in land, labour and GHG emissions management.

For example, we can say that a $1 million increase in policy outlays translated into a 2.39 kilotons reduction in slack per GHG. Given that the average policy outlay across the study group was $780 million and the average slack on GHG was −9873 kilotons, then at the mean values, we can say that a 0.13% increase in policy outlays led to a 0.024% decrease in slack. The slack response is therefore not very strong, as slacks also change under the influence of other factors.

As we can see, there are no trade-offs but synergies between the effects of environmental spending on these three types of inefficiency, which is an optimistic conclusion from a market perspective. It can be said that the policy of reducing GHG emissions does not prevent an increasingly efficient management of land and labour resources in agriculture, and moreover stimulates progress in this area. On the other hand, however, we should consider what the more and more effective management of land and labour in agriculture consists of. The experience of highly developed countries shows that it comes down to decreasing employment and increasing the productivity (partly intensity of use) of agricultural land. Thus, the conclusion that environmental policy, in particular in terms of reducing GHG emissions, promotes the growth of market efficiency may also have negative social connotations in less developed countries, where agriculture plays the role of a reservoir for the labour force, alleviating the problem of urban unemployment. It should also be added that no significant effect of environmental expenditures was found for any other type of inefficiency (slacks), although a positive effect would have been desirable for the dietary variable, which in our analysis addresses the issue of food security. The estimated regression coefficients in the global model can be contrasted with the models for the individual clusters (Tables 5.2, 5.3 and 5.4).

5.2.2 Cluster Perspective

Regression analysis of environmental spending on slacks by cluster shows a more detailed picture of the policy effectiveness condition across country groups. The most important observation concerns the dominant synergy between policy impacts on different slacks. Most of the regression coefficients are positive, except for energy slacks in cluster 1. Let us clarify that the negative sign for production slacks also implies a pro-efficiency impact of public spending in this area, since outputs slacks have positive signs (i.e. they show inefficiency in terms of the possibility to increase production).

In **cluster 1** (see Table 5.2), environmental policies contribute relatively little to allocative efficiency and in this sense cost-effectiveness is relatively lower than in the other clusters. In addition, the energy trade-offs mentioned above occur, which means that the pro-efficiency effects of environmental policies in terms of employment and output value are to some extent undermined by decreasing the efficiency of energy use management. Thus, this is an aspect that policymakers should pay attention to, in particular for the dominant economies in this cluster, i.e. the US and China (we will return to this thread further on). Potentially low environmental effectiveness in this cluster may also be evidenced by the fact that it does not influence (no significant impact) the improvement of economic efficiency in terms of basic factors of environmental pressure, such as GHG, LSU, nitrogen, or pesticides. This may indicate a lack of coordination of environmental policy *sensu largo* with agricultural policy, or a failure to take into account pressures from agriculture in the design of environmental policy expenditures. From the data of Chap. 3 we know that despite relatively high productivity of agricultural inputs in Cluster 1, the potential room for improvement (Pareto's slacks) are relatively high (Table 3.5). Therefore, the postulate to include multidimensional farming efficiency in the design of environmental schemes in this cluster is most reasonable.

In **cluster 2** (see Table 5.3), environmental spending has a positive effect on most slacks (no trade-offs noticed). This shows that in less developed countries with relatively low resource productivity in agriculture it may be paradoxically easier to implement cost-effective environmental and agri-environmental policies. This is confirmed, inter alia, by the many times higher coefficient on the land, employment and GHG slacks than in the global model. In contrast, all cluster-models show no significant effect of environmental policy on pesticides and dietary slacks. The latter

Table 5.2 Cluster 1: Random-effects tobit panel regression model for environmental policy effect on slacks (2005–2018, 21 countries, 246 obs)

VARIABLES	land^r (thous. of ha)	LSU^r (units)	employment^{nr} (thous. of employees)	energy^r (terajoule)	nitrogen^{nr} (tonnes)	pesticides^{nr} (tonnes)	production^r const. Thous. USD	dietary^{nr} (adequacy percentage)	GHG^r (COeq, kilotonnes)
env. Expenditures (central gov. const. million USD)	−0.974	348.760	1.041**	−8.394*	54.349	−0.488	−368.147***	−0.001	1.147
	[2.368]	[631.006]	[0.500]	[4.743]	[51.938]	[1.950]	[127.925]	[0.002]	[1.725]
Constant	−14,245.737	−4294510.085	309.130	−15,156.500	−14,849.869	12,882.502**	−4679284.243***	−17.667***	−18,343.989**
	[12,978.505]	[2795777.682]	[1015.424]	[22,941.724]	[141,087.880]	[6018.212]	[55,117.536]	[4.521]	[8333.971]
RHO	0.851	0.765	0.833	0.798	0.844	0.700	0.998	0.887	0.796

Notes: Standard errors in brackets; *** $p < 0.01$, ** $p < 0.05$, * $p < 0.1$; r radial, nr non-radial; Stata syntax, respectively for the input and output slacks: [xttobit 'slack var.' 'expenditures', ul(0); xttobit production S, ll(0)]

Table 5.3 Cluster 2: Random-effects tobit panel regression model for environmental policy effect on slacks (2005–2018, 33 countries, 362 obs)

VARIABLES	land[r] (thous. of ha)	LSU[r] (units)	employment[nr] (thous. of employees)	energy[r] (terajoule)	nitrogen[nr] (tonnes)	pesticides[nr] (tonnes)	dietary[nr] (adequacy percentage)	GHG[r] (COeq$_2$, kilotonnes)
env. Expenditures (central gov. const. million USD)	**19.672*****	**1647.789****	**5.279****	**6.285****	79.949	0.602	−0.004	**5.613****
	[4.227]	[741.900]	[2.452]	[2.721]	[82.799]	[1.050]	[0.005]	[2.358]
Constant	−19,004.966***	−4721121.311***	1695.670	−9794.059**	277,884.570***	942.418	−4.898	−14,427.626***
	[6280.688]	[1230744.638]	[1148.262]	[3942.557]	[97,393.191]	[2129.254]	[3.497]	[3285.277]
RHO	0.62	0.69	0.58	0.64	0.65	0.86	0.83	0.58

Notes: see Table 5.2; no observation for 'production' slacks >0;

Table 5.4 Cluster 3: Random-effects tobit panel regression model for environmental policy effect on slacks (2005–2018, 27 countries, 377 obs)

VARIABLES	land^r (thous. of ha)	LSU^r (units)	employment^{nr} (thous. of employees)	energy^r (terajoule)	nitrogen^{nr} (tonnes)	pesticides^{nr} (tonnes)	production^r const. Thous. USD	dietary^{nr} (adequacy percentage)	GHG^r (COeq$_2$, kilotonnes)
env. Expenditures (central gov const. Million USD)	**0.581*****	**456.364*****	0.016	1.002	**51.286*****	−0.190	−5.281	−0.000	**1.968*****
	[0.148]	[117.016]	[0.023]	[1.006]	[9.749]	[0.190]	[76.675]	[0.001]	[0.367]
Constant	−2103.530***	−1231791.728***	121.444	−6220.269	−150,512.228***	924.713	314,822.633	−8.756***	−4788.255***
	[540.639]	[331,644.780]	[82.854]	[4091.339]	[37,056.517]	[645.442]	[307,879.991]	[2.525]	[1128.235]
RHO	0.69	0.66	0.77	0.78	0.75	0.74	0.80	0.80	0.72

Notes: see Table 5.2

in particular suggests that environmental policy should pay more attention to food security spillover effects (Table 5.1).

In **cluster 3** (see Table 5.4), the positive effects of environmental policies on slacks are lower than cluster 2, but also no trade-offs between the two can be seen. Of note, this is the only cluster where environmental spending had a significant effect on the nitrogen slack, indicating that EU policies were more effective in this regard than in other countries around the world. We discuss the detailed analysis of the effect of agri-environmental schemes on slacks (under the EU CAP) in the next section (Table 5.5).

We then turned our discussion to the factor analysis (Figs. 5.2, 5.3 and 5.4) conducted using percentage slacks averaged over the entire analysis period (2005–2018). For **cluster 1,** we identified two principal factors explaining a total of 71% of Pareto-type inefficiencies.

In the first cluster there are highly correlated LSU, employment, nitrogen, GHG, dietary and pesticides slacks, with the factor loadings at LSU having the opposite sign; this is significant, since we have the same situation in cluster 3. It follows that in clusters characterized by higher resource productivity in agriculture there may usually occur this type of slack trade-off. Thus, potential Pareto-type progress on LSU input is negatively correlated with improvement on all other inputs and undesirable output. This can be interpreted as follows: the tendency to decrease the intensity of crop production in clusters 1 and 3 is compensated by an increase in the intensity of livestock production. We have a peculiar dual development—on the one hand there is a progressive extensification of field crops, on the other hand there is an intensification of animal production: confined cattle, and poultry and pig farms are displacing extensive, small-scale livestock production. While slacks in the modified production function are being systematically reduced in crop production and a progress favouring sustainability is being introduced (largely through public policy incentives), little is being done in the area of animal production. Therefore, this area would require a particular concentration of agricultural policy efforts.

The second principal component is formed by land and energy slacks. This principal factor seems to be characteristic for cluster 1, as it shows that inefficiencies (and potential room for improvement) in land management and energy consumption are strongly related to each other and at the same time independent of other aspects of agricultural production. This indicates the high energy intensity of primarily field crops and the potential for reducing environmental pressures in this area without interaction with other production aspects. In cluster 3, energy intensity is more integrated with both crop and livestock production.

In **cluster 2,** factor analysis shows a more complex situation (Fig. 5.3).

As in previous steps of the analysis, energy slack is a separate factor (factor 3), suggesting the need to implement independent policy measures in this area. Factor 1 is composed of pesticides, production, dietary, nitrogen and employment slacks, which are strongly positively correlated with each other; LSU and GHG slacks make up principal component 2. This division suggests in which directions (i.e. following three principal factors) agri-environmantal policy schemes should be developed in less developed countries with lower agricultural productivity.

In **cluster 3** all inputs slacks except employment are highly correlated and form factor 1 explaining almost 50% of the inefficiencies variation. As mentioned, the

Table 5.5 Cluster 3: CAP policy options: effect of a subsidy share change on integrated efficiency slacks (random-effects tobit panel regression separate models 2005–2018, 27 countries, 366 obs)

CAP structure change options/IE slacks	land^r (thous. of ha)	LSU^r (units)	employment^{nr} (thous. of employees)	energy^r (terajoule)	nitrogen^{nr} (tonnes)	pesticides^{nr} (tonnes)	production^r const. Thous. USD	dietary^{nr} (adequacy percentage)	GHG^r (COeq$_2$, kilotonnes)
Crops	−5570.450***	−5038777.574***	9.803	−48,239.182***	−181,084.830**	4649.997***	−2282069.587***	−1.945	−13,821.499***
subs. Share	[1011.776]	[865,108.165]	[210.000]	[7720.289]	[85,461.183]	[1754.761]	[686,974.479]	[5.796]	[2718.593]
Livestock	−2297.869*	−3399489.597***	515.331**	−56,864.362***	−236,318.237**	1049.985	−584,605.868	1.683	−9385.247***
subs. Share	[1316.416]	[973,788.612]	[261.648]	[8691.829]	[95,193.824]	[2017.796]	[608,696.574]	[8.555]	[3016.290]
Environment	822.920	1050896.764	168.186	7076.459	35,315.614	841.485	1469216.608	19.912*	3221.278
subs. Share	[1811.478]	[1380112.618]	[336.964]	[14,537.600]	[144,693.852]	[2631.922]	[1024939.929]	[11.123]	[4138.558]
LFA	860.139	1953766.140	−84.858	−14,008.772	301,749.989	−2343.558	173,995.183	−8.662	2244.037
subs. Share	[3042.652]	[20008379.366]	[610.835]	[24,441.052]	[231,423.202]	[4173.293]	[1805514.749]	[16.568]	[5888.575]
Decoupled	2054.093***	2092124.565***	−47.239	32,473.564***	45,873.762	−1730.811*	1175003.044***	1.516	5506.664***
subs. Share	[614.207]	[499,426.660]	[118.078]	[4329.633]	[48,141.743]	[976.039]	[387,367.878]	[3.532]	[1515.425]
Investment	493.089	935,219.121	−111.938	9124.625	100,570.591	1282.457	−1383461.910	−29.951**	1736.843
subs. Share	[1916.263]	[1508428.802]	[344.338]	[15,314.642]	[151,841.016]	[2585.029]	[1093242.899]	[13.212]	[4540.525]
average RHO	0.75	0.40	0.79	0.79	0.62	0.72	0.82	0.80	0.33

Notes: see Table X

factor load of the LSU slack has a negative sign, which indicates a trade-off with other input slacks. This indicates a potential difficulty for the design of agri-environmental policies in the EU that reduce the pressure of inputs. For example, the effects of policy measures in reducing nitrogen and energy use, and GHG emissions, may be undermined due to deteriorating conditions for livestock production, which will reduce the propensity of farmers to adopt sustainable practices. In terms of outputs—the slacks of agricultural production value and meeting caloric needs (dietary) form a separate factor 2 and have opposite signs. This indicates a problem which may be faced also by highly developed countries, that on the one hand the value of agricultural production increases, thanks to specialization, but on the other hand such an action negatively influences the satisfying of food needs. Therefore, policy schemes stimulating the value of agricultural production have to take into account a negative side effect in terms of the level of satisfying food needs.

5.2.3   EU FADN Analysis

We devoted a separate section to the analysis of slacks in the EU (Cluster 3). Here we focused on the shares of individual instruments in the total of agricultural subsidies. Absolute values of subsidies are of less interest, because the discussion of the CAP is more about its structure than about the absolute value of individual programmes (see Chap. 4). As a rule, CAP programming periods last seven years. The total pool of payments for EU agriculture changes relatively infrequently and each time there is political pressure to reduce it. We estimated the matrix of the regression coefficients of models with one variable, which reflects the different types of payments (Table 5.5). The explanatory variables are Pareto's slacks as before. It is worth noting that both in the model with several share variables and with one share variable the omitted variable bias applies only to variables outside the public policy. In a model with a single input variable, e.g. regressing environmental subsidies on dietary income, the effect of other subsidies is also included, because an increase in the share of environmental subsidies means a proportional decrease in the share of the sum of other subsidies, i.e. as the share of "env.subs." increases, then the share of "1-env.subs." decreases. This means that "1-env.susb.", will have the same significance and identical coefficient, just the opposite sign and a different constant term than "env.subs." The only limitation of this approach consists in making only disjoint policy recommendations, but our primary focus here is on the likely direction of the response.

A matrix of regression coefficients of different types of subsidies on slacks reveals a number of synergies and trade-offs of policy options (Fig. 5.5). As a rule, production subsidies (crops and livestock subsidies) negatively affect most slacks under the environmentally and socially adjusted production function (i.e., they increase inefficiencies), and decoupled subsidies have a positive effect that does not, however, offset the negative effect. Only for the pesticides slack is the situation reversed (see Table 5.5). Against this background, environmental subsidies exacerbate inefficiencies in food security (dietary), as noted earlier. In contrast, the Less-

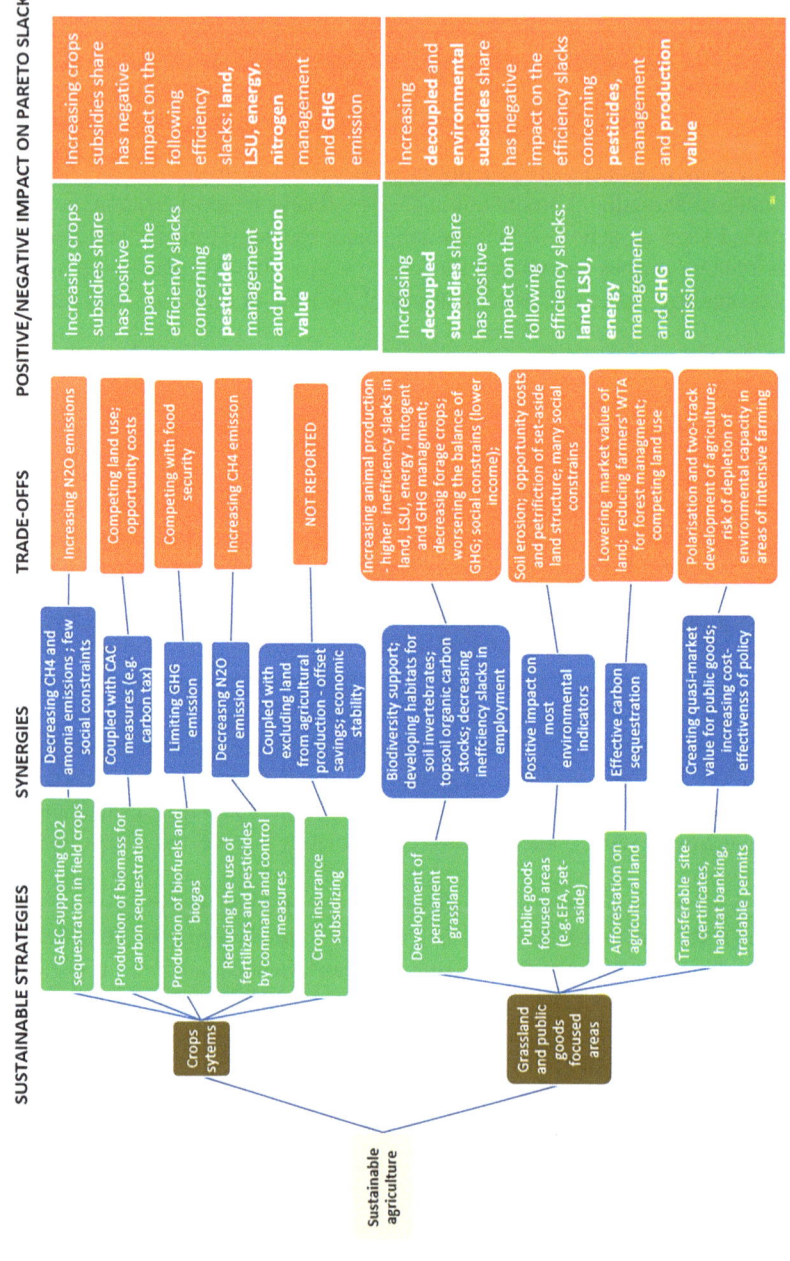

Fig. 5.5 Synergies and trade-offs of sustainable agriculture strategies based on literature review and slacks analysis in EU agriculture

favoured area payment (LFA) have no statistically significant effect on allocative efficiency.

An analysis of the coefficients for crops and livestock subsidies leads to the conclusion that an increasing share of all subsidies except crops (i.e. "1-crops") and all subsidies except livestock (i.e. "1-livestock") have a positive effect on slacks. This confirms that dual development has a *raison d'être*: the current level of production support could be partially left if only a support for extensification would be developed offsetting the negative production impact. Assuming that the starting point is the current support shares, the following (disjoint) scenarios of CAP development beneficial to sustainable agriculture can be identified from Table 5.5:

1. decreasing share of crop subsidies with livestock subsidies unchanged → lower land, LSU, energy, nitrogen and GHG slacks; higher pesticides and production slacks (inefficiencies); policy measures will be designed to compensate for these inefficiencies;
2. decreasing livestock subsidies share with crops subsidies unchanged → lower land, LSU, energy, nitrogen and GHG slacks; higher employment slack; policy measures will be designed to compensate for this inefficiency;
3. Decreasing all production subsides share → lower land, LSU, energy, nitrogen and GHG slacks; larger employment, pesticides and production slacks;
4. Increasing share of environmental subsides but on condition of maintaining the level of food security
5. Increasing the decoupled subsidies share, providing food security (dietary) level is maintained.

5.2.4 Impact of Uncontrollable Policy Variable on the Integrated Efficiency Score

Slacks analysis showed that the public policy expenditures are likely to impact the IE score in each of the clusters studied: Slacks of non-radial input/output data that were significantly affected by changes in the level of environmental expenditures were found in clusters 1 and 2. In cluster 3, however, significant radial variable slacks form together with non-radial ones' principal factors of inefficiency. In addition, detailed analysis of FADN payments revealed significant effects of specific policy measures on non-radial variable slacks (these measures contribute to the central government environmental expenditures).

As we indicated in Fig. 5.1 the third stage of the holistic cost-effectiveness analysis is to compare the IE score while including and not including uncontrollable variables using an approach well established in the literature (Yang & Pollitt, 2009; Mamoon, 2012; Cooper et al., 2007).

We confirmed the significance of differences in rankings with non-parametric tests (Rank sum test, Tables 5.6 and 5.7). The analysis was performed in consecutive years (Table 5.6) and by country (for time series 2015–2018, Table 5.7).

Table 5.6 Impact of uncontrollable variable—environmental spending (central gov.) on change in the efficiency score, constant prices in year

	2005	2006	2007	2008	2009	2010	2011	2012	2013	2014	2015	2016	2017	2018	general
Cluster 1—efficiency scores including and not including policy spending															
included	1.380	1.052	1.029	1.013	1.029	1.023	1.005	1.036	1.051	0.949	1.005	0.980	0.977	1.021	1.031
not included	1.159	0.963	0.869	0.867	0.915	0.855	0.867	0.903	0.958	0.877	0.927	0.883	0.921	0.941	0.917
Rank sum test	0.693	0.689	1.608	2.054	1.016	0.947	1.602	1.598	0.803	0.87	0.965	1.930*	1.013	1.257	**4.523*****
N	12	14	14	15	17	17	17	18	19	18	18	18	18	17	232
Cluster 2—efficiency scores including and not including policy spending															
included	1.380	1.114	1.028	1.134	1.066	1.047	1.005	0.983	0.990	0.994	1.005	0.955	0.953	0.950	1.033
not included	1.207	0.945	0.911	1.001	0.951	0.922	0.890	0.848	0.881	0.898	0.906	0.877	0.846	0.826	0.914
Rank sum test	1.723*	**2.394****	1.488	1.373	1.361	1.992	**2.233****	**2.206*****	1.851*	**1.983****	1.614	1.032	1.116	1.299	**5.893*****
N	19	20	20	22	24	26	26	27	27	28	28	28	27	26	348
Cluster 3—efficiency scores including and not including policy spending															
included	1.558	1.022	0.934	0.950	0.981	0.949	0.930	0.962	0.933	0.936	0.909	0.928	0.926	0.967	0.989
not included	1.131	0.831	0.806	0.829	0.827	0.795	0.816	0.776	0.783	0.792	0.771	0.776	0.792	0.786	0.821
Rank sum test	**1.960****	**1.843***	**1.695***	1.358	**1.894***	**2.434****	**1.698***	**2.569*****	**2.188*****	**1.739***	**1.946***	**2.448****	**2.015****	**1.912***	**7.351*****
N	25	27	27	27	27	26	25	27	27	27	28	26	27	27	

Table 5.7 Impact of uncontrollable variable—environmental spending (central gov.) on change in the efficiency score, constant prices at country level

Country	included	not included	rank sum test	Country	included	not included	rank sum test
Cluster 1—efficiency scores including and not including policy spending				**Cluster 2 cont.—efficiency scores including and not including policy spending**			
Albania	1.105	1.050	0.486	Mongolia	1.09	1.04	−0.218
Algeria	1.354	1.258	0.866	Morocco	1.03	0.83	1.214
Argentina	1.110	1.102	0.528	Namibia	0.68	0.51	1.543
Armenia	1.047	1.021	0.340	Nepal	1.07	1.05	0.265
Canada	1.029	1.027	0.189	New Zealand	1.06	1.05	0.361
Chile	1.043	0.961	1.310	Nicaragua	1.01	0.88	1.567
China	1.042	1.042	0.404	Pakistan	1.10	0.79	**2.849**[***]
Côte d'Ivoire	0.968	0.677	**2.594**[***]	Paraguay	1.01	0.53	**3.130**[***]
Egypt	1.042	0.928	**2.128**[**]	Peru	0.79	0.72	1.155
Israel	1.080	1.078	0.368	South Africa	0.85	0.68	**2.534**[**]
Japan	1.049	1.039	1.149	Tunisia	1.02	1.02	0.334
Jordan	1.150	1.040	1.424	Yemen	1.18	1.15	0.34
Malawi	2.336	1.782	1.000	Zambia	1.02	0.68	**3.63**[***]
Mozambique	1.022	1.000	0.064	**Cluster 3—efficiency scores including and not including policy spending**			
Norway	1.038	1.036	0.023				
Philippines	1.025	1.001	1.057	Austria	0.982	0.976	0.552
Russia	1.021	0.672	**4.319**[***]	Belgium	1.224	1.060	0.827
Switzerland	1.052	1.037	0.506	Bulgaria	0.990	0.434	**4.333**[***]
Thailand	0.869	0.400	**3.906**[***]	Croatia	1.077	1.063	0.781
Turkey	1.061	1.019	**1.667**[*]	Cyprus	1.341	0.361	**4.503**[***]
USA	1.052	1.040	−1.322				
Ukraine	0.680	0.369	**4.457**[***]	Bohemia	1.316	1.296	0.391
Cluster 2—efficiency scores including and not including policy spending				Denmark	1.047	1.037	0.942
Australia	1.04	1.02	0.299	Estonia	0.972	0.587	4.157
Azerbaijan	1.09	0.94	**1.839**[*]	Finland	0.676	0.634	**2.573**[**]
Bangladesh	1.06	0.81	**3.695**[***]	France	1.034	1.024	0.23
Belarus	0.97	0.72	**3.906**[***]	Germany	1.004	0.995	0.69
Brazil	1.07	1.05	1.545	Greece	1.109	1.095	0.333
Burkina Faso	1.14	0.91	0.94	Hungary	0.831	0.668	**3.814**[***]
Costa Rica	1.08	1.07	0.988	Ireland	1.025	1.015	1.241
Ecuador	0.79	0.69	1.359	Italy	1.095	1.044	**2.183**[**]
El Salvador	1.12	1.07	1.218	Latvia	0.733	0.357	**4.503**[***]
Ethiopia	1.03	0.71	**2.117**[**]	Lithuania	1.192	0.685	**4.282**[***]
Georgia	1.07	1.01	1.057	Luxembourg	1.068	1.057	0.643
Guyana	1.15	1.05	**2.108**[**]	Netherlands	1.083	1.062	0.965
Iceland	1.17	1.17	−0.322	Poland	0.858	0.586	**4.319**[***]
Indonesia	1.10	1.09	0.391	Portugal	0.908	0.634	**3.592**[***]

(continued)

Table 5.7 (continued)

Country	included	not included	rank sum test	Country	included	not included	rank sum test
Iran	1.15	1.10	0.104	Romania	1.138	1.037	0.827
Kazakhstan	1.05	1.01	1.424	Slovakia	0.666	0.406	3.768***
Kenya	1.00	0.97	0.333	Slovenia	0.445	0.312	4.503***
Kyrgyzstan	1.21	1.09	0.846	Spain	1.095	0.917	3.814***
Madagascar	0.97	0.88	1.058	Sweden	0.768	0.716	2.849***
Mexico	0.83	0.62	2.096**	United Kingdom	1.046	1.046	0.023

Notes: Data on environmental policy expenditures in China and the US were not available. Therefore, we used data obtained from the U.S. Office of Management and Budget (Outlays of the U.-S. Environmental Protection Agency) for the U.S. and from CEICDATA.COM for China. All amounts were converted to dollars at constant 2015 prices

As expected in **cluster 1,** the effect of the uncontrollable policy variable on the IE ranking is positive in every year, but very weak (mostly statistically insignificant). From the slacks analysis, we know that this impact affected only two non-radial variables slacks. Therefore, in the most holistic view, it only became apparent for one year (i.e. 2016). This confirms the conclusion of the low cost-effectiveness of environmental expenditures in this cluster from the point of view of their impact on improving the resource allocative efficiency in agriculture. The cross-country analysis confirms this conclusion, showing that the significant differences in rankings were only for the less developed countries in this cluster (see Table 5.7), and in the case of the dominant economies of the US and China there was little and no statistical significance (in the case of China no change in the score at all).

In **cluster 2,** the interaction was more pronounced and proved to be statistically significant in six periods, based on which a slight downward dynamic can be detected (i.e. the difference in rankings declined by about 3% on average per year). The cross-country analysis showed that the significant differences in rankings were substantially larger than in cluster 3 (see Azerbaijan, Bangladesh, Belarus, Ethiopia, Guyana, Mexico, Pakistan, Paraguay, South Africa, and Zambia in Table 5.7).

The analysis of the rankings with and without uncontrollable variables in **cluster 3** confirms this last conclusion. In this case the positive impact of the policy was visible in almost all years. However, its dynamics was found to be decreasing—the difference in rankings decreased on average by about 6% per year. Interestingly, by country, this difference was the largest and statistically significant mainly in the New Member States (Central and Eastern European countries) accessed to the EU in 2004. When ordering these countries by decreasing ranking differences (significant and greater than 0.1), the listing is as follows: Cyprus, Bulgaria, Lithuania, Latvia, Portugal, Poland, Slovakia, Spain, Hungary, and Slovenia (see Table 5.7).

In this chapter, we presented the results of the holistic cost-effectiveness analysis. A detailed discussion of these results and recommendations for policymakers related to them can be found in Chap. 7.

Chapter 6
Evolution of Agri-Environmental Schemes Worldwide. Comparing the Agricultural Policy of the EU, the US and the People's Republic of China

6.1 A General Description of Agricultural Policies

6.1.1 The EU Common Agricultural Policy

In this sub-chapter, we focus on the evolution and main directions of agricultural policy in the EU, the USA and the Peoples Republic of China (hence forward—China). These three constitute ca. 45% of global agricultural output (EU—10.4%, USA—9.6% and China—24.9%) and 27% of global arable land (EU—7.1%, USA—11.4% and China—8.6%) (FAOStatat). These figures clearly show their crucial role in the global agricultural system. Although policymakers in all of these states seem to acknowledge all three dimensions of sustainability, the relative roles of each dimension differ, as do the tools used.

The origins of EU agricultural policy (common agricultural policy—CAP) date back to late 1950s and 1960s. Initially, the main aim of the CAP was to increase food production to ensure food safety for all EU citizens. A second intention was to increase and stabilize agricultural income by introducing market price support. The food safety goal was achieved quickly, and output surpluses linked with high costs of policy became problematic. Starting from Mansholt's plan (1968), the CAP has been gradually redesigned and now addresses different aspects of agricultural activity that are considered important but were ignored earlier. The MacSharry's reform (1992), Agenda 2000, Fischler's reform and the introduction of so-called 'greening' in 2013 were clear signs that EU agricultural policy has been gradually changing into an agri-environmental policy with a high priority on the development of rural areas.

This important shift can be justified for two reasons. First, agriculture is responsible for 10.3% of the EU's GHG emissions and nearly 70% of those come from the animal sector (EEA, 2021). Second, EU citizens no longer want to pay for increasing farmers' income. They rather condition their support on providing public goods and

B. Czyżewski, Łu. Kryszak, *Sustainable Agriculture Policies for Human Well-Being*, Human Well-Being Research and Policy Making, https://doi.org/10.1007/978-3-031-09796-6_6

are ready to pay for policies contributing to climate change mitigation and to improving the state of the natural environment.

In the survey undertaken by ECORYS (2017), 92% of non-farmer respondents agreed that agricultural policy should incorporate practices that provide climate and environmental benefits, but only 53% agreed that farmers need direct income support. This opinion was especially held by the citizens of the so-called 'old' EU Member States. Below we focus on the new green architecture of EU agricultural policy which is inseparable from the announcement of the EU Green Deal (EC, 2019) and "Fit for 55" package (EC, 2021a).

The EU Green Deal was announced in December 2019 and is going to be financed by 1/3 of the EUR 1.8 trillion sum set asides for the EU post-pandemic recovery plan called "Next Generation EU" and from the EU seven-year budget. The Green Deal introduces several different initiatives that are linked to UN Sustainable Development Goals. However, two initiatives (or strategies) are particularly important for the agricultural sector. These are the Biodiversity strategy for 2030 (EC, 2020a) and Farm to Fork strategy (EC, 2020b) announced in 2020. Under these strategies, there are number of specific actions that are intended to be undertaken, among which are following: the Organic Action Plan, Welfare of Farmed Animals, sustainable use of pesticides, improved nutrition labelling, EU agri-food promotion policy. The "Fit for 55" package introduces specific actions that will be implemented to attain the EU's ambitious goal of reducing GHG emissions by 55% by 2030, in comparison with the 1990 level. From the agricultural point of view, the most important actions could be those designed to regenerate soil quality to ensure food security.

The most recent CAP reform (agreed in 2021 and to come into force in 2023) is also related to the Green Deal, and its main principles are in line with the general development strategy for the whole EU economy. The CAP has a budget EUR 378 bln for seven years and 40% of this amount need to be spent on climate objectives. The main idea in this CAP is that there are some basic environmental standards that all farmers need to comply to, and some further voluntary actions that farmers can choose to follow for which they receive extra payments. These basic obligatory standards used to be called "cross-compliance rules", but in the new architecture they are strengthened and termed "Good Agricultural and Environmental Conditions" (GAEC). There are 9 GAEC norms at the EU level, for example: appropriate protection of wetlands and peatlands; establishment of buffer zones along watercourses; or crop rotation or other practices to preserve soil potential such as crop diversification. Furthermore, certain statutory management requirements (SMR) based on non-CAP directives are also to be introduced. These are related and are called the "Water Framework Directive" and the "Directive on the Sustainable Use of Pesticide".

The previous CAP was often criticized for its excessive complexity and for the lack of elasticity regarding different needs and conditions across EU Members. It could be questionable whether there is significant simplicity in the new CAP, but higher elasticity is clearly evident in the way the new CAP is formulated at the Member State level. In the new CAP, Member States are obliged to submit their national strategic plans. In these plans, countries should propose specific goals of action undertaken under CAP, but also indicate how some general CAP guidelines will be realized in practice. These make reference to GAEC norms, as well, and the countries must declare what their farmers will need to do (or what is going to be prohibited) for the countries to meet the new CAP. The advantages of elasticity seem to be clear, but there are also some possible drawbacks. For example, it has been negotiated that countries can make some GAEC requirements obligatory only for farms that operate on 10 ha of land or more. This proviso can undermine the achieving of environmental goals in some countries where these rules were introduced and where small-scale farming predominate.

Beyond the obligatory GAEC and SMR standards, the new CAP introduces a new type of I pillar direct payments—the eco-schemes. These subsidies are paid for each hectare of land on which at least one eco-practice is realized. It is not obligatory for the individual farmer to choose any eco-scheme, but it is mandatory for member states to spend at least 20% of their national envelopes for direct payments on eco-schemes payments. In their strategic plans, EU Member States should also propose a list of eco-schemes that will be available for farmers to enroll in. Still, there are some even more ambitious agri-environment climate interventions financed from the second pillar. Obviously, these are also voluntary for farmers.

One can say that new CAP proposes a wide range of eco-practices, and many farms can find their own niche. However, at least in the beginning, farmers may confuse eco-schemes with II pillar interventions.

Perhaps, the most important part of the EU Green Deal with respect to agriculture is the new **Farm to Fork (F2F) strategy** (EC, 2020b). Although EU agriculture is the only major agricultural system that significantly reduced GHG emissions (by 20% since 1990), still much progress needs to be done to approach true sustainability in the sector. The aim of the strategy is to make food production climate-neutral or even climate-friendly, while ensuring that this food is healthy and affordable. Therefore, it is clear that achieving this goal requires not only enacting agricultural reforms, but restructuring of the entire agri-food chain (including food processing and consumption) (EC, 2020b).

There are many particular actions and goals in the F2F strategy. Among them, the reduction of chemical pesticides and hazardous pesticides by 50% by 2030 seems to be very ambitious and challenging. Our analyses in Chap. 3 have shown that the average slack on pesticides in EU countries can be estimated as 20%, meaning that EU countries could relatively easily reduce pesticides use by 20% without changes in production technology. In some new Member States, this slack is much larger, exceeding 50%, but in some other countries, this slack is positive, showing that these countries could even increase pesticides use without depleting efficiency levels. From this perspective, it could be questionable whether this aim is not too ambitious. However, from a purely environmental point of view, a large reduction in pesticide application would be beneficial.

F2F also includes the demand to bring about significant reduction of nutrient loss (50% by 2030), fertilizer use (by 20%) and sales of antimicrobials (by 50%). When it comes to fertilizers, our previous calculations have shown that elimination of slack on nitrogen could save ca. 28% of this input use. In the case of nitrogen, all countries experienced negative slack. Among new Member States, this was often even larger than 50%, meaning that these countries could potentially achieve F2F targets only by slack reduction. In some Member States, where slack is much smaller, further actions need to be undertaken. However, the EU goal is designed to be met at the EU level, so the reduction level can differ between Member States and some countries could make greater or lesser efforts. Achieving these goals would be linked with integrated pest management and integrated nutrient management plans.

Reducing the sales of antimicrobials is related to the problem of food safety. It is estimated that 950,000 deaths (20% of the total number) can be attributed to unhealthy diet. Therefore, actions aiming at enhancing food quality and safety are urgent. Among other undertakings related to these problems are the introduction of a unified system of food labelling that would contain information about the environmental impact of the product, and promotion of EU-sourced high-quality food.

Another big problem is that 68% of UAA is related to animal production—a large portion of crop production is used as feed material. There is a need to reduce the dependency of critical feed material such as soya grown on deforested land, and EU-grown plant protein production need to fostered. In addition, the EU goal is to popularize the use of alternative feed material such as insects, marine feed stocks and by-products such as fish waste. The EU also recognizes the problem of food waste. Since 20% of all food is wasted, the EU aims to halve per capita food waste by 2030.

From the scientific point of view, a very important step forward is the replacing of the Farm Accountancy Data Network (FADN) by the Farm Sustainability Data Network (FSDN). The FADN is a widely used database containing detailed information about the economic and financial aspects of farm functioning that is representative for a vast proportion of the EU agricultural sector (except for the smallest entities). However, it does not contain information about farm social and environmental aspects. Therefore, the use of this database does not directly allow for sustainability and eco-efficiency calculations.

F2F also addresses certain important ethical and life-style issues. It puts great emphasis on animal welfare and intensive livestock production. It also recognizes the evolution of EU diets towards a more plant-based diet with reduced meat consumption. There are two competing views on these matters of course. The first is that the EU should put effort to reduce intensive livestock production, which harms the environment and violates animal welfare. Furthermore, this view holds that meat produced in this model of husbandry has lower nutritional and taste value and can be even dangerous for consumption. In this view, production of meat is not a major problem itself, but rather certain ways of producing it are problematic. However, it is clear that evolution towards reducing animal products in the common diet should be supported. F2F seems to follow this kind of thinking.

Another ambitious plan of the EU is related to **Biodiversity strategy** (EC, 2020a). An important part of this is the extension of protected areas. The goal is that 30% of all of EU's land areas and 30% of its marine areas should be protected. This is a minimum of an additional 4% of land and 19% of marine areas to those currently protected. What is more, 10% of all land areas (instead of 3% at present) should be under strict protection. These goals can impact agriculture by limitation of intensive agricultural methods, especially in these areas or in their close neighborhood. However, the strategy clearly indicates that higher biodiversity pays off also in financial terms. It is estimated that global losses resulting from land cover change are about EUR 3.5 bln up to even 18.5 bln annually, while losses related to land degradation are as high as EUR 5.5–10.5 bln.

The other specific part of EU Green Deal that is related to agriculture is **the Organic Action Plan** (OAP) (EC, 2021b). Despite yields possibly being lower under OA, it seems it is not a very big problem in view of the high level of food security EU-wide. Organic food and agriculture production have benefits both for the natural environment and human health. It is estimated that land farmed organic has ca. 30% more biodiversity than land under conventional agriculture practices (Rundlöf et al., 2016; Tuck et al., 2014). From a health perspective, there are some well-documented examples of the positive impact of organic food consumption, however, it is not very easy to study this impact since people consuming organic food usually follow generally healthier lifestyles (Mie et al., 2017).

The growing trend in OA in the EU is apparent. The area under OA has increased by almost 66% between 2009 and 2019, however, this is still 8.5% of all UAA at the EU-27 level and differs between MS. In countries such as Netherlands, it is below 5%, while in Austria, Estonia and Sweden it is more than 20%. The ambitious aim is to obtain 25% of UAA under OA by 2030. This may be possible thanks to the growing demand for organic food. Sales of certified organic foods have more than doubled in one decade—increasing from EUR 18 bln in 2010, to EUR 41 bln in 2019. There are still big differences between organic food consumption between some "old" and "new" EU Members, but a positive trend is also clear among the latter group (Chiciudean et al., 2019). The number of organic producers and processors is also significant—343,858 and 78,240, respectively. It is worth highlighting that OAP has a multidimensional perspective—it aims both to stimulate demand and convert the entire value chain. Several specific actions have been introduced or will be introduced in the near future. The most important include: promotion of organic food canteens, lower taxation (value added tax) on organic products, promotion of shortening of food value chains, and the previously mentioned changes in animal feeding patterns and husbandry practices.

6.1.2   Farm Bills in the United States

The agricultural sector is heavily supported in the United States of America. However, the way the sector is subsidized differs when compared to the EU. The role of national or state environmental programs and especially rural development programs is much less apparent. The main feature of the agricultural policy is to decrease the level of risk and to stabilize and increase farmer's revenue. The rationale for the policy is built on some conventional wisdom assumptions which are, however, not really true (Goodwin & Smith, 2015). For example, there is an assumption that farm households have lower income and less wealth than other households but in reality the US Department of Agriculture estimated (in 2012) that the median total income of farm households was 34% higher than the median income for all households, and 98% of US farms have "high wealth", where high wealth means wealth above the median level for total economy. It seems that actual design of policy support resulted from the bargaining power of farmers and their groups rather than from real income problems in the sector.

Agricultural policy in the USA is organized in the form of so-called **"farm bills"** (**Agricultural Improvement Acts**) and could be dated back to the reforms of the New Deal in 1930. Farm bills are complex legal acts containing several "titles" to deal with different aspects of the sector. They used to be introduced every six years, but recently a new Act came into force in 2018 (with the budget equal to USD 867$ bln), after the previous one in 2014. However, one should note that the 2018 bill mostly just reinforced practices put forward in the 2014 Act. To understand the basic features of US agricultural support, descriptions of two reforms (in 1996 and in 2014) are particularly important.

Until the 1996 act, the subsidies for farmers had the form of typical market price support. Farmers received payments if the market price of commodity were lower than so-called "target price". The subsidy was calculated as difference between real market price and target price multiplied by the yields. It seems easy, but the system had two specific features. First, the yield used for calculation was not an actual yield in a given year but historically observed yield from 1981 to 1985. This meant that farmers who were more productive in those base years, received higher payments. Second, farmers could not decide what to produce. They were obliged to produce commodities that were assigned to a given acre of land (so called base acres) (Orden & Zulauf, 2015).

In the 1996 reform, **direct payments** were introduced and subsidies did not depend on prices. Under this reform, payments were calculated as the product of historical yields and commodity specific amounts that were not related to actual prices. Therefore, the amount of payments was fixed and known in advance. The reform also introduced production flexibility contracts. This meant that payments were still based on base acres, but farmers could produce another commodity. However, this could lead to situation in which farmers produced corn in a particular year, but received payment based on historical yield and commodity amount assigned for their base acre that could have been soybean, for example. This system appears to be illogical at first, but the logic behind was to decouple subsidies from planting decisions. Since farmers had no impact on the payments they are going to receive, they could make their planting decision based on market signals rather than on the basis of what may be more profitable in terms of the support received.

It should be noted that some evidence from the literature have shown that subsidies do not have an influence on production decisions in fact, but other authors claimed that they have an impact on entry-exit decision (Boussios et al., 2021). If the subsidy system has no (or minimal) impact on planting decision and one can assume that money collected from the taxes, which is spent on agricultural subsidies, would be spent on another program rather than to reduce budgetary outlays then it could be said that so-called "deadweight losses" are minimal or they do not exist at all (Babcock, 2015).

In the 2002 bill, **countercyclical payments** were backed, and in 2008, a new revenue program (Average Crop Revenue Election—ACRE) was introduced. Under the latter, farmers could receive payments even when market prices were higher than target price, as the subsidy was made when farmer revenues were below a moving average of past revenues constructed upon market prices from the previous two crop years and Olympic average[1] yields from the previous 5 years. In practice, this meant that payments were paid even when revenues were quite high, but lower than average. What is more, despite introducing these two programs, direct payments were still in power and the system was criticized for being too generous for farmers. However, it should be noted that enrollment to ACRE program meant renouncing 20% of the amount of direct payments (Orden & Zulauf, 2015).

As mentioned, a very important reform was introduced in 2014. This ended the direct payment approach which was criticized since in this system farmers received money even when their market situation was very good. In the 2014 reform, the existence of base acres as the foundation for payments calculation were retained but farmers could update it to their current preferences at the beginning of the program and they could decide freely on their actual production, so this important feature of the 1996 reform was still maintained. This was a compromise between those who were in favor of base acres and those who claimed that a system built upon planting acres is more rationale.

There were two main programs supporting the income of US farmers introduced in 2014, and farmers could choose between them (Boussios et al., 2021). The first option was the "**Price Loss Coverage Program**" (**PLC**), which was the revised target price program. In this program, the target price was now renamed "PLC Reference Price". This program included a Marketing Loan (ML) that was ca. half of the PLC price. If the market price was below the reference price (but above ML), then payment was calculated as difference between market and reference price multiplied by farm-specific reference yield (PLC yield). However, the based PLC yields were updated and calculated for the 2008–2012 period. If the price was below ML, then payments were calculated using difference between PLC price and ML— meaning that the subsidy was capped at ML level.[2] What is more, the total amount of subsidies for a farm calculated using the rules described above was multiplied by 0.85 and this was the final amount of money that individual farmers received. It need not to be mentioned that all of the rules apply to the updated base acres and not to planted acres.

Instead of PLC, farmers could chose to enroll in another program called "**Agricultural Risk Coverage Program**" (**ARC**). Under this program, the payment was

---

[1] Olympic average is an average calculated for five periods with dropping the highest and the lowest value.

[2] Boussios and O'Donoghue (2019) have shown that prices were never below ML level between 2014 and 2018.

made when the farm revenue was below 86% of benchmark revenue. If actual revenue were higher than 86% of benchmark, then there were no payments. When revenues were between 76% and 86%, the payment just topped-up to 86% of benchmark. If they were below 76% of benchmark, then farmers received 10% of benchmark revenue. It could be said, therefore, that ARC payments were capped at 10% of benchmark. The key thing is to understand how these benchmark revenue were calculated. These were the product of five-year Olympic moving average of county or farm crop yields multiplied by five-year Olympic moving average of US crop year prices. Similarly to PLC, the ARC payments were based on base rather than planted acres and they were paid on 85% of eligible base acres if farmers chose calculations of average yields on county level or 65% if yields on farm level were chosen. It is therefore obvious that the vast majority of farmers (around 99%) who enrolled to ARC chose county-based on schemes.

The key differences between PLC and ARC from farmer perspective were that in ARC farmers could receive payment even if market prices were quite high. In PLC, farmers received money only when prices were below reference level, but if prices were far below this level, then payment could be significant. The reality has shown that in 2015–2017, ARC was way more popular, but this was due to the fact that ARC was chosen particularly often in the Midwest region which is the leading crop region in USA. In other parts of the country, the popularity of PLC and ARC was similar. Between 2015 and 2017, seventy percent of all payments on the national level were paid from ARC (county-based) (Boussios et al., 2021).

When compared to EU's I pillar payments, the system of subsidies from title I in the US farm bill has important consequences from the budgetary perspective. In the EU, the total amount of payments that will be paid is known in advance in the beginning of the seven-year budgetary period. In the US system, the amount of money that will be needed is unknown and can vary a lot. This depends on enrollment rate to the two main programs and on the relation between market price and reference price in PLC or the benchmark price in ARC. The estimates for 2014–2018 showed that total amount of subsidies paid to farmers could be USD 7.4$ bln in a high market prices scenario or USD 27.3$ bln in a low prices scenario (Orden & Zulauf, 2015).

Environmental aspects of agriculture in USA were traditionally supported by the **Conservation Reserve Program (CRP)** first introduced in 1985. The program consisted of paying farmers for land diversion in non-crop uses. Interestingly, from the very beginning, the program was not focused on improving environmental quality for itself, but it aimed at avoiding the soil erosion that could lower agricultural productivity. Until 2002, land conversion programs constituted more than 75% of US spending on conservation schemes. After 2002, working farmland conservation became more important and now around half of expenditures is assigned for sharing the cost of environmental friendly practices on land which is still in agricultural use. This is being done mostly by two programs—**Environmental Quality Incentives Program (EQIP) and Conservation Stewardship Program (CSP)**. The former supports producers who would like to implement pro-environmental practices such as irrigation efficiency, restoring pasture, or nutrient and pest management. Under the EQIP it is possible to receive up to 75 percent of the costs of certain conservation practices (some special groups of farmers can receive 90%). Contracts can be signed for 10 years period but usually last one to three years. CSP is the largest conservation program in the USA. Contracts are signed for five years but they can be prolonged if the initial contract was successful. In CSP participants receive annual payments for the environmental benefits which they provide. The main difference is that EQIP is rather short-term and narrow program which is focused on a specific environmental problem while CSP are longer and wider-scale program which applies to the entire activity. The effectiveness of CRP has been seriously questioned (Lichtenberg, 2015). The CRP enrollment rate was higher in regions where land was cheaper and it constituted good possibility to earn additional money in a relatively easy way, while this rate should have been higher in regions with the biggest environmental problems. It changed to some extent in 1991, when the Environmental Benefit Index (EBI) was introduced. This is a tool for assessing the quality of application for funds. However, the correlation between EBI and the state share of land under CRP (with the latter ranging between 1% and 19%, but usually below 4%) is still close to zero. In contrast, there is clear correlation between share of CRP land and the state share of total US cropland. This clearly indicates that CRP has been treated as an extra-money program rather than a true environmental program.

All three main programs (CRP, EQIP and CSP) are prone to three important problems, namely leakage, additionality and inadequacy. Leakage means that positive outcomes of policy are partially offset by other negative changes. When it comes to CRP, an example of leakage is the estimation that for each 100 acres that are withdrawn from crop usage, another 20 acres are converted to crop uses. The working farmland conservation programs (as CSP) can also be subjected to this problem. Payments from these programs raise farmer income and make it profitable to convert some non-crop land to crop uses because conservation practices reduce soil erosion. The problem is, however, that from environmental point of view, it would be better if this land had not been converted and stayed as grazing land, for example.

Another well-known problem is that in agri-environmental programs there is always a risk that taxpayers pay for activities that farmers would have taken anyway (cf. Pates & Hendricks, 2020; Howard, 2020; Fleming et al., 2018). In other words, there is a question as to what extent additionality is associated with a given practice. In EQIP, additionality should be mitigated because only projects approved by technicians are eligible for funding. In CSP, in turn, it is possible to obtain funding for actions that farmers already do. It can be said that, at least in some cases, under CSP farmers receive support not for taking new actions, but rather for not stopping the current actions. Obviously, this is not exclusively a problem of US conservation schemes. The low level of additionality is a feature of at least some of actions under the second pillar of EU CAP (Chabé-Ferret & Subervie, 2013). The last problem, here referred to as inadequacy, was also discussed in the previous chapter. Simple mechanisms and programs that are designed to deal with complex objectives are usually not objective, unless these partial objectives are highly correlated, but as we know from analyses in this book, there are many possible trade-offs between different environmental goals (see Chaps. 4 and 5).

In this subchapter, we focused on subsidies paid for farmers to increase (stabilize) their income that correspond to economic and social aspects of agricultural sustainability and on conservation schemes related to environmental dimension. However, it should be noted that the highest share of outlays in the current US farm bill (around 80%) is assigned to the Supplemental Nutrition Assistance Program (SNAP) (previously known as food stamps). This is designed to support the food security of low-income US families (Swinnen, 2018). Starting from 2008, it has changed and now improved nutrition is considered more important than just food availability. What is more, stricter eligibility criteria were introduced. However, the idea that only employed people could receive support under SNAP has not yet been introduced.

6.1.3 Agricultural Policy in China

It could be said that China's agricultural policy had two very important goals: to stabilize and increase farmer income, and to ensure food security for a large and dynamically growing population. These goals are further confirmed by the 14th Five Years Plan (2021–2025), which aims to improve food availability through increasing yields and introducing new technologies. The system of agricultural subsidies in China was traditionally oriented mainly toward market price support (MPS). In China, agricultural supports constitute ca. 12.5% of total farm receipts, but 2/3 of all spending is made thorough MPS and 80% of all commodities are subject to price support (OECD, 2021). In the years 2018–2020, farmers received prices that were 10% higher than world prices. Moreover, their market share was and is protected through tariffs. However, the role of direct payments is gradually increasing (cf. Table 6.1). In addition, as it will be described below, in recent years other goals related to social and environmental issues are becoming more and more important. Nevertheless, before we move on to recent policy changes, we will briefly describe the evolution of agricultural policy in the last seventy years.

When the communists took power in China and proclaimed the People's Republic of China, they introduced several reforms. One of the most important was the parceling of agricultural land. They gave around 43% of land to the peasantry (Perkins, 1969). After that, farmer communes were created but they were ineffective which caused the tragedy of mass famine. In addition, in the 1950s and 1960s, the agricultural sector was taxed to support industrialization. Still, it experienced high growth in production, albeit at the cost of chemical fertilizer import.

Between 1978 and 1984, the China's government introduced a very important reform called the "Household Responsibility System" (HRS). Under this reform the village commune was still the owner of land, but individual farmers signed an agreement to cultivate a particular plot. Admittedly, the farmers were subject to a production contract at the beginning of tenure, but they could decide on production and it was in their own interest to increase production level. It is not a surprise that

Table 6.1 Comparison of financial support for agricultural sector in China, USA and EU in 2005 and 2020

Indicator	2005			2020		
	EU	USA	China	EU	USA	China
Total value of production (at farm gate), in bln of USD	337.4	234.7	401.8	452.7	331.5	1619.8
Market Price support, in mln of USD	46.7	7.4	14.6	16.6	2.5	140.9
Market Price support as % of total value of production	13.8%	3.1%	3.6%	3.7%	0.8%	8.7%
PSE as a share of GFR (%)	28.58	14.65	8.02	19.33	11.03	12.17
Most distorting support as % of PSE	50.62	41.97	66.72	22.51	25.14	72.26
Share of GSSE, in total TSE (%)	11.03	9.67	29.19	10.57	12.36	12.61
Share of agricultural knowledge and innovation system, in total GSSE (%)	39.43	26.14	16.15	52.90	29.41	23.91
Consumer support estimate (CSE), in mln of USD	−43.2	17.0	−16.3	−14.7	38.3	−172.2
Percentage TSE (% of GDP)	0.94	0.55	2.09	0.66	0.46	1.59
Payments based on commodity output (share of GFR, %)	13.06	5.09	3.47	3.13	2.47	8.51
Payments based on input use (share of GFR, %)	3.07	3.48	3.14	2.93	2.78	1.14
Payments based on current A/an/R/I, production required (share of GFR, %)	7.25	1.09	0.75	5.38	3.82	1.60
Payments not requiring commodity production (share of GFR, %)	5.32	5.00	0.66	7.78	1.96	0.91
Other payments (share of GFR, %)	−0.12	0.00	0.00	0.12	0.00	0.00

Note: *PSE* Producer Support Estimate, *GFR* Gross Farm Receipts, *GSSE* General Service Support Estimate, *TSE* Total Support Estimate, *CSE* Consumer Support Estimate, *A/An/R/I* area, animal number, farm receipts, farm income
Source: Own elaboration based on OECD.Stat (downloaded on 25 Jan 2022)

after introducing the reform, productivity increased dramatically but the downside of the reform was that it resulted in high land fragmentation. Even nowadays, the majority of crop output is provided by 250 mln farm households with an average land size of only 0.6 ha (Wilkes & Zhang, 2016).

Despite the reform, agricultural production of small farm households in China ensured, however, only 15% of farm family caloric intake, since a large part of the output was sold to buy necessary food. Moreover, the income from farming constituted only 1/3 of farm family total incomes (Wilkes & Zhang, 2016). The rest of the income came from other sources since many farmers were also engaged in different non-farming activities. Furthermore, starting from 1990s, mass migration to cities came about. At the same time, agricultural policy was concentrated on output growth by subsidizing fertilizers and other inputs. This aim was achieved since production grew significantly, but support for inputs has led to excessive use of fertilizers and pesticides and mass environmental degradation.

As discussed in Chap. 3, fertilizer and agrichemical use in China's agriculture is now even much larger, not only when compared to developing countries but also to the developed parts of the world. Indeed, it is estimated that fertilizer use is too high by at least 30–50% and is resulting in mass eutrophication of water bodies in and around China. It should be noted here that input use in China is highly diversified at regional level and the most intensive production is located in eastern and south-eastern regions.

After joining WTO in 2001, China had to decrease tariffs on agricultural commodities and they went down from 42% to 12%, on average (OECD, 2021). As a result, one of the main features of agricultural support was to reduce the income gap between farmers and workers in other sectors by offering minimum prices on commodities and intervention stocks. At the same time, the first agri-environmental program was introduced. This was officially called "Returning Farmland to Forest Programme", but was better known as "Grain for Green". The program resembles the US CRP, as under this scheme farmers receive subsidies to convert marginal degraded or abandoned arable land to forest or grassland. As of 2008, 26.7 mln of ha have been enrolled in this program (Wilkes & Zhang, 2016).

Since 2015, the market price support has gradually become reduced but for wheat and rice it remains high. For cotton a similar support system to the Price Loss Coverage system in the USA exists. The subsidy is dependent on differences between target prices and market prices. In general, MPS for crops is being replaced by agricultural support and protection subsidy payments. These combine direct payments for grain producers, support for agricultural inputs, and subsidies for improved seed variety, and are all paid per unit of land so could be called direct payments. The level of payment is high when calculated per ha regarding average size of farm which is very low. Still, the subsidies usually do not constitute a large portion of rural household income. Existing market price support is operationalized by the grain purchases by the state. These acquired reserves are then used to influence market price in order to stabilize farmer incomes. In addition to this there are also many other types of subsidies, for example, payments for land consolidation or for irrigation constructions.

One of the goal of the agricultural support and protection subsidy payments was to support an increase in farm size (Han et al., 2021) since the vast majority of farms in China operate on very small plots. Beyond these, there are 3.41 mln specialized households that cultivate on areas over 3.3 ha and 870,000 large farms that cultivate on areas of 13.3 ha, on average. This shows that number of large farms is very small when compared to the number of small-scale farms. Furthermore, from the US or EU perspective, 13.3 ha is a small or medium-scale farm, at maximum. In China, there is also 1.29 mln of cooperatives and more than 350,000 agricultural enterprises, including 125,000 large-scale firms. What is more, these large-scale farms engage 125 mln of households in their supply chain (Wilkes & Zhang, 2016). The present agricultural policy supports these bigger farms (3.33 ha or above), although they only operate on ca. 10% of all arable land. Cooperatives encompass 35% of all small farms, but they are often initiated and supported by local government so they operate rather in a top-down approach. When it comes to agribusinesses, these are sometimes involved in production processes (they have land-use rights), but usually they are active in the upper parts of the agribusiness chain and they purchase output from farmers.[3]

According to conventional wisdom, China's policy is oriented rather to economic goals, while environmental issues are not ranked high in policy agenda. It is true that in China there is no specific GHG reduction target for agriculture, but efforts to mitigate emission are being made through, for example, focusing on fertilizer efficiency or controlling methane emissions from rice fields. Indeed, the agricultural sector was responsible for 6.7% of all GHG emissions in China in 2019, while in OECD countries this sector generated 9.5%. Therefore, other environmental challenges appear to be more urgent.

Among China's most important problems are extensive water and soil pollution, land degradation, land desertification, poor waste management practices and low efficiency of fertilizer use. A specific environmental objective was zero pesticide and fertilizer use increase up to 2020 and it was purportedly met. The high level of fertilizer use was brought about, in part, by relatively low fertilizer prices, which was possibly due to policy support for producers of chemicals.

According to released data, only 27.3% of all agricultural land in China is classified as high fertility soils, while 27% is subjected to desertification (Wilkes & Zhang, 2016). However, thanks to an introduced ecological restoration program, the latter figure is starting to decrease. Another problem is that the application of organic manure is declining. Farmers tend to use it rather on their own land, while on the rented areas they are incentivizes to reduce it. In addition, although the intensive use of chemical fertilizers and pesticides has started to decline, the application rate is still very high. China also faces problems with insufficient irrigation and the management of large scale of waste which comes mostly from livestock production.

The rapid economic development of China has led to some social problems in agriculture and in rural areas. There is a phenomenon of migration of young people to cities, while rural areas experience the problem of an ageing population. In addition, the number of hours devoted to work in agriculture has decreased sharply, this outcome being sourced to lack of workers rather than greater mechanization. Obviously, the shortage of workers in farming has led to increase in farm size. Larger farms are sometimes less productive, but the process is generally desirable since larger farms generate higher income and exhibit better fertilizer use efficiency, as well as better carbon management.

The growing importance of environmental issues in the last few years in China is apparent. In particular, two important documents were announced in 2015. The first

---

[3]To understand China's land tenure system, it is important to distinguish between land ownership, contract rights and use rights. Normally, the farmer has contact and use rights to their land, but a village commune is the owner. A farmer can decide to transfer use rights to the cooperative or agribusiness, but he or she will still have contract rights. China's government supports land-use rights markets because it helps in realizing the aim of increasing farm size.

is of a more general nature and is called "Opinions on Hastening Construction of Ecological Civilization", while the second is strictly related to agriculture and is called: "National Agricultural Sustainable Development Plan 2015–2030". The latter focuses on natural resources protection and high quality and efficient agricultural production (OECD, 2021). It also put emphasis on the social dimension of sustainability by considering the problem of food security and availability through increasing grain production. Moreover, the plan not only considers the issue of food availability but also food safety, which is related to the quality of food. All of these issues are very important with regard to changing consumption patterns in Chinese households.

The Plan also underlines the role of environmentally friendly farming practices such as crop residue management, conservation tillage or application of organic manure to increase soil fertility. It can be said that the plan is comprehensive and it genuinely touches different aspects of sustainability. The idea behind the Chinese approach to agricultural sustainability may therefore be called "sustainable intensification".

Another interesting feature of the Plan is that it distinguishes between regions with different conditions for sustainable development of agriculture. Here, "optimized development areas" are deemed to be in good condition for further development, while "conservation and development areas" are marked out for more actions related to natural environment improvement. The "appropriate development" areas are in between and the necessity for multidimensional improvement activity is indicated.

One of the most important problems the agricultural sector faces in China is water quality. In the 1990s, the government proposed the creation of water user associations to promote better collective management of water resources, but they were not as effective as assumed. In 2008, the Water Pollution Law was introduced and it contained harsher penalties for firms that pollute water resources. In 2010, the "three red lines" policy came into force. Among the main aims of the policy (in the 2030 horizon) are achieving irrigation use efficiency of 60%, significant water consumption reduction and improvement of water quality. Recently, the government has also invested in the consolidation of water reservoirs in rural areas to provide adequate water supply in drought periods.

Better control over the process of food production is yet one more issue of importance. In 2020, the "Regulation for the Prevention and Control of Crop Diseases and Insect Pests" came into force to establish a surveillance system for diseases and pests. Herein, to provide a more effective system of control and traceability, a better coordination between central and local agencies is recognized as being needed. This is especially important because there is a serious risk that while reasonable agricultural policy measures will be implemented, as long as they are not effectively controlled, benefits from them may be small. An example of this is livestock production in which low rates of treatment of waste discharged is still reported, despite the introduction of policy regulations (Wilkes & Zhang, 2016).

In last decade, development of organic agriculture in China can be observed. Chinese government introduced organic product standards in 2005 and the market started to develop quickly. The number of organic producers in China is hard to estimate but the rapid growth in organic production is clear when area under organic farming is taken into consideration. It grew from 464,000 ha in 2005 to 3.135 mln in 2018. The output value of organic products reached USD 4.8$ bln in 2018. China is now the third largest producer of organic food in the world and fourth consumer after the USA, Germany and France (Zhang et al., 2020).

6.2   The Structure and Development of Financial Support for Agricultural Sector

In Table 6.1 we have collected data from OECD.Stat on policy support for agriculture in China, the EU and the USA. To document changes in patterns of support, we present data for 2005 and for 2020. In both periods China was the largest agricultural producer but the difference in 2020 was much higher which shows a very dynamic growing trend in China's agricultural production. EU farms were, on average, supported to the largest extent. However, the intensity of support decreased between two period. Producer support estimate (PSE) constituted 19.33% of the gross farm receipts (GFR) in the EU in 2020 and it declined from 28.58% in 2005. In the USA the share of PSE in GFR was 11.03% (decrease from 14.65%). In China, in turn, the level of support increased and in 2020 it was similar to the USA. The share of PSE in GFR was 12.17% (8.02% in 2005). It could be said that intensity of support converged in the analyzed period. However, in China most distorting payments constitute a very large proportion (72.26% in 2020) of PSE. This is because in China market price support is still important (8.7% of the total value of production) while in the EU and USA it is much lower.

Despite producers are supported in all of the studied areas, the structure of this support is much different. China's system is oriented mainly towards subsidies based on output level—these payments constituted 8.51% of GFR in 2020, while in 2005 it was 3.47%. In USA, payments based on current area, animal numbers, farm receipts or income (A/An/R/I) play particularly important role and they constituted 3.82% of GFR. It changed from 2005 when the most important were direct payments (not requiring commodity production) and payments based on commodity output. In the EU support for farmer is now paid mainly through payments not requiring production (7.78% of GFR) and payments based on current A/An/R/I for which production is required (5.38%). These are, for example, some of agri-environmental payments. Subsidies based on commodity output are now negligible after Luxembourg reform but previously they were much higher.

While relative level of PSE was the lowest in China, the total support estimate (TSE) as % of GDP was the highest in the country. Total support to agricultural sector (including producers, consumer and general services support) "costs" 1.59% of GDP in China (decline from 2.09%) while in the USA and the EU in 2020 it was 0.46% and 0.66%, respectively, and in both cases it decreased when compared with

2005. The share of general services support for agricultural sector was on similar level in the EU, China and the USA in 2020. It ranged from 10.57% of TSE in the EU to 12.61% in China. However, in China it declined from 29.19% in 2005 while in the USA and the EU it was relatively stable. Perhaps, more important is the structure of spending on services. In the EU, more than a half (52.9%) of these outlays went to agricultural knowledge and innovation systems while in China and the USA this share was below 25%. This shows that the EU puts a big emphasis on investing and dissemination of innovation in agriculture in its contemporary agricultural policy.

Consumer support in the EU and China was negative both in 2005 and 2020. This is common in many countries since usually consumers pay more in taxes for supporting producers than they receive from the state. An important exception is obviously the USA where the Supplemental Nutrition Assistance Program exist and, as it was mentioned before, outlays on this program constitute the majority of spending under farm bill.

6.3   Major Problems of Implementing Agricultural Policy: Capitalization of Subsidies, Distributional Inequalities, and Ageing

A generous agricultural policy support creates opportunities for farming sector but it could also generate some problems that need to be addressed. As with other policies, the most important thing is the assessment of its effectiveness. This problem was extensively undertaken in previous chapters. Agricultural policy also creates some specific problems among which there are the phenomenon of capitalisation of subsidies in land (or rental) prices, inequalities in subsidies distribution and ageing of farming and rural population.

6.3.1   Capitalisation of Subsidies in Farmland Prices

The capitalisation of subsidies means that some portion of subsidies is captured by landowners and not the farmers. It is because landowners know that farmers receive subsidies so they are incentivized to increase rental prices. Under support system, especially when payments are paid on hectare basis, land becomes more valuable asset. Although there are some studies that indicate that capitalisation may be even negative (subsidies decrease rental prices and/or land value), the vast majority of research reports positive capitalisation. A recent meta-analysis by Baldoni et al. (2021) shows that average capitalisation rate (share of subsidies in the price growth) is 33% for rental price and 12% for land value. However, the results depend on lots of different factors, including type of subsidies, area under study, estimation approach or type of data.

The authors of the cited meta-analysis concluded that higher capitalisation level is associated with decoupled subsidies while instruments design to deal with market imperfections and providing public goods (e.g. support for environmental public goods) can mitigate this effect and it also applies to redistributive payments. On the example of Italy Guastella et al. (2018) approved that capitalisation of coupled payments is higher but in this particular study the level of capitalisation was small at all. However, in other study, Guastella et al. (2021) found that in the years 2006–2008, 28 cents of each euro received by the farmers capitalised into land prices in those EU Member States that chose hybrid model of decoupled payments and 52 cents for countries with regional model. Only historical model did not lead to capitalisation.

According to neoclassical models, capitalisation of subsidies in rental percent should be full if markets are perfectly competitive. Therefore, the paradox is that from farmers perspective (and policymakers) less efficient markets are beneficial since in such circumstances the degree of capitalisation is usually lower. This is confirmed by Valenti et al. (2021) who found that capitalization into land values is smaller when land market is more concentrated. This conclusion is also supported by Van Herck et al. (2013) who have shown that capitalisation is lower in countries where more land is used by corporate farms. The cited study was carried out for the new EU Member States which experienced a significant and quick change from low support for agriculture to a considerable level of support right after accession to the EU. Therefore, the results of this study may be especially interesting as the research was close to quasi-natural experiment. The authors found that each euro of direct payments increased land rent by 13–25 cents.

The problem of capitalisation (sometimes called subsidy incidences) was also extensively studied in the USA. Using data for 1992 and 1997 Kirwan (2009) found that the capitalisation rate in US agriculture was 25%, i.e. 25 cents from the dollar received by a farmer was captured by landowner. More recently, Boussios et al. (2021) found that capitalisation may be much higher. They estimated that every dollar received by the farmers increases rental price by $0.45 to $0.65.

6.3.2   Inequalities

The systems of subsidies is often criticised as large proportion of payments go to the largest farms which usually have relatively good economic situation. In the USA it is estimated that top 20% recipients of subsidies received 89% of total payments in 1995–2012. What is even more striking, 1% of recipients obtained 25% of payments. Similarly, in the EU 20% of largest entities absorb 80% of subsidies (EC, 2018). However, this criticism is not entirely accurate, because it is obvious that as long as payments are calculated on the basis of land area, larger operators will naturally receive more subsidies. For example, in Slovakia where there are relatively few farms but they are large in terms of area size, the proportion of subsidies received by 20% biggest beneficiaries is obviously higher than in Netherlands or Belgium where distribution of land between farms is more equal. We would have a serious problem only if payments represented a larger proportion of the income (or revenue) of the largest farms compared to smaller units. However, in reality, this is not the case.

Based on the example of Croatia, Očić et al. (2018) have shown that subsidies constitute largest proportion of income among medium-sized farms. Similarly, Kirwan (2012) proved on the example of the USA that subsidies to wealthier farms are smaller in percentage terms of their net cash income than in the case of smaller farms with smaller revenues. In the EU there are several mechanisms such as redistributive payments, capping (a maximum amount of subsidies that a farm can receive) and II pillar payments that cause the fact that inequalities in incomes between farms are usually smaller than inequalities in terms of production or land distribution. It could be said that, to some extent, subsidies work as progressive income tax. In fact, subsidies directly contribute to reducing income inequalities as evidenced by Hansen and Offermann (2016), Severini and Tantari (2013) or more recently by Espinosa et al. (2020) and Hanson (2021). The latter author concluded that from the two main CAP mechanism, the Redistributive Payment Scheme was more effective than capping.

In the EU, the distribution of payments between member states is a much debated issue. The payment rates per hectare remain differentiated. The question arises: what distribution of payments between countries is fair? For some discussant the distribution will only be fair when in all countries subsidies rates will be equal (so called flat rate) or if these rate would not differ much from the average for all Member States. Others insist that subsidy rates should reflect the costs of agricultural inputs and the opportunity costs of land and labour not only on the state level but also within country (EC, 2018). Nevertheless, the trend towards a convergence of payment rates and the strengthening of redistributive mechanisms seems inevitable and beneficial from the sustainability point of view. Obviously, this discussion is only legitimate if direct payments remain the main tool of agricultural policy. The example of the USA shows that this need not to be the case. In our opinion, the future may lie rather in income stabilisation mechanisms which are also more easily accepted by the public.

6.3.3 Ageing

Another important problem related to agricultural policy is the ageing of farmers (and rural) population. However, it is different to previously discussed problems since ageing is not a direct result of policy. It is rather a side effect of rapid economic growth in cities which leaves agriculture and rural areas lagging behind. Farming appears to be unattractive place of employment. Effective farming requires technological improvement and high level of motivation but this is domain of young farmers (Hamilton et al., 2015). The policy is usually aware of this problem but the thing is whether policy supporting of young farmers is effective and what kind of policy it should be.

The problem of ageing in China is visible for both urban and rural areas but it is much more severe in rural areas. The rate of ageing (share of population over 65 years old) between 2000 and 2019 increased from 6% to 11% while in rural areas it grew from 7% to 15% (Gu et al., 2022). These authors have shown that ageing has negative effect on many aspects of farming sector functioning, such as: fertilizers use efficiency, farm size, output level or productivity. Interestingly, Gu et al. (2022) found that new farming models (mostly bigger entities) perform better in many aspects that traditional farms and only these forms of farming can address problem of ageing since in larger modern farms are less labour-intensive. Development of these models is supported by agricultural policy which seems reasonable in this context.

The problem of ageing is even more severe in the EU. On the EU-28 level, 33.9% of family farm managers are 65 years old and over. This share is lower among non-family farms (13.8%) but these farms constitute only 4.8% of the total number of farms (Popescu et al., 2021). The tendency is that older farmers manage smaller farms. The motivation to enter farming sectors for young people could be different but it is usually wish to continue farming from the parents or willingness to work with animals and nature (Šimpachová Pechrová et al., 2018). Young farmers in the EU are supported by agricultural policy through increased direct payments for farmers under 40 and through special mechanism in II pillar (mostly grants for investments). May et al. (2019) found that these special payments positively affected young farmers to stay in the sector but they also show that there are many other factors that may discourage young people from working in agriculture. These factors include, among others, problems with acquiring new land, social opinion on farming, possibilities to earn higher salary outside the sector (especially for well-educated people) or lower level of participation in decision-making problems. It seems, therefore, that reversing the negative trend in population ageing will require a number of different public policies as well as stimulating social change to increase the prestige of working in agriculture.

Chapter 7
Building Effective Sustainable Agriculture Policy: Conclusions

7.1 Lessons Learned for Different Agricultural Models: European Union, the USA and China

Our analysis in the previous chapter has shown that it is not possible to construct an ideal model for supporting agricultural development. First, the expectations of farmers, taxpayers and environmental organizations tend to diverge. Second, an ideal agricultural policy would have to pursue economic, social and environmental goals simultaneously. Additionally, it should be relatively cheap, effective, efficient and adjusted to the current economic situation. Third, any justification for the introduced solutions should be based on objectively diagnosed premises and not result from pressure of well-organized agricultural circles. Repeating certain statements that are not entirely true (e.g., about the lower income of farmers in relation to other groups in countries such as the USA) results in the loss of trust in politicians but also in farmers.

A justification for an active agri-environmental policy should be based on three fundamental premises, the last of which must be adequately interpreted depending on the situation. The first premise is that farmers bear a particularly high production risk due to the unpredictability not only of market circumstances (this type of uncertainty also occurs in other industries), but also of climate and weather conditions. The second premise concerns the provision of public goods which agriculture creates and which the market does not fully value. This is a market failure which is quite easy to understand for most taxpayers, not only in highly developed countries. The third rationale may be the need to ensure food security, where food is considered a strategic resource. However, this third premise should be interpreted in a specific context.

The first doubt that concerns the first premise in favor of interventionism in agriculture, i.e. the increased level of risk. Price support present e.g. in China or in the USA (Price Loss Coverage Program) does not fully solve this problem, because it works only when market prices are lower than target prices. In practice, however, a situation may also occur in which prices of agricultural products are high but yields are low, e.g. due to unfavorable weather conditions. Thus, in spite of high prices, agricultural income may be even lower than average. Therefore, the optimal solutions seem to be those which insure not so much prices but incomes or, at least, revenues. Examples of such solutions include Income Stabilization Tools and the American Agricultural Risk Coverage Program. The latter seems to be particularly interesting, as it is not only an auxiliary tool, but the main channel for supporting agriculture in the country that is significant agricultural producer. At the same time, calculating the payment on the basis of base acres instead of planted acres limits the phenomenon of adjusting production decisions depending on the possibility of receiving the subsidy.

An alternative system for supporting agricultural income and revenue is direct payments, such as in the EU. Payments in the EU are also not supposed to influence production decisions, but their rates are predetermined and independent of the current situation in the agricultural sector. On the one hand, this arrangement of support can facilitate planning on a farm, as the farmers know in advance what they will receive. On the other hand, payment rates are often not adapted to the current needs of the farm. Moreover, in a situation of prosperity in agriculture, maintaining a system of subsidies based on a fixed per-hectare rate becomes problematic from the society's point of view. Many taxpayers do not understand why farmers are paid significant amounts when their farms have high incomes. This is particularly the case for large farms which also have significant productive assets at their disposal. However, we have to admit that the system of direct payments has one serious advantage for policymakers—it is easy to estimate its final cost in advance. A middle ground solution could be a mixed (hybrid) system. In such a system, smaller farms, whose incomes are usually lower than in other sectors of the economy and which perform mainly social functions, could receive direct payments based on a specific per-hectare rate, while large farms would participate in an income support system, e.g. in a system similar to the American ARC. Such a hybrid system could be implemented in countries with different levels of development. The lack of guarantee of receiving payments in the ARC system could also help curb the phenomenon of capitalization of subsidies in land prices.

A second rationale for interventionism in agriculture is the need to pay for the public goods that agriculture provides. In the EU's agricultural policy, it usually takes the form of additional payments for undertaking above-standard farming practices. In the new CAP, these practices will be contained in eco-schemes and in Pillar II measures. Since the reforms were initiated in 2002, the American system of support for pro-environmental measures in agriculture has become more and more similar to the European one. In programs such as EQIP or CSP, payments are related to the adoption of specific practices on cultivated agricultural land. In contrast, traditional CRP, i.e. land diversion in non-crop uses, is declining in importance. In China, environmentally friendly farming practices are also promoted but it should be noted that in the case of China, as well as in the USA at the time of the peak popularity of CRPs, the aim of environmental programs was not so much to provide payments for public goods but rather to avoid land degradation and low soil fertility. Thus, the objectives had, in the end, an economic rather than an environmental character. In modern schemes, designed to support eco-friendly practices, the risk of leakage is rather low. The fact that these payments often represent additional income for the farming family is also not necessarily undesirable, considering that these subsidies are payments for a specific service to society (provision of public good). It seems that the biggest challenge is the issue of additionality. The risk that some simple pro-environmental measures would be taken by farmers anyway is significant. Indeed, professional farmers take some actions because they know that they can bring serious long-term benefits for the productivity. Some other actions (such as ploughing manure shortly after application) are simply rational (reducing odor) and do not need to be specifically encouraged. An optimal system of support for environmental practices can therefore be based on payments for specific practices to reduce their cost-intensity and limit the risk of income losses, but the extent to which practices are subsidized should be carefully considered and regularly reviewed. Although, especially in developed countries, payments for eco-friendly practices are easier to justify to taxpayers than traditional income support, there may be pressure for certain farming practices to be treated as mandatory rather than additional.

The third premise concerns food security. Until recently, this may have seemed a premise that should appeal more to people in developing countries who are struggling with famine and malnutrition. However, the COVID-19 pandemic made rich Western societies realize that a system, in which the supply of certain basic commodities relies heavily on international trade rather than domestic production, poses serious risks. This risk comes to the fore in situations of supply disruptions. However, apart from the availability of food itself, the issue of food safety and quality remains a problem which is already much clearer for people in Europe, for example. Similarly, in the USA the Supplemental Nutrition Assistance Program (SNAP) is gradually evolving towards improved nutrition rather than just supporting food security. However, the American system, even after reforms, is very expensive and rather difficult to implement in Europe. Perhaps it would be easier to implement in countries with particularly problematic income disparities. It seems, however, that a two-track approach is optimal. First of all, there is support for local food production, even when the production conditions are slightly worse, in order to ensure food security, especially in crisis situations. Secondly, action is needed to improve food quality—here a good solution may be the promotion of organic farming or action for the compulsory and uniform labelling of products. The Chinese example shows that organic farming can develop successfully not only in the wealthiest societies, where demand is already at a high level. Indeed, organic production can also be an important component of exports.

As for the main problem with the existing support mechanisms discussed in the previous chapter, it seems that the problem of inequality in the distribution of subsidies is exaggerated, at least in the EU. The largest farmers receive nominally more support, but the share of support in their income tends to be smaller. In countries where the problem is an excessively fragmented agrarian structure, it is logical to support larger farms or, at least, to support the creation of larger entities. Chinese actions are good examples here. Taking into account the objectives of sustainable agriculture, one should strive for such an agrarian structure in which medium-sized and medium-large entities prevail. However, the issue of effective support for the process of succession in agricultural holdings remains a very difficult problem. The existing mechanisms, even those applied in the EU, may constitute a valuable financial aid. The question arises, however, whether these solutions really encourage involvement in agricultural activity or are only an additional bonus for those already convinced. The reluctance to work in agriculture is to some extent rational, especially taking into account the high level of risk and the possibility of easier earnings in other areas of economy. Moreover, it results from the previously discussed low prestige of this work, so there is also a cultural factor at play. Therefore, the policy should not only be oriented towards financial support of young farmers but also towards intensive promotion activities which would build a positive image of agriculture in the society.

7.2　Measuring Policy Effectiveness: Integrated Efficiency Gains

Agricultural contributes ca. 5.8 $GtCO_2eq$ per year of GHG emissions which constitutes ~11% of the total anthropogenic emissions (Wollenberg et al., 2016). But the most problematic issue is that global emissions from agriculture are rising and are expected to rise in the future, while projections to other sectors indicate reduction in emissions level. This is mostly due to developing countries, where agricultural emissions are already at high level (Smith et al., 2014).

In Chap. 3 we have shown that total non-CO_2 emission for our sample was 4912 $MtCO_2e$ per year (4.912 Gt) on average in 2005–2018. Therefore, we may say that agriculture of our sampled countries is responsible for ca. 85% of total emissions from global agriculture. We have also claimed that only by reducing slacks it would be possible to decrease agricultural emissions in the sample by as much as 14.6%. However, the question arises how this reduction could contribute to the achievement of global targets?

Wollenberg et al. (2016) presented different models for agricultural emissions. They showed that baseline scenarios for agricultural GHG emission are between 7.52 and 8.97 $GtCO_2eq$ per year in 2030 which indicates significant increase in emissions. To limit global warming in 2100 to 2 °C above pre-industrial level it would be necessary to decrease emissions by 1.37 Gt to 0.92 $GtCO_2eq$ per year by 2030. It means that 11–18% reduction in GHG emission is needed. Our calculations from the Chap. 3 revealed that GHG emission could have been lower between 2005–2018 by 0.72 $GtCO_2eq$ per year (14.6%), on average, if inefficiencies had been removed. **If we extrapolate our results to the near future (2030) and assume that inefficiencies will still exists (baseline scenario), then we can say that vast majority of the GHG reduction problem could be solved just by removing**

slacks! These results are more optimistic than those by (Smith et al., 2008, 2013) who estimated that emission reduction by 0.4 $GtCO_2eq$ per year is possible if widespread dissemination of technical practices occured, or by Havlik et al. (2014) who estimated even smaller reduction (0.21 Gt) when assuming intensification in livestock and crop production with increasing economic efficiency.

More recent report of IPCC (2018) contains even more demanding scenarios regarding emission reduction. In the most ambitious scenario, it is assumed that limiting global warming to 1.5 °C requires reducing agricultural emissions by 16–41% in 2050 relative to 2010, while baseline emissions increase by 24–54% (Leahy et al., 2020). It means that to reach the target, non-CO_2 emissions in 2050 should be roughly half of what they would be under the baseline scenario (emission reduction by 4.8 $GtCO_2eq$). In this context, the gains from removing inefficiencies are not that high. **But even under this very ambitious scenario, reducing emissions by ca. 15% by removing slack would account for one third of the required emissions reduction.**

In our integrated efficiency approach, we concentrate not only on the GHG emission problem but we also present possibilities for the inputs use reduction. Among these inputs, possible reduction of fertilizers and pesticides is particularly often discussed. As indicated in Chap. 6, the ambitious EU targets include the reduction of fertilizers use by 20% and pesticides use by 30% by 2030. Perhaps, these targets may seem to be too ambitious for implementation in other parts of the world, but the EU goals are taken as an example.

Globally, the elimination of slacks would reduce pesticide use by 7.6%. Thus, it can be concluded that efficiency improvements would help to achieve about 15.2% of the target (assuming that the EU targets are treated as global targets).

However, on the country level increasing efficiency by removing slacks can bring much more significant gains. For example, our analyses in Chap. 3 have shown that an average slack on pesticides in the EU countries was estimated as 20%, meaning that EU countries could relatively easily reduce pesticides use by 20% without change in production technology but in some new Member States this slack was much larger, exceeding 50%. It means that these countries could reach EU targets just by removing inefficiencies. Nevertheless, it should be also added that in some other countries slack on pesticides was positive—showing that these countries could even increase pesticides use without depleting efficiency level. However, from purely environmental point of view, a large reduction in pesticides would be beneficial and it is obvious that efficiency cannot be the only one criterion that is taken into account.

When it comes to fertilizers (based on the example of nitrogen), our previous calculations have shown that elimination of slack on nitrogen could save 9.2% of this input use on the global scale. **Therefore, if we treat EU goals as an example, it turns out that elimination of slack on nitrogen use could contribute to achieving ca. 46% of the reduction target.** Similarly to pesticides, slacks on nitrogen are different between countries. In 38 countries from our sample slack on nitrogen was larger than 20% meaning that reduction targets (if introduced on a national scale) in these countries could be achieved just by removing slacks.

Table 7.1 The potential effect of slack reduction on nitrogen equivalent use for countries that exceed 170 kg/ha limit

Country	Nitrogen equivalent/ha	Nitrogen equivalent/ha (after reducing slack)
Brunei Darussalem	1060	967
Egypt	538	462
Netherlands	479	435
Belgium	387	366
Bangladesh	314	174
Japan	261	252
Luxembourg	235	230
Viet Nam	234	224
Denmark	231	252
Norway	218	209
Ireland	213	196
Israel	209	200
Pakistan	206	127
Czechia	204	192
Germany	197	173
India	174	183
Nepal	172	148
Average	**314**	**282**

We can also look on the problem of input use from another perspective. It is clear that in some countries the need to decrease input use is much more urgent that in others. The actual environmental capacity may be hard to estimate but there are some general rules that could be applied. The maximum amount of nitrogen that could be applied depends on a specific crop. Relying on aggregated data we may assume (based on average for different type of crops) that relatively safe limit would be 170 kg/ha per year from all sources. One livestock unit (a one cow) excretes ca. 85 kg of nitrogen per year. Therefore, we calculate nitrogen equivalent as the sum of nitrogen use and nitrogen excreted by LSU divided per ha to see whether the 170 kg/ha limit is exceeded or not. In Table 7.1 data for the countries that exceeded the limit are presented. We also show what the nitrogen use would be if real slacks on nitrogen, LSU and land inputs are considered. Since the theoretical possibility to increase input use (resulting from super-efficiency approach applied in Chap. 3) is not analyzed, all positive real slacks are treated as 0. Calculations results for all countries can be found in Table A12 in appendix. Similarly, we apply 1.75 kg/ha limit of pesticides use. This is because the average use of this input in the EU is 3.5 kg and the Farm to Fork strategy aims to reduce it by half. This means that 1.75 kg/ha limit can be applied, although it seems to be quite ambitious. Data for countries exceeding the limit are included in Table 7.2.

There are 17 countries that exceed the 170 kg/ha limit in our sample. The average level of nitrogen equivalent use for these countries is 314 kg/ha. The reduction of slacks related to nitrogen use (both from fertilizers and excreted by LSU) could lead

Table 7.2 The potential effect of slack reduction on pesticides use for countries that exceed 1.75 kg/ha limit

Country	Pesticides use per ha	Pesticides use per ha (after reducing slack)
Japan	12.33	12.46
Israel	11.86	12.08
Costa Rica	6.76	7.05
Malaysia	6.62	3.42
Czechia	6.24	6.46
Netherlands	5.65	5.62
Italy	4.93	5.12
Belgium	4.89	4.97
Ecuador	3.74	1.54
Portugal	3.47	1.39
China	3.26	3.36
Egypt	2.80	2.80
Germany	2.49	2.43
France	2.45	2.71
Thailand	2.29	1.22
El Salvador	2.19	1.92
Slovenia	1.94	0.78
Honduras	1.92	1.58
Spain	1.83	2.91
Viet Nam	1.76	1.76
Average	4.47	4.08

to a decrease in nitrogen use by 10%. It does not seem like much progress, but in some countries the change is more significant. For example, in Bangladesh and Pakistan the progress could be 45% and 38%, respectively. When it comes to pesticides, the 1.75 kg/ha limit is exceeded by 20 countries. The average level of pesticides use for this group is 4.47 kg/ha. If these countries removed relevant slacks, the pesticides use would be 4.08 kg/ha, meaning that it would decrease by 8.8%. However, in some countries this decline would be much higher. For example, in Portugal, Ecuador and Slovenia the decline would be ca. 60%. One may be interested why in some countries the pesticides or nitrogen use after slack deduction is even higher than before. This is because we also considered the change of land input (denominator used for limit calculations) and in some cases the inclusion slacks for land resulted in the fact that decline in nitrogen or pesticide use was lower than for land and the resulting variable has increased.

7.3 Directional Solutions to Sustainable Agriculture Problems and Their Side Effects

Building an effective system of policy schemes for sustainable agriculture is not an easy task. We propose a problem-based approach, in which we first identify specific solutions (general strategies) for the main problems faced by sustainable agriculture based on the considerations of the previous chapters. Then, we identify the main side effects of these solutions in order to build a systemic solution that could mitigate the observed trade-offs, creating a coherent and internally complementary design of sustainable agriculture policy instruments for farming systems of different intensity.

In most countries, but also globally, priority is given to social problems in agriculture such as food insecurity, land abandonment and ageing of the rural population. In a democratic system, social problems directly translate into votes, therefore the authorities cannot remain indifferent to them. However, these problems are closely related to economic issues, which we described in Chap. 1 as the market treadmill theorem. As a reminder, this refers to the secular mechanism that causes agricultural incomes not to rise, despite increases in technical productivity, or to rise well below the expectations with regard to outlays spent on the productivity growth. In highly developed countries, Cochrane's treadmill disappears in favour of the income growth rate treadmill, which consists in an asymmetry in the business cycle. Agriculture, as we know, suffers from an agricultural price flexibility—the main cause of the treadmill, which results in the commonly known King's law—high agricultural output translates paradoxically into a decrease of farmers' revenue (and income), and in the case of low production the opposite. However, the problem of the treadmill lies in the fact that the mentioned income decrease is not compensated by an income increase in the next swing of the agricultural business cycle. In a more detailed formulation of the above problem (referring to total factor productivity—TFP), a farm earns less than it had been expected by raising the TFP, and these forgone benefits are unfortunately not compensated by a favourable price change under conditions of decreasing productivity. Moreover, in the case of small entities, it can be observed that they lose even a larger part of their income than would result from the decrease in the TFP alone (i.e. the King's effect works only to the disadvantage but it may no longer work to the advantage of small-scale farmers (Czyżewski et al., 2019a, 2019b). In summary, the desire to grow and the aspiration to catch up with other sectors of the national economy in terms of income makes the farm more and more dependent on the treadmill, because to meet the expectations of a higher income growth it falls into a vicious cycle of productivity increases and decreases.

The negative impact of rising productivity on the income growth rate concerns mainly the economically weak firms, whereas high productivity may cushion income decline to some extent. Thus, exceeding a certain threshold of the productivity ratio gives benefits in the conditions of deteriorating price relations, weakening their negative impact on income. The result of this situation is the polarization of the agricultural sector and relative deprivation of farmers' income.

For example, Fig. 7.1 compares the regression of entrepreneurial income growth over productivity in the agricultural sector of the EU12 (Central and Easter European Countries—so called New Member States), EU15 and in the national economy as a whole (EU15), showing the relative position of the mentioned productivity threshold (i.e. function minimum), above which the function becomes increasing.

We can see that this threshold is further located in highly developed countries, from which it can be inferred that with economic development the problems of the

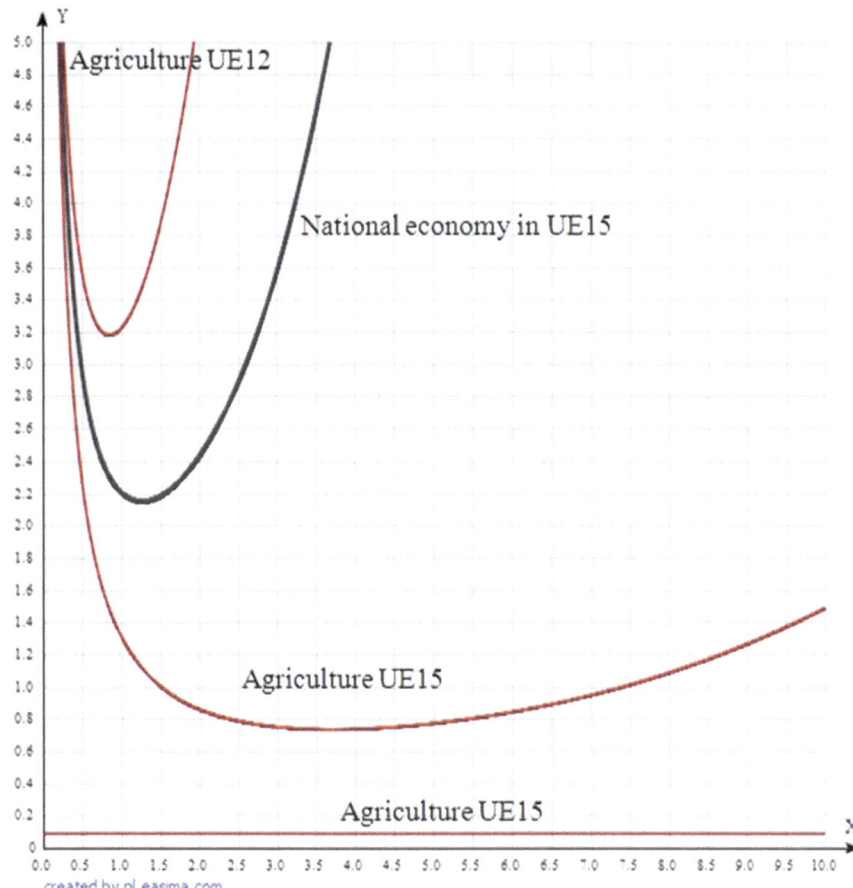

Fig. 7.1 Marginal changes (growth rate) of entrepreneurial income (Y) in agriculture (red lines) and in the entire national economy (black lines) regressed over the productivity coefficient (X). Notes: Y—marginal changes (growth rate) of entrepreneurial income including subsidies (current prices in Million euro); functions are calculated as the derivatives from the log-linear and linear income function estimated on the basis of EU countries panel regression (2005–2015); X— productivity coefficient defined as a ratio of agricultural or total economy output at producer prices, constant prices to total intermediate consumption at constant, basic prices. Source: own elaboration based on Economic Accounts for Agriculture and Czyżewski et al. (2019b)

growth rate treadmill and relative income deprivation in agriculture may become worse in these countries.

The analysis of the phenomenon of the relative deprivation of agricultural income in different EU countries shows that the level of this deprivation in the "Old" EU is actually higher than in the New Member States. Referring to Fig. 7.1, in 2015 the average level of deprivation in the EU-15 was reflected by the fact that agricultural income accounted for about 43% and 53% of income in non-agricultural sectors in the EU-15 and EU-12, respectively. In most cases the deprivation index remained constant, but in some Western European countries (Belgium, Ireland, Greece, Italy and the UK) an increase can be observed in the last decade (Czyżewski & Poczta-Wajda, 2017). One could therefore conclude that the CAP, taking a decoupled path of development, strengthens the income deprivation of farmers in the EU-15 countries.

The recipe for a treadmill of income growth rate could be a two-way impact of agricultural policy that on the one hand (1) facilitates the achievement of the optimal productivity threshold and on the other (2) reduces the effects of price flexibility by financing public goods.

In the first case, instruments can be applied to counterbalance the impact of agricultural price flexibility and, at the same time, protect producers from the cyclical situation in agriculture. A typical response to the business cycle could be intervention (floor) prices, and intervention buying tested in the agricultural policy of the EU (used before decoupling reform in 2003) and operating currently in China. Another solution could be counter-cyclical coupled payments, which still operate in the EU for cereal production and some livestock production, but are flat-rate (not cyclically related). There is ample evidence (cited earlier, see Chap. 4) that flat-rate coupled payments reduce the technical productivity in agriculture, induce inefficiencies slacks, and thus may mitigate the problem of market direction to a very limited extent. Cyclical coupled payments could theoretically improve this mechanism, although their other drawbacks would still remain. At the same time, investment support linked to human capital building through training and education could also counteract the treadmill by facilitating the implementation of technological progress in agriculture. In this way, the productivity threshold above which the treadmill of the income growth becomes irrelevant could be approximated.

A completely different way of fighting market orientation in agriculture is decoupling, the path the EU has chosen. In principle, decoupling linked to GAEC and agri-environmental support breaks the link between subsidies and production, while increasing the heterogeneity of agricultural products and creating the value of environmental public goods (see Fig. 5.5). This results in some decoupling of farmers from agricultural price flexibility and an increase in farmer participation in the food value-chain through the ability to better diversify the quality of delivered products. However, this results in a land value gap (or "land rent gap"). This phenomenon consists in the fact that the productive utilities of land (including the value of the subsidies that can be obtained) explains a smaller and smaller part of its value. On the one hand, this hinders the land sales among farmers and land concentration process. Petrification of fragmented and inefficient agrarian structure takes place. On the other hand, under the conditions of land market regulation, farmers cannot capitalize on the new land utilities (amenities) in any way, for example by selling agricultural land for non-agricultural purposes (e.g. housing and recreation). In many countries, the sale of agricultural land cannot be fully market-based because of the possibility of agricultural land transactions only by farmers or the pre-emptive right of a government agency (Czyżewski & Majchrzak, 2018). However, it is possible to consider relaxing these restrictions for business stakeholders under the condition of acquiring a certain number of site-certificates under a tradable permits scheme or habitat banking system, or charging individual stakeholders an environmental tax on agricultural land purchased for residential purposes. To sum up, land market regulations negatively influence the value of farms and their investment opportunities (e.g. farmers cannot take a loan secured by agricultural land or sell a plot to obtain investment funds).

Payments for public goods should create the conditions for these goods to be indirectly capitalized on by the market, in the form of various services and products offered by rural residents (not necessarily by farms). It is necessary to allow the consumer to see the new utility of the land factor. For instance, "bird payments", as the farmer's saying goes, popular on farmland in the bends of rivers, make these rivers frequented by canoeists, which creates demand for a whole range of services. From a theoretical point of view, the idea is to negate to some extent Samuelson's thesis that a market system will never provide an optimal allocation of public goods (Samuelson 1967) because there is no market valorization of these goods. The agricultural land's new utilities are no longer public goods in the classical sense. We are discussing increasingly "rival" goods, where too much demand leads to their destruction. Environmental utilities in rural areas are close to the definition of common goods, although the lack of money still excludes their consumption to some extent. Certainly, these are so-called merit goods, the supply of which are ensured by the state, even if individuals do not accept this. This is the only thing that distinguishes them from private goods. Thus, ensuring the durability of common goods in rural areas is the role of the state, but their valuation may be an institutional (e.g. subsidies, reserve systems) or quasi-market one through transactions of accompanying services and products (e.g. agritourism, organic food) and the sale of entitlements to the agricultural land of a high nature value. The condition of the effectiveness of the decoupling and valuing of the public goods strategy is, therefore, that the gap in land value resulting from the perception of non-agricultural utility is somehow capitalized on in agriculture, e.g. through the quasi-market systems of tradable permits.

Above, we have considered potential mitigations of social and economic issues concerning sustainable agriculture. There is also the issue of rising environmental pressure including pollution, biodiversity loss, and GHG emissions. It is generally accepted that the instruments described above for valuing public goods and decoupling of payments, which encourage extensification of agricultural production, also serve to address environmental concerns (see Fig. 5.5). Attempts to marketize public goods while reducing pressure on environmental capacity make it possible to maintain a socially desirable supply of these goods.

Nevertheless, choosing one of the paths described above to address social and economic issues involves inevitable side effects, as the analyses in the previous chapters have shown. The counter-cyclical and pro-efficiency agricultural policy pathway produces common side effects that have been described in the literature (see Chap. 5) and arise from our research:

- Inefficiency slacks in: GHG, land, LSU, energy, and nitrogen management, which are mainly associated with the development of intensive crop systems, but also partly due to the increase in livestock production and GHG emissions as a result of promoting the grassland system;
- Increasing N2O emissions due to the intensification of crops systems;
- Exciding environmental capacity by intensive crops farming;
- Meanwhile, small farms face the market treadmill and become laggards.

At the same time, decoupling and public goods valuing produces specific side effects consisting of:

- Inefficiency slacks in production and pesticide use;
- Potential food insecurity due to extensification and decline in agricultural productivity;
- Social constraints in protected areas due to lower incomes in grassland systems and resource allocation constraints in protected areas(lower allocative efficiency); constraints on business and residential functions in rural areas, which results in slower economic development of these regions.

Therefore, an attempt should be made to look for systemic solutions that combine the advantages of both the described mitigation paths and at the same time avoid the side effects. This is not an easy task and requires the integration of various command & control (CAC) and market-based policy measures.

7.4   Systemic Solution

One could think that the postulated dualism, i.e. the combination of the counter-cyclical & pro-efficiency path with decoupling & public goods valuing is a marriage of "fire" and "water". Indeed, it attempts to combine support for productivity with payments for public goods, while in the popular perception the latter are alternative to production and even constitute the "antithesis" of productivity as we are supposed to pay for abstaining from productive activities.

Such a marriage seems necessary if we try to avoid the side effects described in the previous section (see Fig. 7.2). In a systemic solution we should combine counter-cyclical and pro-efficiency instruments with CAC measures introducing mechanisms of environmental capacity protection. We refer here to the GAECs, which we described in more detail in Chap. 5. Their fulfillment would be a condition for receiving counter-cyclical coupled payments and investment subsidies. GAECs should set acceptable maximum limits for nutrient, pesticide use, animal stocking rates and introduce eco-schemes to ensure a certain level of biodiversity. This type of solution has long worked well in the EU, and is being extended in the new strategy from 2023 (European Green Deal and F2F Strategy), as we wrote in Chap. 5. The purpose of the additional conditions for receiving coupled and pro-efficiency support is to prevent over-exploitation of the natural environment, which is the case with the

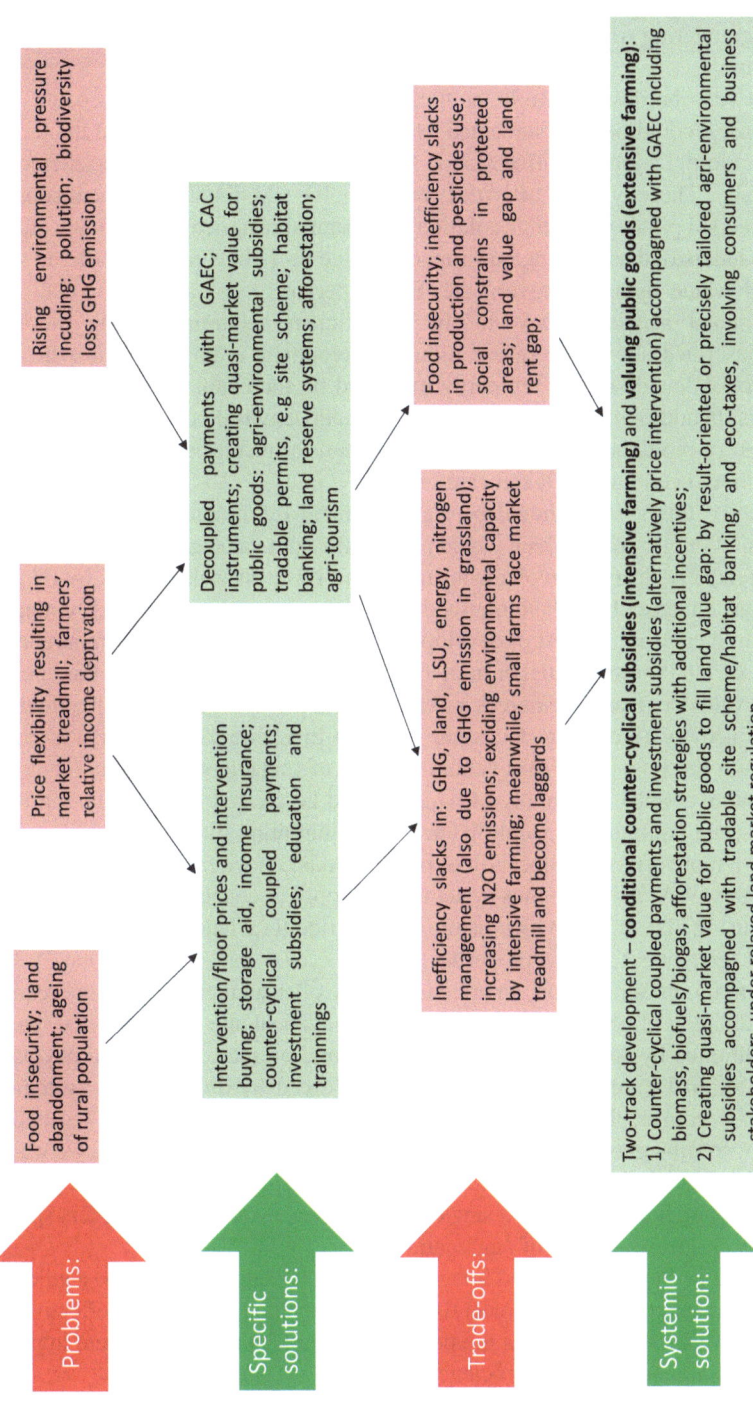

Fig. 7.2 Specific and systemic solutions for main problems of agricultural policy

US' and China's agricultural policy. The experience of these countries shows that simply countering the treadmill is not sufficient for the goals of sustainable agriculture.

GAEC can be designed to be as flexible as possible, taking into account local conditions. It could be complemented by an alternative (at the choice of the farmer) biomass, biofuel, biogas or afforestation strategy that would replace eco-schemes for biodiversity. These types of activities could involve additional investment support.

The second pillar of systemic solution for sustainable agriculture policy would be broadly understood as creating quasi-market value for public goods in order to fill in the land value gap. However, the EU experience so far shows that agri-environmental subsidies may not be a sufficient solution, causing the side-effects described above. Therefore, a dual development is important, in which the socio-economic side-effects are partly mitigated by the coupled and pro-efficiency payments. In addition, more precise tailoring of agri-environmental subsidies for the protection of biodiversity or even the introduction of result-based measures in this respect should certainly be taken care of.

However, this system should be complemented by an additional quasi-market valuation of public goods which would prevent the land value gap. For this system to work effectively, it must involve consumers and business stakeholders by relaxing restrictions on agricultural land transactions. On the example of some EU countries, there is an increasing interest of developers and individual consumers in purchasing agricultural land for residential or rural tourism investments. Obviously, this causes a threat to valuable natural resources and the rural landscape. However, if conducting business activity (including developer activity) in nature-value areas was conditioned with the possession of a certain number of tradable site-permits (or credits from a habitat bank), the supply of which would be created by individual farmers, then a certain compromise would be achieved, inhibiting biodiveristy leakage and enabling a more effective allocation of public goods. As we know, similar solutions work in the USA. An analogous system could apply to consumers who, desiring to make housing investment in high nature value agricultural land, would decide to pay a special 'green tax' on such properties.

To sum up, in our opinion, an effective policy of sustainable agriculture requires the coordination of actions in several areas: anti-cyclical and pro-efficiency support but conditioned by specific limits of environmental pressure; decoupled payments, and institutional public goods valuing but reinforced by quasi-market systems of pricing. Of course, these are only directional recommendations that need to be refined in specific geographical, national and local contexts. In this book we wanted to convey the message that it is not possible to achieve the complex goals of sustainable agriculture through a single path, because each strategy triggers specific trade-offs at different levels of aggregation.

We do hope that the proposed Integrated Efficiency approach may be a guideline to identify these trade-offs and meet the necessary precondition for an effective and persistent policy which says that the policy, above all, shall trigger Pareto improvement and stimulate allocative efficiency.

Appendix

Table A.1 List of developing countries surveyed

Least developed countries	Land locked developing countries	Small Island developing states	Low income and food deficit countries	Net food importing developing countries
Afghanistan	Afghanistan	Antigua and Barbuda	Afghanistan	Afghanistan
Angola	Armenia	Bahrain	Bangladesh	Angola
Bangladesh	Azerbaijan	Barbados	Benin	Antigua and Barbuda
Benin	Bhutan	Belize	Burkina Faso	Bangladesh
Bhutan	Bolivia	Cabo Verde	Burundi	Barbados
Burkina Faso	Botswana	Cook Islands	Cameroon	Benin
Burundi	Burkina Faso	Cuba	Central African Republic	Bhutan
Cambodia	Burundi	Dominican Republic	Chad	Botswana
Central African Republic	Central African Republic	Fiji	Congo	Burkina Faso
Chad	Chad	Grenada	Côte d'Ivoire	Burundi
Eritrea	Ethiopia	Guinea-Bissau	Eritrea	Cambodia
Ethiopia	Kazakhstan	Guyana	Ethiopia	Central African Republic
Gambia	Kyrgyzstan	Jamaica	Gambia	Chad
Guinea	Lao People's Democratic Republic	Maldives	Ghana	Côte d'Ivoire
Guinea-Bissau	Malawi	Mauritius	Guinea	Cuba
Lao People's Democratic Republic	Mali	New Caledonia	Guinea-Bissau	Dominican Republic
Madagascar	Mongolia	Puerto Rico	India	Egypt
Malawi	Nepal	Saint Kitts and Nevis	Kenya	El Salvador
Mali	Niger	Saint Lucia	Kyrgyzstan	Eritrea
Mozambique	North Macedonia	Saint Vincent and the Grenadines	Madagascar	Ethiopia

(continued)

B. Czyżewski, Łu. Kryszak, *Sustainable Agriculture Policies for Human Well-Being*,
Human Well-Being Research and Policy Making,
https://doi.org/10.1007/978-3-031-09796-6

Table A.1 (continued)

Least developed countries	Land locked developing countries	Small Island developing states	Low income and food deficit countries	Net food importing developing countries
Myanmar	Paraguay	Samoa	Malawi	Gambia
Nepal	Republic of Moldova	Seychelles	Mali	Grenada
Niger	Rwanda	Singapore	Mozambique	Guinea
Rwanda	Tajikistan	Suriname	Nepal	Guinea-Bissau
Senegal	Turkmenistan	Timor-Leste	Nicaragua	Honduras
Sierra Leone	Zambia	Trinidad and Tobago	Niger	Jamaica
Timor-Leste	Zimbabwe	Vanuatu	Rwanda	Jordan
Togo			Senegal	Kenya
United Republic of Tanzania			Sierra Leone	Lao People's Democratic Republic
Vanuatu			Syrian Arab Republic	Madagascar
Yemen			Tajikistan	Malawi
Zambia			Togo	Maldives
			United Republic of Tanzania	Mali
			Viet Nam	Mauritius
			Yemen	Mongolia
			Zimbabwe	Morocco
				Mozambique
				Myanmar
				Namibia
				Nepal
				Niger
				Pakistan
				Peru
				Rwanda
				Saint Kitts and Nevis
				Saint Lucia
				Saint Vincent and the Grenadines
				Senegal
				Sierra Leone
				Sri Lanka
				Timor-Leste
				Togo

(continued)

Table A.1 (continued)

Least developed countries	Land locked developing countries	Small Island developing states	Low income and food deficit countries	Net food importing developing countries
				Trinidad and Tobago
				Tunisia
				United Republic of Tanzania
				Vanuatu
				Yemen
				Zambia

Table A.2 The average value of slack in the whole analysed period (in %) in countries of cluster 1—production in international dollars

DMU	PM land	PM LSU	PM energy	PM GHG	Land	LSU	Employment	Energy	Nitrogen	Pesticides	Production	Dietary	GHG
Albania	0.00	0.00	0.00	0.00	0.00	-7.11	-0.61	-5.89	0.16	-1.31	-1.64	-3.57	-15.81
Algeria	0.00	0.00	0.00	0.00	-8.67	-9.29	20.51	-0.54	0.92	1.90	-7.92	-2.26	-5.27
Argentina	0.00	0.00	0.00	0.00	-16.92	-15.45	6.31	-12.80	1.48	0.21	-14.90	-0.40	-17.48
Armenia	-3.99	-4.21	-3.41	-4.39	-14.60	-8.96	-20.03	-17.37	-40.52	-2.04	-2.89	-1.05	0.00
Canada	-8.11	-8.10	-7.90	-8.13	-13.45	-2.49	-13.91	-48.83	-39.57	-26.63	0.00	11.11	-13.68
Chile	-1.25	-1.24	-2.54	-1.13	-1.67	-5.13	-0.35	0.00	-1.62	-1.50	-5.33	2.35	-4.15
China	0.00	0.00	0.00	0.00	-2.86	-5.18	0.05	-6.03	0.04	0.00	-7.69	-0.45	-5.13
Côte d'Ivoire	-0.12	-0.13	-0.12	-0.14	-0.71	-0.22	-1.09	-1.54	-1.62	-0.44	-6.97	0.27	-2.92
Egypt	-0.18	-0.18	-0.18	-0.17	0.00	-7.28	-0.32	-9.56	0.37	-1.16	-10.22	-0.88	-4.72
Ghana	0.00	0.00	0.00	0.00	-7.68	-8.90	0.00	-5.90	0.69	0.08	-12.07	-1.98	-7.23
Israel	0.00	0.00	0.00	0.00	-1.26	-8.07	1.92	-6.68	2.43	0.21	-8.23	-5.07	-3.65
Japan	0.00	0.00	0.00	0.00	0.00	-7.38	0.07	-10.66	0.54	0.00	-6.43	0.00	-5.61
Jordan	-0.74	-0.80	-0.86	-0.82	-4.80	-10.19	-0.84	-9.43	-3.49	4.18	-4.72	-4.01	-9.24
Malawi	-0.08	-0.06	-0.09	-0.07	-0.63	-19.47	-0.17	0.00	-1.55	-0.07	-13.11	-1.99	-9.36
Malaysia	-0.19	-0.20	-0.19	-0.18	0.00	-0.69	-0.16	-4.86	-6.39	-22.65	-4.94	5.69	-7.49
Mozambique	-23.68	-25.05	-51.24	-23.18	-38.06	-7.16	-34.79	-51.19	-24.99	-32.53	-5.62	-5.72	-30.35
North Macedonia	-0.07	-0.07	-0.06	-0.07	-11.14	-7.08	0.09	-6.57	1.05	1.75	-11.14	-12.78	-8.88
Norway	-0.15	-0.16	-0.10	-0.16	0.00	-8.09	5.37	-9.28	0.02	1.55	-1.31	-3.50	-10.89
Philippines	-0.27	-0.28	-0.35	-0.27	-0.37	-5.68	-0.55	0.00	-0.48	-0.64	-6.86	-0.03	-9.25
Russian Federation	-1.29	-1.29	-1.18	-1.28	-11.33	-4.09	-0.77	-7.82	-0.92	-1.59	-6.09	-0.07	-6.74
Saudi Arabia	-51.30	-51.30	-51.67	-51.84	-80.11	-17.82	-51.59	-60.75	-60.88	-7.09	0.00	11.67	0.00
Sri Lanka	-0.66	-0.61	-0.61	-0.51	0.00	0.00	-2.34	0.00	-4.58	-0.09	-7.05	0.15	-14.24
Switzerland	-1.30	-1.29	-1.26	-1.28	-0.39	-8.21	-0.84	-0.08	-1.73	-0.33	-4.95	-0.43	-6.73

Thailand	0.00	0.00	0.00	0.00	−6.13	0.00	−0.29	−6.64	0.00	−0.14	−6.76	0.13	−17.30
Turkey	0.00	0.00	0.00	0.00	−0.56	−4.14	−0.15	−12.24	0.50	−3.64	−8.85	−1.01	−1.83
Ukraine	−0.53	−0.57	−0.52	−0.53	−5.63	−0.62	−0.37	−3.10	−3.74	−7.71	−12.52	0.30	−4.16
USA	−0.37	−0.38	−0.39	−0.38	−8.28	−8.89	−0.01	−8.13	−0.65	−0.41	−9.24	−0.06	−8.54
Viet Nam	0.00	0.00	0.00	0.00	−0.46	−6.52	0.00	−0.22	0.00	0.00	−6.21	−0.75	−5.00
Average	**−3.37**	**−3.43**	**−4.38**	**−3.38**	**−8.42**	**−6.93**	**−3.39**	**−10.93**	**−6.59**	**−3.57**	**−6.92**	**−0.51**	**−8.42**

Table A.3 The average value of slack in the whole analysed period (in %) in countries of cluster 2—production in international dollars

DMU	PM land	PM LSU	PM energy	PM GHG	Land	LSU	Employment	Energy	Nitrogen	Pesticides	Production	Dietary	GHG
Australia	-1.37	-1.35	-1.37	-1.40	-12.13	-7.52	-1.17	-10.44	-0.57	-1.33	-9.03	-1.30	-10.02
Azerbaijan	-2.74	-2.88	-3.24	-2.95	-1.65	-4.26	-7.48	-9.40	-2.28	-10.21	-6.90	-0.80	-0.74
Bangladesh	0.00	0.00	0.00	0.00	0.00	-5.75	0.00	-6.50	0.06	0.00	-7.55	-0.27	-6.35
Belarus	-0.18	-0.18	-0.18	-0.19	-0.69	-0.07	-0.29	-4.68	-0.48	-0.86	-6.53	-0.90	-9.91
Bolivia	-4.41	-4.56	-4.19	-4.56	0.00	-14.82	1.01	-0.04	0.16	-10.30	-5.30	0.49	-13.46
Brazil	0.00	0.00	0.00	0.00	-4.98	-8.03	2.03	-7.71	0.59	0.00	-9.61	-0.38	-8.15
Brunei Darussalam	0.00	0.00	0.00	0.00	-6.28	-14.39	1.18	-24.03	4.49	-0.29	-14.21	-7.97	-9.07
Burkina Faso	-18.26	-19.54	-26.28	-19.42	-8.09	-29.39	1.38	0.00	-0.74	-7.12	-4.30	-0.17	-25.31
Cameroon	0.00	0.00	0.00	0.00	-1.97	-4.28	0.29	0.00	0.40	0.00	-11.16	-0.91	-2.80
Colombia	-17.77	-17.97	-18.52	-17.97	-19.51	-3.38	-54.94	0.00	-13.65	-54.80	0.00	5.85	-3.71
Costa Rica	0.00	0.00	0.00	0.00	-1.38	-3.94	0.40	-1.01	0.60	0.00	-10.32	-0.52	-4.07
Ecuador	-4.34	-4.40	-4.88	-4.41	0.00	-9.49	-62.55	-21.51	-7.19	-43.41	-0.37	11.33	0.00
El Salvador	-18.88	-19.14	-18.26	-19.18	-2.33	-0.12	-62.30	-4.28	-42.05	-7.69	-0.12	8.01	-5.53
Ethiopia	-13.07	-14.45	-18.50	-14.40	-2.86	-16.58	-17.09	0.00	0.00	-5.75	-3.95	1.37	-13.46
Georgia	-26.22	-25.88	-30.80	-25.84	-6.98	-0.48	-65.37	-11.39	-38.08	-60.80	0.00	4.27	-17.74
Guyana	-5.82	-6.00	-5.87	-5.82	-25.22	0.00	-8.49	-56.95	-22.25	-38.29	-0.02	4.69	-69.91
Honduras	-41.47	-41.72	-42.59	-41.58	0.00	-10.48	-52.92	0.00	-36.63	-6.58	0.00	8.26	-1.25
Iceland	0.00	0.00	0.00	0.00	-14.94	-2.31	2.02	-12.96	0.12	4.78	-11.35	-1.97	-12.90
India	-0.02	-0.02	-0.02	-0.02	-8.70	-11.40	-0.02	-15.37	-0.46	-1.06	-12.95	-0.34	-11.14
Indonesia	0.00	0.00	0.00	0.00	-12.56	-14.62	0.94	-23.71	0.33	0.00	-16.53	-1.21	-13.54
Iran	-0.74	-0.74	-0.87	-0.62	-9.04	-9.83	-0.73	-18.06	4.08	0.86	-9.82	0.07	-3.68
Iraq	-25.75	-26.50	-25.54	-26.20	-15.51	-0.73	-51.18	-45.64	-46.98	-9.03	0.00	24.40	-4.94
Kazakhstan	-2.45	-2.28	-2.31	-2.40	-21.20	-6.77	-0.78	-11.14	-7.79	-0.45	-11.38	-1.13	-12.54
Kenya	-7.25	-7.77	-6.73	-7.68	-8.92	-12.94	-8.03	-0.25	0.00	0.00	-4.59	3.89	-7.67

Kyrgyzstan	-0.63	-0.68	-0.64	-0.68	-6.16	-3.60	-1.16	0.00	-2.26	-1.84	-9.90	-0.05	-7.87
Madagascar	-5.52	-5.79	-7.34	-5.24	-32.72	0.00	-7.52	0.00	-62.13	-49.54	-4.50	8.31	-8.33
Mexico	-2.41	-2.44	-2.37	-2.43	-27.62	-5.84	-19.34	-1.28	-8.05	-5.18	-1.26	-0.94	0.00
Mongolia	-6.05	-4.95	-6.69	-5.13	-19.25	-19.79	0.15	0.00	-8.68	-2.84	-10.18	0.74	-19.56
Morocco	-1.96	-1.83	-2.00	-1.86	-7.89	-8.28	-12.04	-1.84	-4.87	-12.71	-4.76	-1.31	0.00
Namibia	-52.42	-51.68	-60.02	-53.52	-44.53	-9.72	-13.96	-23.34	-67.52	-17.82	-4.70	15.16	-34.46
Nepal	0.00	0.00	0.00	0.00	-1.14	-10.67	0.00	-9.64	0.06	0.00	-9.35	-1.41	-9.74
New Zealand	0.00	0.00	0.00	0.00	-1.71	-6.98	-0.31	-4.91	0.07	-0.03	-8.98	-0.19	-7.84
Nicaragua	-4.92	-5.78	-6.18	-5.70	-10.82	-29.00	0.00	0.00	-20.41	-19.84	-7.91	0.16	-28.84
Niger	0.00	0.00	0.00	0.00	-15.65	-20.17	0.00	-8.65	0.32	1.87	-15.69	-1.64	-19.79
Pakistan	0.00	0.00	0.00	0.00	-2.75	-18.39	0.33	-4.84	0.00	2.03	-8.28	-0.05	-12.33
Paraguay	-12.63	-12.86	-12.72	-12.70	-31.25	-23.76	-9.82	0.00	-2.55	-15.20	-1.44	2.90	-28.10
Peru	-1.25	-1.25	-1.32	-1.26	-34.87	-7.62	0.00	0.00	-2.42	-23.03	-1.57	3.04	-0.32
Senegal	-40.17	-40.68	-42.41	-40.43	-2.17	-11.62	-2.22	0.00	-29.49	-30.95	-2.45	8.24	-15.75
South Africa	-4.11	-3.98	-5.89	-3.96	-74.83	-16.51	-2.04	-5.52	-3.70	-53.48	-1.06	4.82	0.00
Tajikistan	-0.55	-0.52	-0.64	-0.52	-4.70	-4.59	-1.73	-23.48	-3.99	-6.91	-9.46	6.90	-11.73
Tunisia	0.00	0.00	0.00	0.00	-3.71	-3.15	0.00	-4.25	1.56	0.45	-5.56	-1.94	-0.97
Uruguay	-0.86	-0.87	-0.86	-0.87	-41.56	-34.79	-2.72	0.00	-5.79	-11.37	-3.03	-0.89	-32.21
Yemen	-13.80	-13.18	-20.50	-13.19	-41.74	-26.71	-3.80	-35.85	-11.50	-40.20	-5.16	11.98	-5.92
Zambia	-36.82	-37.82	-37.93	-37.07	-67.34	0.00	-54.12	0.00	-57.73	-83.87	0.00	38.33	-59.97
Average	-8.52	-8.63	-9.49	-8.62	-14.94	-10.29	-11.69	-9.29	-11.30	-14.15	-6.16	3.34	-12.99

Table A.4 The average value of slack in the whole analysed period (in %) in the EU countries—production in international dollars

DMU	PM land	PM LSU	PM energy	PM nutrient	PM production	PM GHG	Land	LSU	Employment	Energy	Nitrogen	Pesticides	Production	Dietary	GHG
Austria	-0.54	-0.54	-0.54	-0.64	0.55	-0.55	-8.06	-2.59	-4.65	-0.32	-4.27	-0.77	0.50	-1.53	-5.18
Belgium	1.49	1.50	1.43	1.53	-1.52	1.50	-1.79	-5.47	15.82	-6.96	-6.04	3.33	0.03	-2.30	-5.92
Bulgaria	-38.95	-40.14	-40.53	-37.24	37.75	-38.65	-69.10	0.00	-18.39	-2.84	-79.86	-40.54	0.00	1.33	-47.33
Croatia	-0.11	-0.11	-0.08	-0.10	0.11	-0.11	-17.23	-18.37	3.82	-19.08	-22.05	32.31	0.59	-2.73	-15.31
Cyprus	-71.42	-71.21	-71.12	-70.73	70.94	-70.94	-42.30	0.00	-4.00	-33.58	-40.24	-88.20	0.00	5.00	-15.73
Czechia	12.03	10.31	12.03	9.58	-10.55	10.23	-6.01	-12.27	0.55	-15.82	-10.94	-0.30	0.97	-5.02	-6.54
Denmark	0.26	0.25	0.25	0.25	-0.28	0.26	-9.21	-4.20	13.22	-0.81	-7.67	10.34	0.00	0.58	-2.01
Estonia	-27.25	-27.29	-26.99	-26.92	26.97	-27.28	-80.32	0.00	-3.65	-61.27	-45.91	-58.55	0.00	1.32	-53.69
Finland	-30.04	-29.91	-29.66	-29.54	29.93	-29.91	-42.95	0.00	-2.27	-49.40	-62.86	-24.96	0.00	0.00	-43.84
France	1.15	1.14	1.20	1.11	-1.19	1.16	-8.31	-3.91	0.28	-4.41	-7.86	-4.52	0.00	0.02	-4.04
Germany	1.25	1.25	1.26	1.26	-1.26	1.27	-0.91	-6.21	3.99	-1.48	-6.68	5.37	0.00	-0.20	-5.01
Greece	-0.73	-0.77	-0.34	-0.91	0.56	-0.74	-19.99	-0.59	-2.89	-16.32	-11.09	-4.40	0.00	-7.85	-6.75
Hungary	-3.49	-3.46	-3.22	-3.37	3.25	-3.45	-35.68	0.00	0.72	-12.54	-48.06	-14.16	0.00	-0.53	-15.25
Ireland	-0.05	-0.05	-0.06	-0.05	0.04	-0.05	-14.38	-18.28	0.66	0.00	-17.23	0.00	0.00	-3.24	-20.70
Italy	2.80	2.64	2.88	2.98	-2.89	2.69	-1.79	-2.78	-3.95	-7.09	-4.49	1.52	0.00	-0.06	-5.30
Latvia	-48.24	-48.33	-48.20	-48.08	48.14	-48.25	-75.96	0.00	-33.19	-44.99	-56.28	-73.45	0.00	0.00	-51.65
Lithuania	-1.08	-1.05	-1.07	-1.13	1.07	-1.08	-31.91	-0.03	-3.74	-0.08	-43.57	-7.59	0.00	-0.66	-31.39
Luxembourg	-0.10	-0.10	-0.10	-0.10	0.10	-0.10	-2.97	-5.23	6.72	-2.06	-5.49	1.47	0.36	-9.62	-6.56
Netherlands	4.09	3.95	4.17	4.36	-3.89	3.98	-0.75	-9.17	3.25	-11.89	-6.23	1.76	0.00	-0.32	-8.80
Poland	0.90	0.88	0.90	0.88	-0.87	0.89	-5.61	0.00	-15.50	-22.91	-17.69	3.67	0.00	-0.27	-6.91
Portugal	-10.31	-10.49	-10.08	-10.91	10.51	-10.43	-6.69	0.00	-40.81	-3.32	-0.56	-55.85	0.00	-0.32	-7.21
Romania	2.30	2.49	1.03	2.20	-2.52	2.47	-14.67	-8.12	-3.12	-1.44	-14.43	3.88	0.00	-2.72	-9.74
Slovakia	-49.91	-49.81	-49.85	-50.00	49.64	-49.90	-69.04	0.00	0.00	-10.61	-61.56	-68.17	0.00	8.07	-31.74
Slovenia	-55.82	-55.93	-55.97	-55.91	55.95	-55.91	-43.53	-0.28	-48.37	-28.67	-3.23	-77.77	0.00	0.00	-21.60
Spain	2.79	2.76	2.71	2.60	-2.92	2.78	-11.17	-6.17	-0.86	-2.41	-4.76	1.77	0.00	-0.12	-2.44

Sweden	−21.85	−21.81	−21.67	−21.95	21.80	−21.86	0.00	−0.04	−46.98	−54.71	−30.96	0.00	0.20	−40.43
United Kingdom	1.73	1.84	1.64	1.83	−1.72	1.86	−8.14	1.66	0.00	−6.63	3.44	0.00	−1.22	−8.18
Average	**−12.19**	**−12.30**	**−12.22**	**−12.18**	**12.14**	**−12.23**	**−4.14**	**−4.99**	**−15.08**	**−24.09**	**−17.83**	**0.09**	**−0.82**	**−17.75**

Table A.5 The change in relative slack value between 2005–2007 and 2016–2018 sub-periods in Cluster 1 (in pp.)—international dollars

DMU	PM land	PM LSU	PM energy	PM GHG	Land	LSU	Employment	Energy	Nitrogen	Pesticides	Production	Dietary	GHG
Albania	0.00	0.00	0.00	0.00	0.00	-22.79	0.00	-18.99	1.58	0.00	3.99	-6.81	-29.29
Algeria	0.00	0.00	0.00	0.00	-23.42	-24.52	-69.94	1.87	-8.29	-2.95	1.13	-1.89	-2.40
Argentina	0.00	0.00	0.00	0.00	-38.94	-36.27	67.90	-27.28	0.00	1.00	-30.99	0.85	-36.17
Armenia	15.71	15.63	15.76	15.71	-1.73	-5.32	54.77	-40.23	72.51	6.09	-10.92	-4.09	0.00
Canada	-4.65	-4.50	-4.97	-4.81	-1.22	-10.46	-9.60	21.27	29.15	27.38	0.00	6.14	-2.18
Chile	4.28	4.33	5.31	4.27	-3.67	-11.98	-0.20	0.00	-7.89	4.58	-13.62	10.36	-15.11
China	0.00	0.00	0.00	0.00	-6.32	-15.59	0.00	-8.87	0.19	0.00	-15.68	0.30	-15.26
Côte d'Ivoire	0.57	0.58	0.57	0.60	0.00	1.01	-2.53	3.27	-4.73	-1.80	1.79	-1.26	0.58
Egypt	0.43	0.46	0.44	0.46	0.00	-16.97	0.19	-22.00	-2.70	1.46	-22.07	-2.15	-9.94
Ghana	0.00	0.00	0.00	0.00	-33.27	-25.40	0.00	-7.07	-5.11	-4.18	-27.42	-2.43	-29.12
Israel	0.00	0.00	0.00	0.00	3.97	-4.31	-0.84	-17.30	-0.49	0.19	-25.19	-20.57	-1.38
Japan	0.00	0.00	0.00	0.00	0.00	-19.10	0.36	-23.30	-0.49	0.00	-15.11	0.00	-14.28
Jordan	2.40	2.42	2.50	2.41	-18.99	-8.89	2.27	-6.15	-54.34	-9.97	0.81	-14.91	3.66
Malawi	0.00	0.00	0.00	0.00	-1.80	0.58	0.00	0.00	0.00	-9.14	-22.91	-5.05	9.81
Malaysia	0.34	0.35	0.21	0.35	0.00	1.30	-0.79	-24.46	-0.15	4.61	-14.03	0.91	1.41
Mozambique	36.67	35.53	53.76	34.64	11.37	-16.61	49.90	62.33	29.91	47.81	-19.72	-27.49	-1.93
North Macedonia	0.00	0.00	0.00	0.00	-26.34	-22.24	1.68	5.17	0.23	0.00	-31.95	-44.08	-23.90
Norway	0.74	0.73	0.75	0.73	0.00	-8.75	-8.36	-25.01	-0.08	-7.87	-0.47	-10.57	-14.56
Philippines	0.00	0.00	0.00	0.00	0.00	-12.81	0.02	0.00	-3.81	0.00	-14.24	0.29	-13.70
Russian Federation	0.61	0.63	0.62	0.61	-23.72	-14.14	0.00	-22.56	1.13	1.39	-10.78	-0.06	-18.35
Saudi Arabia	24.75	24.75	24.00	24.78	-15.84	-11.67	70.13	-21.36	-3.53	20.63	0.00	7.43	0.00
Sri Lanka	2.95	2.98	2.89	2.65	0.00	0.00	5.88	0.00	8.87	0.00	1.38	0.24	6.74
Switzerland	1.72	1.72	1.76	1.73	1.21	-18.58	1.07	-0.33	4.71	-0.48	-15.75	-2.10	-19.48

Thailand	0.00	0.00	0.00	0.00	0.00	0.00	−2.01	0.00	0.00	0.00	−3.19	0.00	−12.51
Turkey	0.00	0.00	0.00	0.00	−1.71	−3.15	−0.78	−35.07	−0.81	−18.61	−13.08	−0.40	0.87
Ukraine	−1.22	−1.24	−1.23	−1.19	1.65	−1.11	−0.70	3.12	−15.40	−16.81	7.74	−0.20	10.43
USA	0.62	0.63	0.60	0.62	−30.03	−27.30	−0.06	−26.93	−2.93	0.18	−28.89	0.14	−29.42
Viet Nam	0.00	0.00	0.00	0.00	0.55	−11.86	0.00	0.80	0.00	0.00	−8.69	−0.43	−6.73
Average	3.07	3.04	3.68	2.98	−7.44	−12.39	5.73	−8.25	1.34	1.55	−11.71	−4.21	−9.36

Table A.6 The change in relative slack value between 2005–2007 and 2016–2018 sub-periods in Cluster 2 (in pp.)—international dollars

DMU	PM land	PM LSU	PM energy	PM GHG	Land	LSU	Employment	Energy	Nitrogen	Pesticides	Production	Dietary	GHG
Australia	-3.18	-3.18	-3.35	-3.33	-26.62	-22.36	-1.85	-12.08	-0.71	0.45	-14.60	0.04	-22.13
Azerbaijan	6.10	6.13	6.12	6.14	-7.73	-6.82	23.10	-0.40	-1.43	16.23	-26.34	-2.52	-1.69
Bangladesh	0.00	0.00	0.00	0.00	0.00	-15.06	0.00	1.50	0.00	0.00	-14.24	0.95	-17.72
Belarus	0.00	0.00	0.00	0.00	0.00	0.00	1.84	0.52	2.16	3.21	-7.75	-1.11	-15.65
Bolivia	3.00	3.03	3.03	3.00	0.00	-19.17	-1.35	-0.16	0.67	11.40	-11.64	0.00	-20.19
Brazil	0.00	0.00	0.00	0.00	-14.66	-21.92	0.78	-8.63	0.00	0.00	-20.50	-0.09	-23.89
Brunei Darussalam	0.00	0.00	0.00	0.00	0.24	-33.05	14.05	-35.53	-17.26	4.09	-31.86	-34.92	-29.85
Burkina Faso	46.01	45.92	44.98	46.04	0.00	-6.93	-0.93	0.00	-4.97	-23.22	-15.47	0.39	-5.69
Cameroon	0.00	0.00	0.00	0.00	-2.06	-3.13	0.00	0.00	0.00	0.00	-19.73	0.25	-1.88
Colombia	-1.29	-1.33	-1.40	-1.35	-32.45	-5.80	-38.25	0.00	-18.95	-44.26	0.00	9.82	-8.38
Costa Rica	0.00	0.00	0.00	0.00	-4.64	0.70	-1.08	3.16	0.00	0.00	-17.60	1.11	1.24
Ecuador	9.98	9.95	9.99	9.97	0.00	7.95	42.64	2.60	21.96	65.14	0.00	-10.27	0.00
El Salvador	-5.55	-5.56	-5.60	-5.56	0.00	0.00	-63.98	5.47	-41.74	0.00	0.56	-4.47	-1.60
Ethiopia	48.32	48.48	48.69	48.39	-8.14	-0.99	59.89	0.00	0.00	18.40	-11.20	6.05	-8.13
Georgia	-1.83	-1.21	-1.81	-1.10	10.68	0.00	-28.45	-7.10	68.24	31.22	0.00	7.42	-9.46
Guyana	8.94	8.77	8.72	8.78	-17.63	0.00	-0.52	-18.01	16.91	3.10	-0.12	2.34	-5.22
Honduras	24.30	24.34	25.25	24.45	0.00	-2.08	40.83	0.00	15.63	18.93	0.00	-6.12	3.00
Iceland	0.00	0.00	0.00	0.00	-32.68	-7.09	-0.87	-36.83	0.51	8.60	-14.83	0.57	-24.50
India	0.00	0.00	0.00	0.00	-25.64	-33.59	0.00	-30.66	0.20	0.00	-29.52	0.00	-31.53
Indonesia	0.00	0.00	0.00	0.00	-28.10	-7.01	1.97	-3.68	0.00	0.00	-25.05	0.66	-25.47
Iran	3.48	3.48	3.54	3.40	-22.04	-10.89	3.37	-19.99	0.00	3.22	-27.28	1.25	-6.98
Iraq	22.45	22.57	24.16	22.61	27.88	2.14	10.74	-22.78	27.15	-4.50	0.00	6.21	-7.72
Kazakhstan	0.00	0.00	0.00	0.00	-42.95	-6.52	5.50	-26.92	0.00	0.00	-25.09	-3.37	-13.77
Kenya	18.46	18.28	18.81	18.25	-28.81	-17.29	17.46	-1.03	0.00	0.00	-9.27	9.53	-19.33

Kyrgyzstan	0.00	0.00	0.00	0.00	0.94	7.14	−3.33	0.00	0.00	−1.96	16.65	0.22	−2.81
Madagascar	17.13	17.13	17.65	16.95	11.99	0.00	24.42	0.00	77.15	52.56	−19.22	19.82	−25.74
Mexico	0.00	0.00	0.00	0.00	10.55	−1.83	13.05	0.00	9.66	3.37	−2.84	−3.25	0.00
Mongolia	0.00	0.00	0.00	0.00	−32.02	−25.63	−0.67	0.00	0.00	0.00	−4.35	1.79	−26.32
Morocco	−4.71	−4.78	−5.49	−4.62	−16.87	−6.17	−24.40	0.00	−11.51	−21.86	−4.40	2.00	0.00
Namibia	78.86	78.86	78.85	78.82	17.80	−19.50	38.01	−16.66	84.46	28.84	−6.06	15.36	−12.15
Nepal	0.00	0.00	0.00	0.00	−5.25	−34.09	0.00	19.31	−3.36	0.00	−33.48	1.48	−35.06
New Zealand	0.00	0.00	0.00	0.00	3.02	−16.88	0.00	−13.59	−3.62	0.00	−21.61	−0.13	−19.14
Nicaragua	10.70	10.77	10.57	10.76	−33.30	8.84	0.00	0.00	62.39	68.01	−27.85	−1.61	9.60
Niger	0.00	0.00	0.00	0.00	−32.36	−22.90	0.00	13.00	−2.82	−3.68	−21.40	0.25	−20.25
Pakistan	0.00	0.00	0.00	0.00	−3.48	−16.40	0.00	6.40	0.00	31.48	−6.71	0.00	−11.32
Paraguay	14.07	14.05	14.13	14.04	−7.08	7.54	0.00	0.00	0.00	16.09	−7.08	2.39	11.06
Peru	1.45	1.44	1.38	1.48	−23.01	−2.33	0.00	0.00	−0.93	15.34	1.52	−3.31	0.00
Senegal	23.03	23.16	21.29	23.45	0.00	−12.01	5.00	0.00	17.04	29.67	−5.60	−2.74	−14.25
South Africa	9.09	9.09	10.36	8.94	−6.37	−12.68	−8.82	−2.44	4.13	27.17	−2.39	8.01	0.00
Tajikistan	0.00	0.00	0.00	0.00	−13.86	2.33	−13.79	−66.47	−1.08	−17.36	2.51	−23.46	−12.73
Tunisia	0.00	0.00	0.00	0.00	−7.75	−3.67	0.00	4.98	0.00	−3.42	−12.06	−3.42	2.07
Uruguay	1.04	1.05	1.03	1.05	−47.10	−41.12	2.01	0.00	18.01	−9.99	−5.70	0.23	−36.84
Yemen	−33.03	−32.88	−33.17	−32.86	13.39	0.44	−19.64	−44.82	−19.64	−53.10	0.00	20.60	0.00
Zambia	7.11	6.95	6.35	7.59	−22.81	0.00	5.71	0.00	−20.28	15.44	0.00	−16.46	−19.97
Average	6.91	6.92	6.91	6.94	−10.20	−9.04	2.33	−7.06	6.32	6.56	−10.95	0.03	−11.60

Table A.7 The change in relative slack value between 2005–2007 and 2016–2018 sub-periods in Cluster 3 (in pp.)—international dollars

DMU	PM land	PM LSU	PM energy	PM nutrient	PM production	PM GHG	Land	LSU	Employment	Energy	Nitrogen	Pesticides	Production	Dietary	GHG
Austria	0.66	0.67	0.66	0.68	0.66	0.66	6.92	-3.31	20.55	0.00	0.53	2.98	0.00	-3.33	-0.99
Belgium	-0.71	-0.71	-0.65	-0.74	-0.68	-0.71	3.88	-13.65	-30.63	10.18	-10.63	-7.11	0.15	-7.04	-13.39
Bulgaria	-19.01	-19.12	-18.84	-19.70	-18.50	-19.15	16.52	0.00	-0.84	9.43	23.36	8.90	0.00	6.07	27.36
Croatia	0.00	0.00	0.00	0.00	0.00	0.00	-31.43	-26.89	4.84	-12.21	-30.76	-79.60	2.68	-6.85	-31.34
Cyprus	5.51	5.49	5.52	4.33	5.53	5.32	3.08	0.00	4.26	-4.81	-21.79	-4.18	0.00	-4.96	-21.41
Czechia	-34.14	-31.98	-31.90	-34.39	-33.27	-32.23	8.96	-29.68	0.30	-46.11	9.62	2.46	4.47	-2.41	-3.42
Denmark	0.72	0.73	0.69	0.67	0.65	0.70	-9.57	-3.91	-17.31	-2.11	4.92	-46.58	0.00	0.40	-0.02
Estonia	-5.61	-5.72	-5.75	-5.50	-5.76	-5.66	4.51	0.00	-12.52	16.35	33.76	6.51	0.00	1.81	3.11
Finland	18.16	18.15	18.11	18.45	18.18	18.22	12.10	0.00	-9.76	-5.59	-12.15	-13.52	0.00	0.00	-5.71
France	-3.94	-3.88	-4.11	-3.75	-4.18	-3.90	-15.13	-12.05	0.99	-12.70	-10.57	11.12	0.00	-0.21	-11.66
Germany	-1.98	-1.98	-1.98	-2.00	-1.97	-1.99	2.47	-7.53	3.19	0.85	-13.92	-31.30	0.00	-0.67	-7.98
Greece	5.66	5.50	5.49	5.68	4.88	5.46	-41.55	1.73	12.80	-27.79	10.97	27.94	0.00	-16.80	-9.18
Hungary	-1.95	-1.93	-1.81	-2.16	-1.77	-1.94	23.92	0.00	-2.40	24.88	42.42	20.15	0.00	-2.10	16.12
Ireland	-0.22	-0.22	-0.22	-0.22	-0.22	-0.22	-7.78	-15.05	3.35	0.00	-6.36	0.00	0.00	3.26	-16.17
Italy	-7.52	-7.34	-7.50	-7.29	-7.60	-7.33	-0.38	-1.13	11.08	-7.38	6.00	1.94	0.00	-0.17	-1.97
Latvia	-2.92	-2.82	-2.94	-2.86	-2.78	-2.86	2.72	0.00	-40.89	10.83	20.69	9.44	0.00	0.00	6.04
Lithuania	0.00	0.00	0.00	0.00	0.00	0.00	-15.55	0.00	-2.65	0.35	-16.25	-2.49	0.00	-0.32	-14.38
Luxembourg	0.00	0.00	0.00	0.00	0.00	0.00	-6.87	-8.99	6.42	0.49	-12.24	0.00	-1.33	-26.60	-11.21
Netherlands	-8.99	-8.88	-9.53	-9.04	-8.95	-8.90	2.34	-25.86	0.24	-29.86	-15.08	1.44	0.00	0.55	-24.45
Poland	-0.93	-0.92	-0.97	-0.91	-0.93	-0.92	-19.77	0.00	-14.03	-17.07	-6.70	-3.90	0.00	1.44	-2.33
Portugal	12.99	12.93	12.91	12.44	12.88	12.89	-17.34	0.00	0.22	-13.80	0.00	37.88	0.00	-1.45	-0.15
Romania	-9.30	-8.96	-8.09	-9.88	-9.93	-9.19	-33.74	-30.95	0.00	5.09	-1.12	14.17	0.00	-7.40	-22.85
Slovakia	-4.30	-4.25	-3.91	-4.60	-4.31	-4.30	9.52	0.00	0.00	17.15	17.66	-3.52	0.00	3.83	12.04
Slovenia	-1.32	-1.33	-1.33	-1.33	-1.32	-1.32	15.82	0.00	-16.58	4.39	-4.42	-2.20	0.00	0.00	0.95
Spain	1.39	1.28	1.58	1.31	1.52	1.41	-16.52	-7.83	3.73	-1.61	1.88	-19.54	0.00	0.97	-1.12

Sweden	5.88	5.78	5.71	5.73	5.55	5.80	0.52	0.00	-0.20	-3.37	2.94	-2.72	0.00	-0.97	1.81
United Kingdom	-5.71	-6.02	-6.35	-6.14	-5.99	-6.04	-25.81	-21.73	-0.09	0.00	-20.64	-8.11	0.00	-5.22	-22.44
Average	0.66	0.67	0.66	0.68	0.66	0.66	6.92	-3.31	20.55	0.00	0.53	2.98	0.00	-3.33	-0.99

Table A.8 Changes in GHG emissions from agriculture in the EU countries

Country	2005–2019	Projected 2005–2030 with existing policy measures	Projected 2005–2030 with additional measures
Croatia	−17.7	−20.5	−25.4
Malta	−16.6	−27.9	−27.9
Greece	−11.9	2.1	2.1
Italy	−8.7	−13.4	−13.4
Romania	−8.1	5.6	2
France	−4.4	−7.2	−7.2
Belgium	−4	−7.9	−18.1
Cyprus	−3.9	−8.4	−8.9
Denmark	−2.7	−8.2	−8.2
Germany	−2.3	−9.5	−9.5
Spain	−2.3	−5.6	−16.5
EU-27	**−1.8**	**−2**	**−5**
Sweden	−1.4	−13.3	−13.3
Czechia	−0.6	1.4	1.4
Netherlands	−0.2	−0.6	−1.3
Sloveia	0.3	5.1	−0.8
Finland	1.4	−2.1	−9.6
Austria	1.9	3.6	−6.4
Poland	2.5	3	3
Portugal	2.7	−1.8	−4.4
Lithuania	4.7	−0.2	−8.1
Slovakia	5.6	−5.6	−5.6
Ireland	6.2	10.1	−5.1
Luxembourg	10.2	5.8	−18.9
Hungary	16.2	22.4	14.4
Latvia	22.8	54.2	44.4
Bulgaria	24.7	48.1	48.1
Estonia	27	32.8	32.8

Source: European Environment Agency (EEA 2021) https://www.eea.europa.eu/data-and-maps/indicators/greenhouse-gas-emissions-from-agriculture/assessment

Table A.9 Eco-efficiency change and its components (2006–2018 average)—international USD, Cluster 1

DMU	MI	EC	TC	OBTC	IBTC	MATC
Albania	1.033	1.001	1.033	0.967	1.005	1.063
Algeria	0.982	0.962	1.021	0.968	1.012	1.043
Argentina	1.032	0.982	1.051	0.947	1.006	1.103
Armenia	0.979	0.945	1.036	0.985	0.991	1.062
Canada	1.007	0.982	1.026	1.007	1.000	1.018
Chile	0.998	0.982	1.017	0.992	0.994	1.031
China	1.010	0.987	1.023	0.978	0.999	1.046
Côte d'Ivoire	1.029	1.000	1.030	0.969	1.003	1.059
Egypt	1.000	0.981	1.019	0.979	1.003	1.038
Ghana	0.982	0.971	1.011	0.977	1.013	1.022
Israel	0.978	0.955	1.024	0.992	0.984	1.049
Japan	1.004	0.989	1.015	0.990	0.995	1.030
Jordan	1.002	0.983	1.020	0.982	1.001	1.038
Malawi	1.001	0.977	1.024	0.968	1.015	1.043
Malaysia	1.028	0.989	1.040	0.998	0.990	1.052
Mozambique	0.974	0.953	1.022	0.996	0.985	1.041
North Macedonia	0.997	0.960	1.039	0.966	0.997	1.079
Norway	0.995	0.980	1.016	0.992	0.995	1.028
Philippines	1.005	0.988	1.017	0.983	1.001	1.033
Russian Federation	1.004	0.988	1.015	0.978	1.008	1.030
Saudi Arabia	0.982	0.953	1.031	1.001	1.000	1.030
Sri Lanka	1.027	0.998	1.028	0.977	0.997	1.055
Switzerland	1.001	0.988	1.014	0.993	0.993	1.027
Thailand	1.025	0.997	1.028	0.968	1.006	1.055
Turkey	1.007	0.993	1.015	0.982	1.004	1.030
Ukraine	1.020	0.999	1.020	0.977	1.004	1.041
USA	0.992	0.972	1.021	0.989	0.990	1.043
Viet Nam	1.016	0.995	1.021	0.976	1.004	1.041
Average	1.004	0.980	1.024	0.981	1.000	1.044

Note: *MI* Malmquist index (sequential technology), *EC* efficiency change, *TC* technical change, *OBTC* output-based technological change, *IBTC* input-based technological change, *MATC* mixed technological change

Table A.10 Eco-efficiency change and its components (2006-2018 average)—international USD, Cluster 2

DMU	MI	EC	TC	OBTC	IBTC	MATC
Australia	0.997	0.979	1.018	0.991	0.992	1.036
Azerbaijan	0.989	0.974	1.015	0.989	1.001	1.025
Bangladesh	1.004	0.988	1.016	0.978	1.006	1.033
Belarus	1.008	0.991	1.017	0.982	1.002	1.034
Bolivia	1.032	0.998	1.034	0.969	1.005	1.061
Brazil	0.999	0.981	1.018	0.973	1.010	1.036
Brunei Darussalam	0.946	0.928	1.019	0.962	1.020	1.039
Burkina Faso	0.978	0.944	1.036	0.986	0.996	1.055
Cameroon	1.010	0.987	1.023	0.968	1.010	1.047
Colombia	1.045	1.025	1.019	1.000	0.999	1.020
Costa Rica	1.014	0.986	1.028	0.979	0.994	1.057
Ecuador	1.028	0.978	1.051	1.004	0.998	1.049
El Salvador	1.048	1.025	1.022	0.999	0.998	1.026
Ethiopia	0.970	0.934	1.038	0.974	1.009	1.055
Georgia	0.995	0.972	1.024	1.002	0.992	1.029
Guyana	1.026	1.001	1.025	0.999	0.999	1.027
Honduras	0.973	0.945	1.030	1.001	1.000	1.030
Iceland	1.013	0.984	1.029	0.969	1.003	1.059
India	1.018	0.982	1.037	0.965	1.000	1.074
Indonesia	1.013	0.979	1.035	0.961	1.006	1.070
Iran	1.000	0.975	1.026	0.972	1.003	1.052
Iraq	0.973	0.951	1.022	0.999	1.000	1.023
Kazakhstan	1.005	0.984	1.021	0.973	1.009	1.040
Kenya	1.004	0.981	1.024	0.983	0.996	1.046
Kyrgyzstan	1.021	1.001	1.020	0.976	1.010	1.034
Madagascar	0.965	0.950	1.016	0.993	0.997	1.027
Mexico	1.013	0.995	1.018	0.996	1.003	1.019
Mongolia	1.047	0.996	1.052	0.959	0.992	1.105
Morocco	1.027	0.999	1.028	0.969	1.031	1.029
Namibia	0.924	0.891	1.038	0.991	1.001	1.046
Nepal	0.991	0.975	1.017	0.973	1.011	1.033
New Zealand	1.001	0.980	1.021	0.984	0.995	1.042
Nicaragua	0.982	0.956	1.028	0.982	1.008	1.038
Niger	1.025	0.978	1.048	0.957	0.997	1.099
Pakistan	1.043	0.995	1.048	0.963	0.993	1.096
Paraguay	1.011	0.978	1.034	1.002	1.001	1.030
Peru	1.069	1.013	1.055	0.986	1.007	1.064
Senegal	1.059	1.010	1.049	0.992	0.991	1.067
South Africa	1.035	0.983	1.052	0.999	1.015	1.037
Tajikistan	1.031	0.989	1.042	0.966	0.997	1.082
Tunisia	1.008	0.989	1.020	0.981	1.000	1.039
Uruguay	1.020	0.982	1.039	0.988	0.991	1.061

(continued)

Table A.10 (continued)

DMU	MI	EC	TC	OBTC	IBTC	MATC
Yemen	1.055	1.039	1.015	0.975	1.024	1.017
Zambia	1.021	0.991	1.031	1.000	1.000	1.031
Average	1.010	0.981	1.029	0.982	1.003	1.046

Table A.11 Eco-efficiency change and its components (2006-2018 average)—international USD, Cluster 3

DMU	MI	EC	TC	OBTC	IBTC	MATC
Austria	1.001	0.998	1.003	0.996	1.003	1.005
Belgium	1.005	0.984	1.021	0.982	0.998	1.042
Bulgaria	1.042	1.014	1.028	0.999	0.998	1.030
Croatia	0.997	0.975	1.022	0.987	0.993	1.043
Cyprus	1.015	0.994	1.021	0.999	0.999	1.023
Czechia	0.966	0.895	1.079	0.966	0.960	1.163
Denmark	0.991	0.971	1.021	0.987	0.995	1.040
Estonia	1.023	1.007	1.016	0.998	1.005	1.013
Finland	1.015	0.982	1.033	1.000	1.000	1.033
France	0.989	0.975	1.014	0.993	0.995	1.026
Germany	1.001	0.982	1.020	0.989	0.993	1.038
Greece	0.987	0.970	1.018	0.981	1.005	1.032
Hungary	1.011	0.994	1.017	0.997	1.000	1.020
Ireland	1.020	1.000	1.020	0.984	0.998	1.039
Italy	0.996	0.975	1.021	0.988	0.992	1.041
Latvia	1.040	1.008	1.032	1.009	1.000	1.023
Lithuania	1.022	1.000	1.022	0.993	0.998	1.032
Luxembourg	0.998	0.979	1.020	0.990	0.991	1.040
Netherlands	1.005	0.975	1.030	0.981	0.990	1.061
Poland	1.008	0.996	1.013	0.992	0.998	1.022
Portugal	0.998	0.973	1.026	1.000	0.993	1.034
Romania	0.995	0.969	1.027	0.983	0.991	1.054
Slovakia	1.053	1.010	1.042	0.999	0.999	1.044
Slovenia	1.012	1.005	1.007	1.000	1.001	1.007
Spain	1.024	1.001	1.023	0.977	1.002	1.045
Sweden	1.008	0.984	1.024	1.000	1.000	1.024
United Kingdom	0.994	0.970	1.025	0.990	0.986	1.050
Average	1.001	0.998	1.003	0.996	1.003	1.005

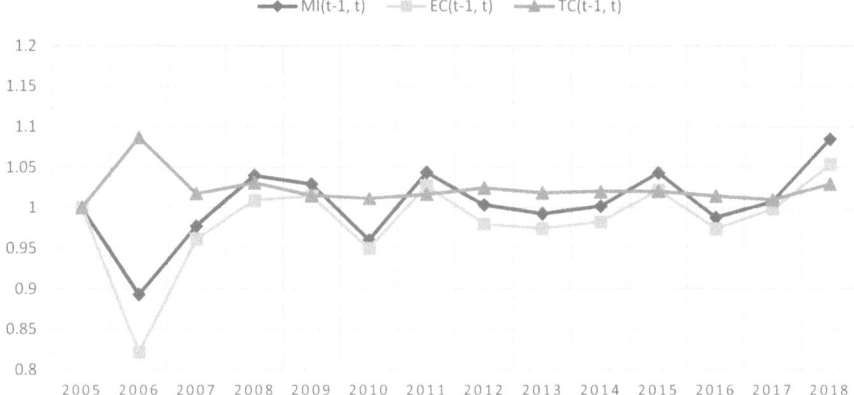

Figure A.1 Evolution of Malmquist Index and its component—cluster 1

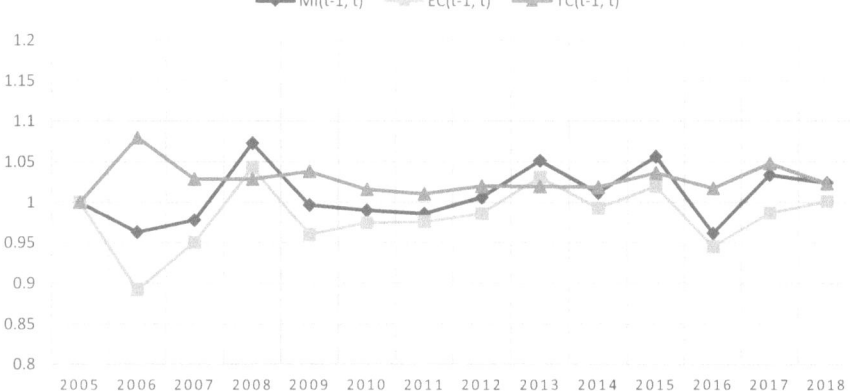

Figure A.2 Evolution of Malmquist Index and its component—cluster 3

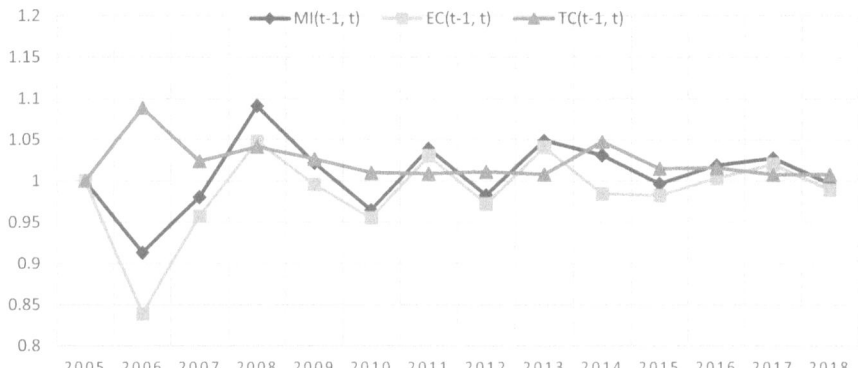

Figure A.3 Evolution of Malmquist Index and its component—cluster 3

Table A.12 The potential effect of slack reduction on nitrogen equivalent use and pesticides use in sampled countries

Country	Nitrogen equivalent/ ha	Nitrogen equivalent/ha (after slack)	Country	Pesticides use per ha	Pesticides use per ha (after slack)
Brunei Dar.	1060	967	Japan	12.33	12.46
Egypt	538	462	Israel	11.86	12.08
Netherlands	479	435	Costa Rica	6.76	7.05
Belgium	387	366	Malaysia	6.62	3.42
Bangladesh	314	174	Czechia	6.24	6.46
Japan	261	252	Netherlands	5.65	5.62
Luxembourg	235	230	Italy	4.93	5.12
Viet Nam	234	224	Belgium	4.89	4.97
Denmark	231	252	Ecuador	3.74	1.54
Norway	218	209	Portugal	3.47	1.39
Ireland	213	196	China	3.26	3.36
Israel	209	200	Egypt	2.80	2.80
Pakistan	206	127	Germany	2.49	2.43
Czechia	204	192	France	2.45	2.71
Germany	197	173	Thailand	2.29	1.22
India	174	183	El Salvador	2.19	1.92
Nepal	172	148	Slovenia	1.94	0.78
France	146	140	Honduras	1.92	1.58
Switzerland	142	134	Spain	1.83	2.91
Cyprus	138	167	Viet Nam	1.76	1.76
United Kingdom	134	132	Hungary	1.74	2.75
New Zealand	129	124	Philippines	1.72	1.70
Austria	126	121	Argentina	1.57	2.16
Poland	121	87	Nicaragua	1.56	1.51
Finland	118	116	Brazil	1.44	1.72
Côte d'Ivoire	118	103	Bangladesh	1.44	1.05
Malaysia	116	82	Poland	1.42	1.44
El Salvador	113	92	Côte d'Ivoire	1.42	2.23
Slovenia	110	189	Austria	1.37	1.38
Sweden	109	233	Cyprus	1.36	1.07
Italy	109	108	Colombia	1.35	0.68
Philippines	109	101	Switzerland	1.34	1.35
Costa Rica	107	108	Ukraine	1.30	0.45
Thailand	104	75	Denmark	1.26	1.47
Indonesia	100	119	Brunei Dar.	1.24	1.29
China	98	99	Greece	1.17	1.34
Ecuador	95	77	United Kingdom	1.15	1.59
Sri Lanka	94	50	Luxembourg	1.15	1.19

(continued)

Table A.12 (continued)

Country	Nitrogen equivalent/ ha	Nitrogen equivalent/ha (after slack)	Country	Pesticides use per ha	Pesticides use per ha (after slack)
Belarus	92	60	Canada	1.12	0.85
Ethiopia	90	63	Turkey	1.10	1.55
Honduras	90	75	Georgia	1.09	1.07
Hungary	86	95	Uruguay	1.07	2.07
Spain	84	109	Jordan	1.03	1.15
Albania	83	78	USA	0.97	1.76
Slovakia	82	163	Paraguay	0.96	1.23
Portugal	81	82	Slovakia	0.84	0.91
Brazil	79	80	Lithuania	0.83	0.90
Lithuania	75	91	Norway	0.70	0.70
Turkey	72	92	Finland	0.69	0.69
Bulgaria	71	76	Latvia	0.65	0.63
Jordan	70	66	Ireland	0.64	0.79
Nicaragua	68	60	Chile	0.62	0.56
Uruguay	67	100	Sweden	0.61	1.24
Canada	64	54	Sri Lanka	0.57	0.35
Burkina Faso	59	38	Estonia	0.56	1.13
USA	57	79	Belarus	0.53	0.16
Azerbaijan	57	53	Romania	0.48	0.81
Estonia	57	209	Mexico	0.48	0.30
Iran	56	59	Morocco	0.45	0.39
Greece	55	61	New Zealand	0.45	0.46
Armenia	55	47	Cameroon	0.44	0.44
Romania	55	65	Bulgaria	0.43	0.54
Colombia	51	70	Albania	0.40	0.39
Kenya	51	43	Bolivia	0.36	0.39
Latvia	51	136	Ghana	0.31	0.13
Iraq	46	46	South Africa	0.28	0.26
Mexico	45	54	Guyana	0.27	0.05
Paraguay	43	54	Peru	0.27	0.12
Cameroon	43	42	India	0.25	0.28
Malawi	42	37	Russia	0.24	1.10
Chile	42	37	Tunisia	0.23	0.25
Peru	41	54	Malawi	0.22	0.23
North Macedonia	39	40	Armenia	0.19	0.28
Georgia	38	56	Zambia	0.15	0.08
Senegal	37	40	Iran	0.13	0.15
Ukraine	35	23	Algeria	0.13	0.17
Guyana	34	95	Australia	0.12	0.27
Tajikistan	33	37	Nepal	0.10	0.10

(continued)

Table A.12 (continued)

Country	Nitrogen equivalent/ ha	Nitrogen equivalent/ha (after slack)	Country	Pesticides use per ha	Pesticides use per ha (after slack)
Argentina	32	35	North Macedonia	0.10	0.10
Tunisia	29	31	Ethiopia	0.10	0.08
Morocco	28	38	Azerbaijan	0.08	0.06
Bolivia	23	22	Pakistan	0.08	0.03
Niger	18	18	Senegal	0.07	0.02
South Africa	17	54	Iraq	0.07	0.05
Ghana	17	18	Kenya	0.06	0.06
Yemen	16	19	Tajikistan	0.05	0.03
Kyrgyzstan	16	27	Kyrgyzstan	0.05	0.09
Russia	15	58	Burkina Faso	0.05	0.02
Zambia	14	42	Kazakhstan	0.04	0.07
Iceland	13	54	Saudi Arabia	0.03	0.25
Algeria	13	13	Indonesia	0.03	0.04
Madagascar	12	26	Mozambique	0.02	0.03
Australia	11	20	Yemen	0.01	0.01
Croatia	8	22	Madagascar	0.01	0.01
Mongolia	6	11	Croatia	0.00	0.05
Namibia	5	6	Iceland	0.00	0.01
Mozambique	4	9	Namibia	0.00	0.00
Kazakhstan	3	6	Mongolia	0.00	0.00
Saudi Arabia	3	16	Niger	0.00	0.00

References

Abdulai, A., & Huffman, W. E. (2005). The diffusion of new agricultural technologies: The case of crossbred-cow technology in Tanzania. *American Journal of Agricultural Economics, 87*(3), 645–659.

Adam, A., & Tsarsitalidou, S. (2019). Environmental policy efficiency: Measurement and determinants. *Economics of Governance, 20*, 1–22.

Adenuga, A. H., Davis, J., Hutchinson, G., Donnellan, T., & Patton, M. (2018). Modelling regional environmental efficiency differentials of dairy farms on the island of Ireland. *Ecological Indicators, 95*, 851–861.

Adenuga, A. H., Davis, J., Hutchinson, G., Donnellan, T., & Patton, M. (2019). Environmental efficiency and pollution costs of nitrogen surplus in dairy farms: A parametric hyperbolic technology distance function approach. *Environmental and Resource Economics, 74*(3), 1273–1298. https://doi.org/10.1007/s10640-019-00367-2

Adenuga, A. H., Davis, J., Hutchinson, G., Patton, M., & Donnellan, T. (2020). Modelling environmental technical efficiency and phosphorus pollution abatement cost in dairy farms. *Science of the Total Environment, 714*, 136690. https://doi.org/10.1016/j.scitotenv.2020.136690

Adrian, T., & Shin, H. S. (2010). Liquidity and leverage, Federal Reserve Bank of New York Staff Report 328, dostępne w Internecie. http://www.newyorkfed.org/research/staff_reports/sr328.pdf (11.05.2013).

Agronomy News. (1989, January). American Society of Agronomy, Crop Science Society of America, Soil Science Society of America, p. 15.

Ajzen, I. (1991). The theory of planned behavior. *Organizational Behavior and Human Decision Processes, 50*, 179–211. https://doi.org/10.1016/0749-5978(91)90020-T

Alene, A. D. (2010). Productivity growth and the effects of R&D in African agriculture. *Agricultural Economics, 41*(3-4), 223–238.

Alesina, A., & Giuliano, P. (2015). Culture and Institutions. *Journal of Economic Literature, American Economic Association, 53*(4), 898–944.

Allen, B., Hart, K., Radley, G., Tucker, G., Keenleyside, C., Oppermann, R., Underwood, E., Menadue, H., Poux, X., Beaufoy, G., Herzon, I., Povellato, A., Vanni, F., Pra˘zan, J., Hudson, T., & Yellachich, N. (2014). *Biodiversity protectionthrough results based remuneration of ecological achievement. Report prepared for the European Commission, DG Environment* (p. 167). Institute for European Environmental Policy.

Alston, J. M., & Pardey, P. G. (2014). Agriculture in the global economy. *The Journal of Economic Perspectives, 28*(1), 121–146.

Andersen, P., & Petersen, N. C. (1993). A procedure for ranking efficient units in data envelopment analysis. *Management Science, 39*(10), 1261–1264.

Ando, A., Camm, J., Polasky, S., & Solow, A. (1998). Species distributions, land values, and efficient conservation. *Science, 279*, 2126–2128.

Arata, L., & Sckokai, P. (2016, February 1). The impact of agri-environmental schemes on farm performance in five E.U. member states: A DID-matching approach. *Land Economics, 92*(1), 167–186. https://doi.org/10.3368/le.92.1.167

Babcock, B. (2015). Welfare effects of title one program. In V. H. Smith (Ed.), *The economic welfare and trade relations implications of the 2014 farm bill*. Emerald Publishing.

Bachev, H. (2017). *An assessment of sustainability of Bulgarian farms*. MPRA Paper No. 77463. https://mpra.ub.uni-muenchen.de/77463/ (Accessed on 20 November 2021).

Backhaus, J. G. (1997). Ingenious tax: A Contemporary Restatement – Special Issue: Commemorating the 100th Anniversary of the death of Henry George. *American Journal of Economics and Sociology, 4*, 4, 6–10. Wiley-Blackwell

Bacon, C., Getz, C., Kraus, S., & Holland, K. (2012). The social dimensions of sustainability in diversified, industrial and hybrid farming systems. *Ecology and Society, 17*, 41. Available online: http://food.berkeley.edu/wp-content/ uploads/2014/09/Social-Dimensions.pdf (Accessed on 2 May 2017).

Badunenko, O., & Tauchmann, H. (2018). *Simar and Wilson two-stage efficiency analysis for Stata* (FAU Discussion Papers in Economics 08/2018). Friedrich-Alexander University Erlangen-Nuremberg, Institute for Economics.

Badunenko, O., & Tauchmann, H. (2019). Simar and Wilson two-stage efficiency analysis for Stata. *The Stata Journal, 19*(4), 950–988.

Bakam, I., Balana, B. B., & Matthews, R. (2012). Cost-effectiveness analysis of policy instruments for greenhouse gas emission mitigation in the agricultural sector. *Journal of Environmental Management, 112*, 33–44. https://doi.org/10.1016/j.jenvman.2012.07.001

Baldoni, E., Kancs, D., Ciaian, P., & Drabik, D. (2021). *The capitalisation of agricultural subsidies into farmland prices– A meta-analysis of empirical literature*. Publications Office of the European Union.

Banga, R. (2014). *Impact of green box subsidies on agricultural productivity, production and international trade*. Unit of Economic Cooperation and Integration Amongst Developing Countries (ECIDC), Background Paper, No. RVC-11: 15–21.

Barett, C. B. (1993). *On price risk and the inverse farm size-productivity relationship*. University of Wisconsin-Madison, Department of Agricultural Economics Staff Paper, Series No. 369, pp. 1–17.

Barnard, C. H. (2000). Agriculture and the rural economy: Urbanization affects a large share of farmland. *Rural Conditions and Trends, 10*(2), 57–63.

Barro, R. J., & Lee, J. W. (2013). A new data set of educational attainment in the world, 1950–2010. *Journal of Development Economics, 104*, 184–198.

Bartolini, F., Brunori, G., Coli, A., Landi, C., & Pacini, B. (2015). *Assessing the causal effect of decoupled payments on farm labour in tuscany using propensity score methods* (No. 1008-2016-80180).

Bartolini, F., Gallerani, V., Raggi, M., & Viaggi, D. (2012). Modelling the linkages between cross-compliance and agri-environmental schemes under asymmetric information. *Journal of Agricultural Economics, 63*, 310–330.

Bartolini, F., Vergamini, D., Longhitano, D., & Povellato, A. (2021). Do differential payments for agri-environment schemes affect the environmental benefits? A case study in the north-eastern italy. *Land Use Policy, 107*, 104862. https://doi.org/10.1016/j.landusepol.2020.104862

Batáry, P., Dicks, L. V., Kleijn, D., & Sutherland, W. J. (2015). The role of agri-environment schemes in conservation and environmental management. *Conservation Biology, 29*, 1006–1016.

Beltrán-Esteve, M., Reig-Martínez, E., & Estruch-Guitart, V. (2017). Assessing eco-efficiency: A metafrontier directional distance function approach using life cycle analysis. *Environmental Impact Assessment Review, 63*, 116–127.

Berlan, A. (2013). Social sustainability in agriculture: An anthropological perspective on child labour in cocoa production in Ghana. *Journal of Development Studies, 49*, 1088–1100.

Bilbao-Terol, C., et al. (2017). Rural tourism accommodation prices by land use-based hedonic approach: First results from the case study of the self-catering cottages in Asturias. *Sustainability, 9*, 1688. https://doi.org/10.3390/su9101688

Bindlish, V., & Evenson, R. E. (1997). The impact of T&V extension in Africa: The experience of Kenya and Burkina Faso. *The World Bank Research Observer, 12*(2), 183–201.

Bingen, J., Serrano, A., & Howard, J. (2003). Linking farmers to markets: Different approaches to human capital development. *Food Policy, 28*(4), 405–419. https://doi.org/10.1016/j.foodpol.2003.08.007

Birge, T., Toivonen, M., Kaljonen, M., & Herzon, I. (2017). Probing the grounds: Developing a payment-by-results agri-environment scheme in finland. *Land Use Policy, 61*, 302–315. https://doi.org/10.1016/j.landusepol.2016.11.028

Bock, B. (2012). Social innovation and sustainability; how to disentangle the buzzword and its application in the field of agriculture and rural development. *Studies in Agricultural Economics, 114*, 57–63.

Boháčková, I. (2014). Some notes to income disparity problems of agriculture. *Agris on-line Papers in Economics and Informatics, 5*(4), 2–11.

Bonfiglio, A., Arzeni, A., & Bodini, A. (2017a). Assessing eco-efficiency of arable farms in rural areas. *Agriculture Systems, 151*, 114–125.

Bonfiglio, A., Camaioni, B., Coderoni, S., Esposti, R., Pagliacci, F., & Sotte, F. (2017b). Are rural regions prioritizing knowledge transfer and innovation? Evidence from Rural Development Policy expenditure across the EU space. *Journal of Rural Studies, 53*, 78–87.

Börjesson, M., & Ahlgren, E. O. (2012). Cost-effective biogas utilization - A modelling assessment of gas infrastructural options in a regional energy system. *Energy, 48*(1), 212–226. https://doi.org/10.1016/j.energy.2012.06.058

Borresch, R., Maas, S., Schmitz, K., & Schmitz, P. M. (2009). Modeling the value of a multifunctional landscape: A discrete choice experiment. In *Paper presented at the International Association of Agricultural Economics Conference*, Beijing, China.

Boussios, D., & O'Donoghue, E. (2019). Potential variability in commodity support: Agriculture risk coverage and price loss coverage programs. *Economic Research Report-Economic Research Service, USDA*, (267).

Boussios, D., Castillo, M., & Brewer, B. (2021). The unintended beneficiaries of farm subsidies. *Land Economics*, 081319-0118R2.

Braat, L., & ten Brink, P. (Eds.). (2008). *The cost of policy inaction, the case of not meeting the 2010 biodiversity target*. Alterra, Alterra-rapport 1718, with J. Bakkes, K. Bolt, I. Braeuer, B. ten Brink, A. Chiabai, H. Ding, H. Gerdes, M. Jeuken, M. Kettunen, U. Kirchholtes, C. Klok, A. Markandya, P. Nunes, M. van Oorschot, N. Peralta-Bezerra, M. Rayment, C. Travisi, M. Walpole.

Breustedt, G., & Habermann, H. (2011). The incidence of EU per-hectare payments on farmland rental rates: A spatial econometric analysis of german farm-level data. *Journal of Agricultural Economics, 62*(1), 225–243.

Brink, C., van Ierland, E., Hordijk, L., & Kroeze, C. (2005). Costeffective emission abatement in agriculture in the presence of interrelations: Cases for The Netherlands and Europe. *Ecological Economics, 53*(1), 59–74.

Buchanan, M. J. (2009). *The demand and supply of public goods* (J. Ma Trans.). Shanghai Renmin Press.

Bureau, J.-C., & Mahé, L. P. (2008). *CAP reform beyond 2017: An idea for a longer time perspective*. EU Parliament.

Cai, W., & Pandey, M. (2015). The agricultural productivity gap in Europe. *Economic Inquiry, 53*, 1807–1817.

Caiado, R. G. G., de Freitas Dias, R., Mattos, L. V., Quelhas, O. L. G., & Leal Filho, W. (2017). Towards sustainable development through the perspective of ecoefficiency—A systematic literature review. *Journal of Cleaner Production, 165*, 890–904.

Caicedo, F. V. (2019). The mission: Human capital transmission, economic persistence, and culture in South America. *The Quarterly Journal of Economics, 134*(1), 507–556. https://doi.org/10.7910/DVN/ML1155

Capraro, V., & Rand, D. G. (2018). Do the right thing: Experimental evidence that preferences for moral behavior, rather than equity or efficiency per se, drive human prosociality. *Forthcoming in Judgment and Decision Making, 1*, 99–111.

Carpenter, S. R., Caraco, N. F., Correl, D. L., Horwarth, R. W., Sharler, A. N., & Smith, V. H. (2012). Nonpoint pollution of surface waters with phosphorus and nitrogen. *Ecological Applications, 8*(3), 559–568.

Carroll, N., Fox, J., & Bayon, R. (Eds.). (2008). *Conservation and biodiversity banking: A guide to setting up and running biodiversity credit trading systems.* Earthscan.

Carson, R. T., & Czajkowski, M. (2014). The discrete choice experiment approach to environmental contingent valuation. In S. Hess & A. Daly (Eds.), *Handbook of choice modelling.* Elgar.

Chabé-Ferret, S., & Subervie, J. (2013). How much green for the buck? Estimating additional and windfall effects of French agro-environmental schemes by DID-matching. *Journal of Environmental Economics and Management, 65*(1), 12–27.

Chagwiza, C., Muradian, R., & Ruben, R. (2016). Cooperative membership and dairy performance among smallholders in Ethiopia. *Food Policy, 59*, 165–173. https://doi.org/10.1016/j.foodpol.2016.01.008

Chamberlain, D. E., Fuller, R. J., Bunce, R. G. H., Duckworth, J. C., & Shrubb, M. (2000). Changes in the abundance of farmland birds in relation to the timing of agricultural intensification in England and Wales. *Journal of Applied Ecology, 37*(5), 771–788.

Chang, H.-J. (2009). Rethinking public policy in agriculture—Lessons from distant and recent. *Journal of Peasant Studies, 36*, 477–515.

Charnes, A., Cooper, W. W., & Rhodes, E. (1978). Measuring the efficiency of decision making units. *European Journal of Operational Research, 2*(6), 429–444.

Chen, Z., Huffman, W. E., & Rozelle, S. (2011). Inverse relationship between productivity and farm size: The case of China. *Contemporary Economic Policy, 29*(4), 580–592.

Chen, L., & Jia, G. (2017). Environmental efficiency analysis of China's regional industry: A Data Envelopment Analysis (DEA) based approach. *Journal of Cleaner Production, 142*, 846–853.

Chen, S., & Ravallion, M. (2013). More relatively-poor people in a less absolutely-poor world. *The Review of Income and Wealth, 1*(59), 1–28.

Cheng, G. (2014). *Data envelopment analysis: Methods and MaxDEA Software* (pp. 208–209). Intellectual Property Publishing House Co. Ltd.

Chèze, B., David, M., & Martinet, V. (2020). Understanding farmers' reluctance to reduce pesticide use: A choice experiment. *Ecological Economics, 167*, 106349.

Chiciudean, G. O., Harun, R., Ilea, M., Chiciudean, D. I., Arion, F. H., Ilies, G., & Muresan, I. C. (2019). Organic food consumers and purchase intention: A case study in Romania. *Agronomy, 9*(3), 145.

Chiron, F., Prince, K., Paracchini, M. L., Bulgheroni, C., & Jiguet, F. (2013). Forecasting the potential impacts of CAP-associated land use changes on farmland birds at the national level. *Agriculture, Ecosystems and Environment, 176*, 17–23.

Chiu, Y. H., Luo, Z., Chen, Y. C., Wang, Z., & Tsai, M. P. (2013). A comparison of operating performance management between Taiwan banks and foreign banks based on the Meta-Hybrid DEA model. *Economic Modelling, 33*, 433–439.

Chiueh, Y. W., & Chen, M. C. (2008). Environmental multifunctionality of paddy fields in Taiwan: An application of contingent valuation method. *Paddy and Water Environment, 6*, 229–236. https://doi.org/10.5539/enrr.v2n4p114

Christensen, T., Pedersen, A. B., Nielsen, H. O., Mørkbak, M. R., Hasler, B., & Denver, S. (2011). Determinants of farmers' willingness to participate in subsidy schemes for pesticide-free buffer zones—A choice experiment study. *Ecological Economics, 70*(8), 1558–1564.

Chung, Y. H., Färe, R., & Grosskopf, S. (1997). Productivity and undesirable outputs: A directional distance function approach. *Journal of Environmental Management, 51*(3), 229–240.

Ciaian, P., & Swinnen, J. (2009). Credit market imperfections and the distribution of policy rents. *American Journal of Agricultural Economics, 91: 1*, 123–124, 140.

Cochrane, W. (1958). *Farm prices: Myth and reality.* University of Minnesota Press.

Cochrane, W. (1979). *The development of American agriculture: A historical analyses* (2nd ed.). University of Minnesota Press.

Cooper, W. W., Seiford, L. M., & Tone, K. (2007). *Data envelopment analysis: A comprehensive text with models, applications, references and DEA-solver software* (Vol. 2). Springer.

Cooper, W. W., Seiford, L. M., & Zhu, J. (Eds.). (2011). *Handbook on data envelopment analysis.* Springer.

Cortignani, R., & Dono, G. (2015). Simulation of the impact of greening measures in an agricultural area of the southern Italy. *Land Use Policy, 48*, 525–533. https://doi.org/10.1016/j.landusepol. 2015.06.028

Czajanow, A. (1931). The socio-economic nature of peasant farm economy. In P. A. Sorokin, C. C. Zimmerman, & C. J. Galpin (red.), *A Systematic Source Book in Rural Sociology, tom 2,* University of Minnesota Press

Czajanow, A. (1966). *The theory of peasant economy.* The University of Wisconsin Press.

Czajanow, A. (1991). *The theory of peasant co-operatives.* I.B. Tauris.

Czajkowski, M., Hanley, N., & La Riviere, J. (2014). The effects of experience on preferences: Theory and empirics for environmental public goods. *American Journal of Agricultural Economics, 97*(1), 333–351.

Czekaj, S., Majewski, E., & Was, A. (2014). The impact of the "greening" of the Common Agricultural Policy on the financial situation of Polish farms. *Problems of Agricultural Economics, 4*, 105–121.

Czyżewski, B. (2009). *The evolution of land rent theory and its significance for the EU agriculture* (pp. 83–90). Jelgawa.

Czyżewski, B. (2017). *Kierat rynkowy w europejskim rolnictwie.* PWN.

Czyżewski, A., & Bak, U. (1995). Makroekonomiczne uwarunkowania przedsiębiorczości w agrobiznesie. In A. Czyżewski (red.), *Rozwój rolnictwa i agrobiznesu w skali krajowej i lokalnej, Wyd.* Ośrodka Doradztwa Rolniczego w Bielinku, p. 47.

Czyżewski, A., & Czyżewski, B. (2014). A new paradigm of development as a modern challenge in agriculture. *Management, 1*, 460–472.

Czyżewski, A., & Czyżewski, B. (2016). Research challenges for agricultural economics in the new paradigm. In: B. Czyżewski (red.), *Political rents of European farmers in the sustainable development paradigm. International, national and regional perspective* (pp. 18–27). Polish Scientific Publishers (PWN).

Czyżewski, B., Czyżewski, A., & Kryszak, Ł. (2017). Macroeconomic models for agricultural incomes in Europe: How the market treadmill has changed since the 60s, Department of Macroeconomics and Agricultural Economics, Working Paper No 4/2017/KMIGŻ. Poznań University of Economics and Business.

Czyżewski, B., Czyżewski, A., & Kryszak, Ł. (2019a). The market treadmill against sustainable income of European Farmers: How the CAP has struggled with Cochrane's curse. *Sustainability, 11*(3), 791.

Czyzewski, B., Matuszczak, A., & Miśkiewicz, R. (2019b). Public goods versus the farm price-cost squeeze: Shaping the sustainability of the EU's common agricultural policy. *Technological and Economic Development of Economy, 25*(1), 82–102.

Czyzewski, B., & Guth, M. (2021). Impact of policy and factor intensity on sustainable value of European agriculture: Exploring trade-offs of environmental, economic and social efficiency at the regional level. *Agriculture, 11*, 78. https://doi.org/10.3390/agriculture11010078

Czyżewski, B., & Kryszak, Ł. (2018). Impact of different models of agriculture on greenhouse gases (GHG) emissions: A sectoral approach. *Outlook on Agriculture, 47*(1), 68–76.

Czyżewski, B., & Majchrzak, A. (2018). Market versus agriculture in Poland–macroeconomic relations of incomes, prices and productivity in terms of the sustainable development paradigm. *Technological and Economic Development of Economy, 24*(2), 318–334.

Czyżewski, B., & Matuszczak, A. (2016a). Interwencjonizm rolny: pogoń za rentą a wybór publiczny lub korygowanie rynku. *Ekonomista, 5*, 674–703.

Czyżewski, B., & Matuszczak, A. (2016b). A new land rent theory for sustainable agriculture. *Land Use Policy, 55*(1–8), 222–229. https://doi.org/10.1016/j.landusepol.2016.04.002

Czyżewski, B., & Matuszczak, A. (2017). Wpływ płatności za dobra publiczne na renty ekonomiczne w rolnictwie w Polsce na tle krajów UE27. In J.St. Zegar, *Z badań nad rolnictwem społecznie zrównoważonym*, IERiGŻ PIB, Warszawa (w druku).

Czyżewski, B., & Matuszczak, A. (2018). Towards measuring political rents in agriculture: Case studies of different agrarian structures in the EU. *"Agricultural Economics" (AGRICECON)*. https://doi.org/10.17221/286/2016-AGRICECON.

Czyżewski, B., Matuszczak, A., Czyżewski, A., & Brelik, A. (2020). Public goods in rural areas as endogenous drivers of income: Developing a framework for country landscape valuation. *Land Use Policy, 107*, 104646.

Czyzewski, B., Matuszczak, A., Czyzewski, A., & Brelik, A. (2021). Public goods in rural areas as endogenous drivers of income: Developing a framework for country landscape valuation. *Land Use Policy, 107*, 104646. https://doi.org/10.1016/j.landusepol.2020.104646

Czyzewski, B., Matuszczak, A., Grzelak, A., Guth, M., & Majchrzak, A. (2020). Environmental sustainability value in agriculture revisited: How does Common Agricultural Policy contribute to ecoefficiency-? *Sustainability Science, 16*(2021), 137–152. https://doi.org/10.1007/s11625-020-00834-6

Czyżewski, B., & Poczta-Wajda, A. (2017). Effects of policy and market on relative income deprivation of agricultural labour. *Wieś i Rolnictwo, 3*(176), 53–70.

Czyzewski, B., Polcyn, J., & Brelik, A. (2022). Political orientations, economic policies, and environmental quality: Multi-valued treatment effects analysis with spatial spillovers in country districts of Poland. *Environmental Science & Policy, 128*(2022), 1–13. https://doi.org/10.1016/j.envsci.2021.11.001

Czyżewski, B., Sapa, A., & Kułyk, P. (2021). Human capital and eco-contractual governance in small farms in Poland: Simultaneous confirmatory factor analysis with ordinal variables. *Agriculture, 11*, 46. https://doi.org/10.3390/agriculture11010046

Czyżewski, B., Smędzik-Ambroży, K., & Mrówczyńska-Kamińska, A. (2020). Impact of environmental policy on eco-efficiency in country districts in Poland: How does the decreasing return to scale change perspectives? *Environmental Impact Assessment Review, 84*, 106431. https://doi.org/10.1016/j.eiar.2020.106431

Czyżewski, B., Trojanek, R., Dzikuć, M., & Czyżewski, A. (2020). Cost effectiveness of the common agricultural policy and environmental policy in country districts: Spatial spillovers of pollution, bio-uniformity and green schemes in Poland. *Science of The Total Environment, 726*(2020), 138254. https://doi.org/10.1016/j.scitotenv.2020.138254

Dakpo, K. H., Jeanneaux, P., & Latruffe, L. (2017). Greenhouse gas emissions and efficiency in French sheep meat farming: A non-parametric framework of pollution-adjusted technologies. *European Review of Agricultural Economics, 44*(1), 33–65.

Daly, H. E., & Cobb, J. B. (1989). *For the common good: Redirecting the economy toward community, the environment, and a sustainable future*. Boston Beacon Press.

David, K., & William, J. (2003, December). Sutherland how effective are European agri-environment schemes in conserving and promoting biodiversity? *Journal of Applied Ecology, 40*(6), 947–969.

Davidova, S., Bailey, A., Dwyer, J., Erjavec, E., Gorton, M., & Thomson, K. (2013). *Semi-subsistence farming: Value and directions of development*. European Parliament.

Davis, J. (1959). A formal interpretation of the theory of relative deprivation. *Sociometry, 4,* 280–296.

De Gorter, H., & Swinnen, J. (2002). Political economy of agricultural policy. *Handbook of Agricultural Economics, 2,* 1893–1943.

Delbecq, B. A., Kuethe, T. H., & Borchers, A. M. (2014). Identifying the extent of the urban fringe and its impact on agricultural land values. *Land Economics, 90*(4), 587–600.

Dessart, F. J., Barreiro-Hurlé, J., & van Bavel, R. (2019). Behavioural factors affecting the adoption of sustainable farming practices: A policy-oriented review. *European Review of Agricultural Economics, 46*(3), 417–471.

Dharmapala, S. P. (2018). Bias-correction in DEA efficiency scores using simulated beta samples: An alternative view of bootstrapping in DEA. *International Journal of Mathematics in Operational Research, 12*(4), 438–456.

Dibb, D. W. (1990). *Sustainable agriculture: The world scene.* Potash and Phosphate Institute.

Dickie, I., & Tucker, G. (2010). *The use of market-based instruments for biodiversity protection - the case of habitat banking, technical report for the European Commission.* DG Environment. http://ec.europa.eu/environment/enveco/pdf/eftec_habitat_technical_report.pdf

Diotallevi, F., Blasi, E., & Franco, S. (2015). Greening as compensation for production of environmental public goods: How do common rules have an influence at local level? The case of durum wheat in Italy. *Agricultural and Food Economics, 3,* 3–17.

Donald, P. F., Green, R., & Heath, M. F. (2001). Agricultural intensification and the collapse of Europe's farmland bird populations. *Proceedings of the Royal Society, Series B, 155,* 39–43.

Dong, G., Wang, Z., & Mao, X. (2018). Production efficiency and GHG emissions reduction potential evaluation in the crop production system based on emergy synthesis and nonseparable undesirable output DEA: A case study in Zhejiang Province, China. *PloS one, 13*(11), e0206680.

Du, J. (2014). Study on the environmental performance of China's agricultural growth. *The Journal of Quantitative & Technical Economics, 11,* 53–69.

Du, J., Wang, R., & Wang, X. H. (2016). Environmental total factor productivity and agricultural growth: A two-phase analysis based on the DEA-GML index and panel Tobit model. *Chinese Rural Economy, 3,* 65–81.

Ducos, G., Dupraz, P., & Bonnieux, F. (2009). Agri-environment contract adoption under fixed and variable compliance costs. *Journal of Environmental Planning and Management, 52*(5), 669–687.

Dudu, H., & Krsitkova, Z. S. (2017). *Impact of CAP Pillar II payments on agricultural productivity. JRC technical report.* Publications Office of the European Union. https://doi.org/10.2760/802100

Durkheim, É., & (edited and translated by François Pizarro Noël and Ronjon Paul Datta). (2020). An unpublished manuscript by Durkheim 'On the general physics of law and morality, 4th year of the course, 1st lecture, December 2, 1899, course outline: On penal sanctions'. *Durkheimian Studies, 24*(2020), 45–56.

Eberhardt, M., & Teal, F. (2012). No mangoes in the Tundra: Spatial heterogeneity in agricultural productivity analysis. *Oxford Bulletin of Economics and Statistics, 75*(6), 914–939. https://doi.org/10.1111/j.1468-0084.2012.00720.x

EC. (2011). Our life insurance, our natural capital: An EU biodiversity strategy to 2020 (Brussels: European Commission) COM (2011) 244 final

EC. (2017a). Science for Environment Policy. Agri-environmental schemes: How to enhance the agriculture-environment relationship. Thematic Issue 57. Issue produced for the European Commission DG Environment by the Science Communication Unit, UWE, Bristol. Available at: http://ec.europa.eu/science-environment-policy

EC. (2017b). https://ec.europa.eu/info/sites/default/files/food-farming-fisheries/key_policies/documents/ext-eval-payment-practices-climate-leaflet_2017_en.pdf

EC. (2019). Communication from the commission "The European green deal", COM/2019/640 final, 11 December 2019.

EC. (2020a). Communication from the commission, EU biodiversity strategy for 2030, "Bringing nature back into our lives", COM/2020/380 final, 20 May 2020.

EC. (2020b). Communication from the Commission "A farm to fork strategy for a fair, healthy and environmentally-friendly food system", COM/2020/381 final, 20 May 2020.

EC. (2020c). https://ec.europa.eu/info/sites/default/files/food-farming-fisheries/key_policies/docu ments/cap-post-2020-environ-benefits-simplification_en.pdf

EC. (2021a). https://ec.europa.eu/eurostat/cache/infographs/energy/bloc-4a.html

EC. (2021b). Communication from the commission, "Fit for 55": Delivering the EU's 2030 climate target on the way to climate neutrality, COM/2021/550 final, 14 July 2021.

EC. (2021c). Communication from the commission, "On an action plan for the development of organic productions", COM/2021/141 final, 19 April 2021

EC. (2021d). https://ec.europa.eu/clima/eu-action/forests-and-agriculture_en

EC (European Commission). (2018). CAP specific objectives explained – Brief No 1. Ensuring viable farm income, Brussels.

ECCP European Climate Change Programme. (2016). The second european climate change programme final report working group ECCP review - topic group agriculture and forestry. https://ec.europa.eu/clima/system/files/2016-11/review_agriculture_en.pdf (Accessed 12.2021)

Ecorys. (2017, July 7). Modernising & simplyfing the common agricultural policy. Summary of the results of the public consultation. Client: European Comission – DG Agri, Brussels.

EEA (European Environment Agency). (2004). *High nature value farmland: Characteristics, trends and policy challenges* (p. 31). European Union Publications Office. isbn:92-9167-664-0. https://www.eea.europa.eu/publications/report_2004_1.

EEA -European Environment Agency. (2021). https://www.eea.europa.eu/data-and-maps/indica tors/greenhouse-gas-emissions-from-agriculture/assessment.

Elstrand, E. (1969). Norwegian experience from extension work in farm management. *International Journal of Agrarian Affairs, 5*(4), 91–95.

EP European Parliament. (2021a). https://www.europarl.europa.eu/news/en/headlines/soci ety/20210303STO99110/carbon-leakage-prevent-firms-from-avoiding-emissions-rules

EP European Parliament. (2021b). https://www.europarl.europa.eu/factsheets/pl/sheet/70/energia ze-zrodel-odnawialnych.

Espinosa, M., Louhichi, K., Perni, A., & Ciaian, P. (2020). EU-wide impacts of the 2013 cap direct payments reform: A farm-level analysis. *Applied Economic Perspectives and Policy, 42*(4), 695–715.

European Climate Change Programme Report. (2006)

European Commision. (2019). *Organic farming in the EU. A fast growing sector*. Directorate-Geenral for Research and Innovation.

European Commission. (2015). *EU agriculture spending - focused on results*. DG Agri Fact-Sheet.

European Commission. (2019). *Organic farming in the EU. A fast growing sector*. Directorate-General for Research and Innovation.

European Court of Auditors. (2011). Is agri-environment support well designed and managed? Special Report No. 7/2011. European Union, p. 75. http://eca.europa.eu/portal/pls/portal/docs/1/8772726.PDF

European Court of Auditors. (2017). *More efforts needed to implement the natura 2000 network to its full potential*. Publications Office of the European Union.

European Network for Rural Development. (2010). Report on the contribution of the European network for rural development to the publicdebate on the common agricultural policy after 2013 (13/07/2010). European Commission. http://enrd.ec.europa.eu/enrd-static/fms/pdf/DAB81B97-9E9B-F50F-6F18-C76EBF6B1A4A.pdf.

Eurostat. (2019). *Sustainable development indicators*. European Commission. Available online: http://ec.europa.eu/eurostat/web/sdi/indicators/socioeconomic-development.15.03.20170 (Accessed on 10 November 2019)

Ewert, F., Rounsevell, M. D. A., Reginster, I. R., Metzger, M., & Leemans, R. (2005). Future scenarios of European agricultural land use. Part I. Estimating changes in crop productivity. *Agriculture, Ecosystems and Environment, 107*, 101–116.

Falconer, K., & Saunders, C. (2002). Transaction costs for SSSIs and policy design. *Land Use Policy, 19*, 157e166.

Falconer, K., & Whitby, M. (2000). Untangling red tape: Scheme administration and the invisible costs of European agri-environmental policy. *European Environment, 10*, 193e203.

Fałkowski, J. (2013). Does it matter how much land your neighbour owns? In *The functioning of land markets in Poland from a social comparison perspective* (pp. 1–20). IDEAS Working Paper Series from RePEc.

FAO. (2021a). http://www.fao.org/sustainable-development-goals/overview/fao-and-the-post-201 5-development-agenda/sustainable-agriculture/en/

FAO. (2021b). http://www.fao.org/sustainability/en/

FAO, IFAD, UNICEF, WFP and WHO. (2020). *The state of food security and nutrition in the world 2020 transforming food systems for affordable healthy diets*. FAO.

Faostat. (2020) Emissions due to agriculture. Global, regional and country trends 2000–2018. Faostat Analytical Brief 18.

Färe, R., & Grosskopf, S. (2003). Nonparametric productivity analysis with undesirable outputs: Comment. *American Journal of Agricultural Economics, 85*(4), 1070–1074.

Farms in Poland. (2021). Simultaneous confirmatory factor analysis with ordinal variables. *Agriculture, 11*, 46. https://doi.org/10.3390/agriculture11010046

Felipe, J., & McCombie, J. (2012a). *Problems with regional production functions and estimates of agglomeration economies: A caveat emptor for regional scientists* (Working Paper No. 725). Levy Economics Institute.

Felipe, J., & McCombie, J. (2012b). Agglomeration economies, regional growth, and the aggregate production function: A caveat emptor for regional scientists. *Spatial Economic Analysis, 7*, 461–484.

Fielding, K. S., Terry, D. J., Masser, B. M., & Hogg, M. A. (2008). Integrating social identity theory and the theory of planned behaviour to explain decisions to engage in sustainable agricultural practices. *British Journal of Social Psychology, 47*(1), 23–48. https://doi.org/10.1348/014466607X206792

Figge, F., & Hahn, T. (2004). Sustainable value added – measuring corporate contributions to sustainability beyond eco-efficiency. *Ecological Economics, 48*(2), 173–187.

Figge, F., & Hahn, T. (2005). The cost of sustainability capital and the creation of sustainable value by companies. *Journal of Industrial Ecology, 9*(4), 47–58.

Finn, J. A., Bartolini, F., Bourke, D., Kurz, I., & Viaggi, D. (2009). Ex post environmental evaluation of agri-environment schemes using experts' judgements and multicriteria analysis. *Journal of Environmental Planning and Management, 52*, 717–737.

Flanders, A., White, F. C., & Escalante, C. L. (2004). Equilibrium of land values from agricultural and general economic factors for cropland and pasture capitalization in Georgia. *Journal of Agribusiness, 22*(1), 49–60.

Fleming, P., Lichtenberg, E., & Newburn, D. A. (2018). Evaluating impacts of agricultural cost sharing on water quality: Additionality, crowding in, and slippage. *Journal of Environmental Economics and Management, 92*, 1–19.

Foley, J. A., Defries, R., Asner, G. P., Barford, C., Bonan, G., Carpenter, S. R., Chapin, F. S., Coe, M. T., Daily, G. C., et al. (2005). Global consequences of land use. *Science, 309*(5734), 570–574.

Food and Agriculture Organization (FAO). (2003). *World Agriculture: Towards 2015/2030: An FAO Perspective* (p. 432). FAO.

Foster, A. D., & Rosenzweig, M. R. (1995). Learning by doing and learning from others: Human capital and technical change in agriculture. *Journal of Political Economy, 103*, 1176–1209.

Freibauer, A., Rounsevell, M. D. A., Smith, P., & Verhagen, J. (2004). carbon sequestration in the agricultural soils of Europe. *Geoderma, 122*, 1e23.

Fretz, T. A. (1992). Sustainable agriculture: Our role as horticulturists. *ASHS Newsletter, 8*(5), 3–4.

Frey, B. S., Luechinger, S., & Stutzer, A. (2009). The life satisfaction approach to valuing public goods: The case of terrorism. *Public Choice, 138*, 317–345.

Gadanakis, Y., Bennett, R., Park, J., & Areal, F. (2015). Evaluating the sustainable intensification of arable farms. *Journal of Environmental Management, 150*, 288–298.

Garcia-Quijano, J. F., Deckmyn, G., Moons, E., Proost, S., Ceulemans, R., & Muys, B. (2005). An integrated decision support framework for the prediction and evaluation of efficiency, environmental impact and total social cost of domestic and international forestry projects for greenhouse gas mitigation: Description and case studies. *Forest Ecology and Management, 207*, 245–262.

Gardner, B. L. (1992). Changing economic perspectives on the farm problem. *Journal of Economic Literature, 30*(1), 62–101.

Garrod, G. D., & Willis, K. G. (1992). Goods' characteristics: an application of the hedonic price method to environmental attributes. *Journal of Environmental Management, 34*, 59–76. https://doi.org/10.1016/SO301-4797(05)80110-0

Ge, P. F., Wang, S. J., & Huang, X. L. (2018). Measurement for China's agricultural green TFP. *China Population, Resources and Environment, 28*(5), 66–74.

Geoghegan, J., Wainger, L. A., & Bockstael, N. E. (1997). Spatial landscape indices in a hedonic framework: An ecological economics analysis using GIS. *Ecological Economics, 23*(3), 251–264.

Ghali, M., Latruffe, L., & Daniel, K. (2016). Efficient use of energy resources on French farms: An analysis through technical efficiency. *Energies, 9*(8), 601.

Gizicki-Neundlinger, M., & Güldner, D. S. (2017). Scarcity and soil fertility in pre-industrial Austrian agriculture—The sustainability costs of inequality. *Sustainability, 9*, 265. Available online: http://www.mdpi.com/2071-1050/9/3/332/htm (Accessed on 1 May 2018)

Gocht, A., Ciaian, P., Bielza, M., Jean-Michel Terres, J. M., Röder, N., Himics, M., & Salputra, G. (2017). EU-wide economic and environmental impacts of CAP greening with high spatial and farm-type detail. *Journal of Agricultural Economics, 68*(3), 651–681. https://doi.org/10.1111/1477-9552.12217

Gocht, A., Ciaian, P., Bielza, M., Terres, J., Roder, N., Himics, M., & Salputra, G. (2017). EU-wide economic and environmental impacts of CAP greening with high spatial and farm-type detail. *Journal of Agricultural Economics, 68*(3), 651–681.

Gómez-Limón, J. A., Picazo-Tadeo, A. J., & Reig-Martínez, E. (2012). Eco-efficiency assessment of olive farms in Andalusia. *Land Use Policy, 29*, 395–406.

Goodwin, B. K., & Smith, V. H. (2015). The 2014 farm bill – an economic welfare disaster or triumph? In V. H. Smith (Ed.), *The economic welfare and trade relations implications of the 2014 farm bill*. Emerald Publishing.

Góral, J., & Kulawik, J. (2015). Problem kapitalizacji subsydiów w rolnictwie. *Zagadnienia ekonomiki rolnej, 342*, 3–24.

Görlach, B., Interwies, E., Newcombe, J., & Johns, H. (2005). *Cost-effectiveness of environmental policies. In An inventory of applied ex-post evaluation studies with a focus on methodologies, guidelines and good practice*. Specific agreement no. 3475/B2004.EEA. https://onlinelibrary.wiley.com/doi/pdf/10.1002/sd.378.

Gostomczyk. (2017). State and prospects for the development of the biogas market in the EU and Poland - Economic Approach. *Zeszyty Naukowe Szkoły Głównej Gospodarstwa Wiejskiego w Warszawie Problemy Rolnictwa Światowego, 17*(XXXII), notebook 2, 48–64. https://doi.org/10.22630/PRS.2017.17.2.26

Grados, D., & Schrevens, E. (2019). Multidimensional analysis of environmental impacts from potato agricultural production in the Peruvian central Andes. *Science of the Total Environment, 663*, 927–934. https://doi.org/10.1016/j.scitotenv.2019.01.414

Granovetter, M. (1985). Economic action and social structure: The problem of embeddedness. *American Journal of Sociology, 91*(3), 481–510.

Grant, P., Abrams, D., Robertson, D., & Garay, J. (2015). Predicting protests by disadvantaged skilled immigrants: A test of an integrated social identity, relative deprivation, collective efficacy (SIRDE) model. *Social Justice Research, 1*(28), 1–6.

Groot, R. S., Wilson, M. A., & Boumans, R. M. J. (2002). A typology for the classification, description and valuation of ecosystem functions, goods and services. *Special Issue: The Dynamics and Value of Ecosystem Services: Integrating Economic and Ecological Perspectives, Ecological Economics, 41*, 393–408.

Grzelak, A. (2015). The problem of complexity in economics on the example of the agricultural sector. *Agricultural Economics Czech, 61*(12), 577–568.

Grzelak, A., Guth, M., Matuszczak, A., Czyżewski, B., & Brelik, A. (2019). Approaching the environmental sustainability value in agriculture: How factor endowments foster the eco-efficiency. *Journal of Cleaner Production, 241*, 118304. https://doi.org/10.1016/j.jclepro.2019.118304

Gu, B., Ren, C., Zhou, X., Wang, C., Guo, Y., Diao, Y., Shen, S., Reis, S., Li, W., & Xu, J. (2022). *Aging threatens sustainability of smallholder farming in China* (Preprint). Research Square.

Guastella, G., Moro, D., Sckokai, P., & Veneziani, M. (2018). The capitalisation of CAP payments into land rental prices: A panel sample selection approach. *Journal of Agricultural Economics, 69*(3), 688–704.

Guastella, G., Moro, D., Sckokai, P., & Veneziani, M. (2021). The capitalisation of decoupled payments in farmland rents among EU regions. *Bio-based and Applied Economics, 10*(1), 7–17.

Guesmi, B., & Serra, T. (2015). Can we improve farm performance? The determinants of farm technical and environmental efficiency. *Applied Economic Perspectives and Policy, 37*(4), 692–717.

Guiso, L., Sapienza, P., & Zingales, L. (2006). Does culture affect economic outcomes? *Journal of Economic Perspectives, 20*, 23–48. https://doi.org/10.2139/ssrn.876601

Guth, M., Smędzik-Ambroży, K., Czyżewski, B., & Stępień, S. (2020). The economic sustainability of farms under common agricultural policy in the European Union countries. *Agriculture, 10*(2), 34. https://doi.org/10.3390/agriculture10020034

Hai, A. T. N., & Speelman, S. (2020). Economic-environmental trade-offs in marine aquaculture: The case of lobster farming in Vietnam. *Aquaculture, 516*, 734593.

Hailu, A., & Veeman, T. S. (2001). Non-parametric productivity analysis with undesirable outputs: An application to the Canadian pulp and paper industry. *American Journal of Agricultural Economics, 83*(3), 605–616.

Halkos, G., & Petrou, K. N. (2019). Treating undesirable outputs in DEA: A critical review. *Economic Analysis and Policy, 62*, 97–104.

Hamilton, W., Bosworth, G., & Ruto, E. (2015). Entrepreneurial younger farmers and the "young farmer problem" in England. *Agriculture and Forestry, 61*(4), 61–69.

Hamuda, H. E. A. F. B., & Patkó, I. (2010). Relationship between environmental impacts and modern agriculture. *Óbuda University e-Bulletin, 1*(1), 87–98.

Han, X., Xue, P., & Zhang, N. (2021). Impact of grain subsidy reform on the land use of smallholder farms: Evidence from Huang-Huai-Hai Plain in China. *Landscape, 10*(9), 929.

Han, H. B., Zhong, Z. Q., Wen, C. C., & Sun, H. G. (2018). Agricultural environmental total factor productivity in china under technological heterogeneity: Characteristics and determinants. *Environmental Science and Pollution Research, 25*, 32096–32111.

Hansen, H., & Offermann, F. (2016). Direct payments in Germany-income and distributional effects of the 2013 CAP reform. *German Journal of Agricultural Economics, 65*(2), 77–93.

Hanson, A. (2021). Assessing the redistributive impact of the 2013 CAP reforms: An EU-wide panel study. *European Review of Agricultural Economics, 48*(2), 338–361.

Hartig, F., & Drechsler, M. (2009). Smart spatial incentives for market-based conservation. *Biological Conservation, 142*(4), 779–788.

Hasund, K. P. (2013). Indicator-based agri-environmental payments: A payment-by-result model for public goods with a Swedish application. *Land Use Policy, 30*(1), 223–233.

Havlík, P., Valin, H., Herrero, M., Obersteiner, M., Schmid, E., Rufino, M. C., Mosnier, A., Thornton, P. K., Böttcher, H., Conant, R. T., Frank, S., Fritz, S., Fuss, S., Kraxner, F., & Notenbaert, A. (2014). Climate change mitigation through livestock system transitions. *Proceedings of the National Academy of Sciences, 111*(10), 3709–3714.

Hayami, Y., & Ruttan, V. W. (1970a). Agricultural productivity differences among countries. *American Economic Review, 60*(5), 895–911.

Hayami, Y., & Ruttan, V. W. (1970b). Factor prices and technical change in agricultural development: The United States and Japan, 1880–1960. *Journal of Political Economy, 78*(5), 1115–1141.

Hayami, Y., & Ruttan, V. W. (1971). *Agricultural development: An international perspective*. Johns Hopkins Press.

Hazel, P., Poulton, C., Wiggins, S., & Dorward, A. (2007, May). The future of small farms for poverty reduction growth, 2020 discussion paper 42, IFPRI, Washington, DC.

Heberton, E. (1967). The law of demand – the roles of Gregory King and Charles Davenant. *The Quarterly Journal of Economics, 81*(3), 483–492.

Hediger, W. (2008). *Agriculture's multifunctionality, sustainability, and social responsibility*. Available online: Ageconsearch.umn.edu/bitstream/36854/2/Hediger.pdf (Accessed on 10 May 2018)

Heinrichsmayer, W., & Witzke, H. P. (1991). Agrarökonomische Grundlagen, Agrarpolitik, Bd. 1, UTB 1651 Stuttgart.

Hendrickson, J. R., Hanson, J. D., Tanaka, D., & Sassenrath, G. (2008). Principles of integrated agricultural systems: Introduction to processes and definition. *Renewable Agriculture and Food Systems., 23*(4), 265–271. https://doi.org/10.1017/S1742170507001718

Hennessy, D. A. (1998). The production effects of agricultural income support polices under uncertainty. *American Journal of Agricultural Economics, 80*, 46–55.

Hermans, F., Horlings, I., Beers, P. J., & Mommaas, H. (2010). The contested redefinition of a sustainable countryside: Revisiting frouws' rurality discourses. *Sociologia Ruralis, 50*, 46–63.

Hermoso, V., Morán-Ordóñez, A., Canessa, S., & Brotons, L. (2019). Four ideas to boost EU conservation policy as 2020 nears. *Environmental Research Letters, 14*(10), 101001. https://doi.org/10.1088/1748-9326/ab48cc

Hill, B., & Bradley, D. (2015). Comparison of farmers' incomes in the EU Member states. In *Study for the European parliament's committee on agriculture and rural development*. European Parliament.

Hoang, V. N., & Coelli, T. (2011). Measurement of agricultural total factor productivity growth incorporating environmental factors: A nutrients balance approach. *Journal of Environmental Economics and Management, 62*(3), 462–474.

Hoang, V.-N., & Rao, D. P. (2010). Measuring and decomposing sustainable efficiency in agricultural production: A cumulative exergy balance approach. *Ecological Economics, 69*, 1765–1776.

Hofstede, G. (2001). *Culture's consequences: Comparing values, behaviors, institutions, and organizations across nations* (2nd ed.). Sage Publications.

Howard, G. (2020). Additionality violations in agricultural payment for service programs: Experimental evidence. *Land Economics, 96*(2), 244–264. https://stars.library.ucf.edu/etd/2093

Huang, W., Bruemmer, B., & Huntsinger, L. (2016). Incorporating measures of grassland productivity into efficiency estimates for livestock grazing on the Qinghai-Tibetan Plateau in China. *Ecological Economics, 122*, 1–11.

Hvid, A. (2015). Increasing natural resource rents from farmland: A curse or a blessing for the rural poor? *Peace Economics, Peace Science, and Public Policy, 21*(1), 59–78.

Hyll, W., & Schneider, L. (2014). Relative deprivation and migration preferences. *Economics Letters, 2*(122), 334–337.

IEEP Report. (2010). *The use of market-based instruments for biodiversity protection -the case of habitat banking - technical report*. http://ec.europa.eu/environment/enveco/index.htm

IPCC. (2018). Global warming of 1.5°C. In V. Masson-Delmotte et al. (Eds.), *An IPCC special report on the impacts of global warming of 1.5 °C above pre-industrial levels and related global greenhouse gas emission pathways, in the context of strengthening the global response to the threat of climate change, sustainable development, and efforts to eradicate poverty*. Intergovernmental Panel on Climate Change.

IPCC. (2021). *The intergovernmental panel on climate change, special report: Global warming of 1.5° C*. https://www.ipcc.ch/sr15/; 19.11.2021

James, H. S., Jr. (2015). Generalized morality, institutions and economic growth, and the intermediating role of generalized trust. *Kyklos, 68*(2), 165–196.

Jayanta, S., & Dipti, P. (2013). Changes in reative deprivation and social well-being. *International Journal of Social Economics, 6*(40), 528–536.

Jitmun, T., Kuwornu, J. K., Datta, A., & Anal, A. K. (2020). Factors influencing membership of dairy cooperatives: Evidence from dairy farmers in Thailand. *Journal of Co-operative Organization and Management, 8*(1), 100109. https://doi.org/10.1016/j.jcom.2020.100109

Johansson, P.-O. (1993). *Cost-benefit analysis of environmental change*. Cambridge University Press.

Johnson, D. G. (2000). *Reducing the urban-rural income disparity*. University of Chicago, Paper No. 00–07.

Johnsrud, M. (1988). Sustainable agriculture. *Better Crops, 73*(Winter), 4–5.

Kahneman, D., & Thaler, R. H. (2006). Anomalies: Utility maximization and experienced utility. *Journal of Economic Perspectives, 20*(1), 221–234.

Kahneman, D., Wakker, P. P., & Sarin, R. (1997). Back to Bentham? Explorations of experienced utility. *Quarterly Journal of Economics, 112*(2), 375–406.

Kaminski, J., Kan, I., & Fleischer, A. (2012). A structural land-use analysis of agricultural adaptation to climate change: A proactive approach. *American Journal of Agricultural Economics, 95*(1), 70–93.

Kettunen, M. (2011). *Assessment of the Natura 2000 co-financing arrangements of the EU financing instrument* (A project for the European Commission - final report). Institute of European Environmental Policy.

Khoshroo, A., Izadikhah, M., & Emrouznejad, A. (2018). Improving energy efficiency considering reduction of CO2 emission of turnip production: A novel data envelopment analysis model with undesirable output approach. *Journal of Cleaner Production, 187*, 605–615.

Kijek, T., Nowak, A., & Domańska, K. (2016). The role of knowledge capital in Total Factor Productivity changes: The case of agriculture in EU countries. *German Journal of Agricultural Economics, 65*(670-2019-669), 171–181.

Kirchner, M., Schönhart, M., & Schmid, E. (2016). Spatial impacts of the CAP post-2013 and climate change scenarios on agricultural intensification and environment in Austria. *Ecological Economics, 123*, 35–56.

Kirwan, B. E. (2009). The incidence of US agricultural subsidies on farmland rental rates. *Journal of Political Economy, 117*(1), 138–164.

Kirwan, B. (2012). *United States farm subsidies: A question of equity and efficiency*. CANRP.

Klassert, C., & Möckel, S. (2013). Improving the policy mix: The scope for market-based instruments in EU biodiversity policy. *Environmental Policy and Governance, 23*(5), 311–322. https://doi.org/10.1002/eet.1623

Kleijn, D., & Sutherland, W. J. (2005). How effective are European agri-environment schemes in conserving and promoting biodiversity? *Journal of Applied Ecology, 40*(6), 947–969. https://doi.org/10.1111/j.1365-2664.2003.00868.x

Kneip, A., Simar, L., & Wilson, P. W. (2008). Asymptotics and consistent bootstraps for DEA estimators in nonparametric frontier models. *Econometric Theory, 24*, 1663–1697.

Knetsch, J., & Borcherding, T. (1979). Expropriation of private property and the basis for compensation. *University of Toronto Law Journal, 29*, 237–252.

Kronbak, L., & Vestergaard, N. (2013). Environmental cost-effectiveness analysis in intertemporal natural resource policy: Evaluation of selective fishing gear. *Journal of Environmental Management, 131*, 270–279. https://doi.org/10.1016/j.jenvman.2013.09.035

Kryszak, Ł., & Herzfeld, T. (2021). One or many European models of agriculture? How heterogeneity influences income creation among farms in the European Union. *Agricultural Economics - Czech, 67*(11), 445–456. https://doi.org/10.17221/154/2021-AGRICECON

Kryszak, Ł., & Matuszczak, A. (2019). Determinants of farm income in the European Union in new and old member states. A regional study. *Roczniki (Annals), 21*(3), 200–211.

Kuosmanen, T. (2005). Weak disposability in nonparametric production analysis with undesirable outputs. *American Journal of Agricultural Economics, 87*(4), 1077–1082.

Kuosmanen, T., & Kuosmanen, N. (2009). How not to measure sustainable value (and how one might). *Ecological Economics, 69*, 235–243.

Lakner, S., & Breustedt, G. (2017). Efficiency analysis of organic farming systems a review of concepts, topics, results and conclusions. *German Journal of Agricultural Economics, 66*(670-2020-978), 85–108.

Latruffe, L. (2010). Competitiveness, productivity and efficiency in the agricultural and agri-food sectors. In *OECD food, agriculture and fisheries working paper*. OECD Publishing.

Latruffe, L., Diazabakana, A., Bockstaller, C., Desjeux, Y., Finn, J., Kelly, E., Ryan, M., & Uthes, S. (2016). Measurement of sustainability in agriculture: A review of indicators. *Studies in Agricultural Economics, 118*, 123–130.

Lauwers, L. (2009). Justifying the incorporation of the materials balance principle into frontier-based eco-efficiency models. *Ecological Economics, 68*, 1605–1614.

Le, T. L., Lee, P. P., Peng, K. C., & Chung, R. H. (2019). Evaluation of total factor productivity and environmental efficiency of agriculture in nine East Asian countries. *Agricultural Economics, 65*(6), 249–258.

Leahy, S., Clark, H., & Reisinger, A. (2020). Challenges and prospects for agricultural greenhouse gas mitigation pathways consistent with the Paris agreement. *Frontiers in Sustainable Food Systems, 4*, 69.

Levidow, L., Birch, K., & Papaioannou, T. (2012). EU agri-innovation policy: Two contending visions of the bio-economy. *Critical Policy Studies, 6*, 40–65.

Levins, R. A., & Cochrane, W. W. (1996). The treadmill revisited. *Land Economics, 72*(4), 550–553.

Levinson, A. (2012). Valuing public goods using happiness data: The case of air quality. *Journal of Public Economics, 96*(9–10), 869–880.

Li, D., Yini Zhao, Y., Sun, Y., & Yin, D. (2017). Corporate environmental performance, environmental information disclosure, and financial performance: Evidence from China. *Human and Ecological Risk Assessment: An International Journal, 23*(2), 323–339.

Liang, L., Lal, R., Ridoutt, B. G., Zhao, G., Du, Z., Li, L., Feng, A., Wang, L., Peng, P., Hang, S., & Wu, W. (2018). Multi-indicator assessment of a water-saving agricultural engineering project in north Beijing. *China, Agricultural Water Management, 200*, 34–46. https://doi.org/10.1016/j.agwat.2018.01.007

Lichtenberg, E. (2015). Conservation, the farm bill and US agri-environmental policy. In V. H. Smith (Ed.), *The economic welfare and trade relations implications of the 2014 farm bill*. Emerald Publishing.

Lin, J. Y. (1991). Education and innovation adoption in agriculture: Evidence from hybrid rice in China. *American Journal of Agricultural Economics, 73*(3), 713–723. https://doi.org/10.2307/1242823

Lipton, M. (2005). *Can small farms survive prosper or be the key channel to cut mass poverty?* Presentation to FAO Symposium on Agricultural Commercialization and the Small Farmer, 4–5, May, Rome.

Liu, Y., & Feng, C. (2019). What drives the fluctuations of "green" productivity in China's agricultural sector? A weight Russell directional distance approach. *Resources, Conservation and Recycling, 147*, 201–213.

Livingston, M., Erickson, K., & Mishra, A. (2010). Standard and Bayesian random coefficient model estimation of US corn – Soybean farmer risk attitudes. In V. E. Ball, R. Fanfani, & L. Gutierrez (Eds.), *The economic impact of public support to agriculture: An international perspective* (pp. 329–343). Springer.

Long, X., Luo, Y., Sun, H., & Tian, G. (2018). Fertilizer using intensity and environmental efficiency for China's agriculture sector from 1997 to 2014. *Natural Hazards, 92*(3), 1573–1591.

Lowe, P., Buller, H., & Ward, N. (2002). Setting the next agenda? British and French approaches to the second pillar of the common agricultural policy. *Journal of Rural Studies, 18*, 1–17.

Lowenberg-DeBoer, J., & Boehlje, M. (1986). The impact of farmland price changes on farm size and financial structure. *American Journal of Agricultural Economics, 68*(4), 838–848.

Luechinger, S. (2009). Valuing air quality using the life satisfaction approach. *Economic Journal, 119*(536), 482–515.

MacDonald, J., Korb, P., & Hoppe, R. (2013). *Farm size and the organization of U.S. crop farming.* U.S. Department of Agriculture, Economic Research Report, No. 152, Washington DC.

Madsen, B., Carroll, N., Kandy, D. & Bennett, G. (2011). *2011 Update: State of biodiversity markets. Report.* Forest Trends, Washington, DC, USA [www document]. URL http://www.ecosystemmarketplace.com/reports/2011_update_sbdm

Mahy, L., Dupeux, B., Van Huylenbroeck, G., & Buysse, J. (2015). Simulating farm level response to crop diversification policy. *Land Use Policy, 45*, 36–42.

Maia, R., Silva, C., & Costa, E. (2016). Eco-efficiency assessment in the agricultural sector: The Monte Novo irrigation perimeter, Portugal. *Journal of Cleaner Production, 138*(2), 217–228.

Majchrzak, A., & Stępień, S. (2016). Flows of rents as an economic barometer for agriculture: The case of the EU-27. In B. Czyżewski (Ed.), *Political rents of European farmers in the sustainable development paradigm. International, national and regional perspective.* Polish Scientific Publishers.

Mamoon, A. (2012). *Assessing the impact of economies of scale and uncontrollable factors on the performance of U.S. cities.*

Marsden, T. (1998). Agriculture beyond the treadmill? Issues for policy, theory and research practice. *Progress in Human Geography, 22*(2), 265–275.

Martey, E., Etwire, P. M., Wiredu, A. N., & Dogbe, W. (2014). Factors influencing willingness to participate in multi-stakeholder platform by smallholder farmers in Northern Ghana: Implication for research and development. *Agric Food Econ, 2*(1), 11. https://doi.org/10.1186/s40100-014-0011-4

Martínez-García, C. G., Dorward, P., & Rehman, T. (2013). Factors influencing adoption of improved grassland management by small-scale dairy farmers in central Mexico and the implications for future research on smallholder adoption in developing countries. *Livestock Science, 152*(2–3), 228–238. https://doi.org/10.1016/j.livsci.2012.10.007

Matuszczak, A. (2021). Ewolucja Kwestii Agrarnej A Środowiskowe Dobra Publiczne, Instytut Ekonomiki Rolnictwa I Gospodarki Żywnościowej – Państwowy Instytut Badawczy, Warszawa 2020.

May, D., Arancibia, S., Behrendt, K., & Adams, J. (2019). Preventing young farmers from leaving the farm: Investigating the effectiveness of the young farmer payment using a behavioural approach. *Land Use Policy, 82*, 317–327.

McCann, L., Colby, B., Easter, K. W., Kasterine, A., & Kuperan, K. V. (2005). Transaction cost measurement for evaluating environmental policies. *Ecological Economics, 52*, 527e542.

McGillivray, D. (2012). Compensating biodiversity loss: The EU Commission's approach to compensation under article 6 of the habitats directive. *Journal of Environmental Law.* https://doi.org/10.1093/jel/eqs007

Menozzi, D., Fioravanzi, M., & Donati, M. (2015). Farmer's motivation to adopt sustainable agricultural practices. *Bio-Based and Applied Economics, 4*, 125–147. https://doi.org/10.13128/bae-14776

Miao, R., Feng, H., Hennessy, D. A., & Du, X. (2016). Assessing cost-effectiveness of the conservation reserve program (CRP) and interactions between the CRP and crop insurance. *Land Economics, 92*(4), 593–617. https://doi.org/10.3368/le.92.4.593

Miceli, T. J., & Minkler, A. P. (1995, April). Willingness-to-accept versus willingness-to-pay measures of value: Implications for rent control, eminent domain, and zoning. *Public Finance Quarterly, 23*(2), 255–270.

Mie, A., Andersen, H. R., Gunnarsson, S., Kahl, J., Kesse-Guyot, E., Rembiałkowska, E., et al. (2017). Human health implications of organic food and organic agriculture: A comprehensive review. *Environmental Health, 16*(1), 1–22.

Mohammadzadeh, A., Damghani, A. M., Vafabakhsh, J., & Deihimfard, R. (2018). Environmental and economic analysis of saffron and canola production systems: In East Azerbaijan province of Iran. *International journal of plant production, 12*(2), 73–83. https://doi.org/10.1016/j.biombioe.2019.105271

Mojo, D., Fischer, C., & Degefa, T. (2017). The determinants and economic impacts of membership in coffee farmer cooperatives: Recent evidence from rural Ethiopia. *Journal of Rural Studies, 50*, 84–94. https://doi.org/10.1016/j.jrurstud.2016.12.010

Moran, D., MacLeod, M., Wall, E., Eory, V., Pajot, G., Matthews, R., McVittie, A., Barnes, A., Rees, B., Moxey, A., Williams, A., & Smith, P. (2008). UK marginal abatement cost curves for the agriculture and land use, land-use change and forestry sectors out to 2022, with qualitative analysis of options to 2050. Final.

Mottet, A., Ladet, S., Coque, N., & Gibon, A. (2006). Agricultural land-use change and its drivers inmountain landscapes: A case study in the Pyrenees. *Agriculture, Ecosystems and Environment, 114*(2-4), 296–310.

Moutinho, V., Robaina, M., & Macedo, P. (2018). Economic-environmental efficiency of European agriculture —a generalized maximum entropy approach. *Agricultural Economics—Czech, 64*, 423–435.

Mouysset, L. (2014). Agricultural public policy: Green or sustainable? *Ecological Economics, 102*, 15–23. https://doi.org/10.1016/j.ecolecon.2014.03.004

Mouysset, L., Doyen, L., & Jiguet, F. (2013). How do the farmer risk preferences affect the biodiversity? *Ecological Applications, 23*, 96–109.

National Statistical Office—Statistical Office in Katowice. (2011). *Indicators of sustainable development in Poland* (pp. 72–111). National Statistical Office—Statistical Office in Katowice.

Nickerson, C., Morehart, M., Kuethe, T., Beckman, J., Ifft, J., & Williams, R. (2012). *Trends in U.S. farmland values and ownership*. Economic Information Bulletin, No. 92.

Nilsson, P., & Johansson, S. (2013). Location determinants of agricultural land prices. *Jahrb Reg wiss, 33*, 1–21.

O'Dea, C. (2013). *Lawmakers look to restrict farmland tax break to working farmers*. NJ Spotlight. dostępne w Internecie. http://www.njspotlight.com/stories/13/03/01/farmland-assessment/ (28.02.2014).

O'Donnell, C. J., Rao, D. P., & Battese, G. E. (2008). Metafrontier frameworks for the study of firm-level efficiencies and technology ratios. *Empirical Economics, 34*(2), 231–255.

Očić, V., Grgić, Z., Batelja Lodeta, K., & Šakić Bobić, B. (2018). The impact of subsidies on agricultural income in The Republic of Croatia. *Poljoprivreda, 24*(2), 57–62.

OECD. (1989). *Economic instruments for environmental protection*. OECD.

OECD. (1998). *Eco-efficiency*. OECD Publishing.

OECD. (2000). *A matrix approach to evaluating policy: Preliminary findings from PEM pilot studies of crop policy in the EU, the US, Canada and Mexico*. OECD Directorate for Food, Agriculture and Fisheries Trade Directorate.

OECD. (2021). *Agricultural policy monitoring and evaluation 2021: Addressing the challenges facing food systems*. OCED Publishing.

Olesen, O. B., & Petersen, N. C. (2016). Stochastic data envelopment analysis—A review. *European Journal of Operational Research, Elsevier, 251*(1), 2–21.

Olesen, J. E., & Bindi, M. (2002). Consequences of climate change for European agricultural productivity, land use and policy. *European Journal of Agronomy, 16*, 239–262.

Olson, M. L. (1965). *The logic of collective action: Public goods and the theory of groups*. Harvard University Press.

Olson, D. (1995). *Decision aids for selection problems.* Springer Verlag.

Orden, D., & Zulauf, C. (2015). The 2014 farm bill in historical perspective. In V. H. Smith (Ed.), *The economic welfare and trade relations implications of the 2014 farm bill.* Emerald Publishing.

Ostrom, E. (1990). *Governing the commons: The evolution of institutions for collective action.* Cambridge University Press.

Pain, D. J., & Pienkowski, M. J. (Eds.). (1997). *Farming and birds in Europe: The common agricultural policy and its implications for bird conservation.* Academic Press.

Pang, J. X., Chen, X. P., Zhang, Z. L., & Li, H. J. (2016). Measuring eco-efficiency of agriculture in China. *Sustainability, 8*(4), 398.

Paper, Installment 6 of Creating a Sustainable Food Future. (n.d.). World Resources Institute. Available online at http://www.worldresourcesreport.org.

Paracchini, M. L., & Britz, W. (2010, March 23–26). Quantifying effects of changed farm practice on biodiversity in policy impact assessment - an application of CAPRI-Spat. In *OECD workshop: Agri-environmental Indicators: Lessons Learned and Future Directions*, Leysin, Switzerland.

Park, Y. S., Egilmez, G., & Kucukvar, M. (2016). Emergy and end-point impact assessment of agricultural and food production in the United States: A supply chain-linked ecologically-based life cycle assessment. *Ecological Indicators, 62,* 117–137. https://doi.org/10.1016/j.ecolind.2015.11.045

Parman, J. (2012). Good schools make good neighbors: Human capital spillovers in early 20th century agriculture. *Explorations in Economic History, 49*(3), 316–334. https://doi.org/10.1016/j.eeh.2012.04.002

Pates, N. J., & Hendricks, N. P. (2020). Additionality from payments for environmental services with technology diffusion. *American Journal of Agricultural Economics, 102*(1), 281–299.

Peacock, W. G., Hoover, G. A., & Kilian, C. D. (1988). Divergence and convergence in international development: A decomposition analysis of inequality in the world system. *American Sociological Review, 53,* 843.

Peña, C. R., Serrano, A. L. M., de Britto, P. A. P., Franco, V. R., Guarnieri, P., & Thomé, K. M. (2018). Environmental preservation costs and eco-efficiency in Amazonian agriculture: Application of hyperbolic distance functions. *Journal of Cleaner Production, 197,* 699–707.

Pérez Urdiales, M., Oude Lansink, A., & Wall, A. (2016). Eco-efficiency among dairy farmers: The importance of socio-economic characteristics and farmer attitudes. *Environmental and Resource Economics, 64*(4), 559–574.

Performance Of U.S. Cities. (2012). Electronic theses and dissertations, 2004–2019. 2093.

Perkins, D. (1969). *Agricultural Development in China, 1368-1968.* Aldine Publishing Company.

Picazo-Tadeo, A. J., Beltrán-Esteve, M., & Gómez-Limón, J. A. (2012). Assessing eco-efficiency with directional distance functions. *European Journal of Operational Research, 220*(3), 798–809.

Picazo-Tadeo, A. J., Gómez-Limón, J. A., & Reig-Martínez, E. (2011). Assessing farming eco-efficiency: A data envelopment analysis approach. *Journal of Environmental Management, 92*(4), 1154–1164.

Pindado, E., Sánchez, M., Verstegen, J. A., & Lans, T. (2018). Searching for the entrepreneurs among new entrants in European Agriculture: The role of human and social capital. *Land Use Policy, 77,* 19–30. https://doi.org/10.1016/j.landusepol.2018.05.014

Plantinga, A. J., & Miller, D. (2001). Agricultural land values and future development. *Land Economics, 77*(1), 56–67.

Platteau, J.-P. (1993). Behind the market stage where real societies exist-part II: The role of moral norms. *The Journal of Development Studies, 4,* 753–817.

Platteau, J. P. (2000). *Institutions, social norms, and economic development.* Academic Publishers.

Plaxico, J., & Kletke, D. (1979). The value of unrealized farm land capital gains. *American Journal of Agricultural Economics, 61,* 327–330.

Poczta-Wajda, A., Sapa, A., Stępień, S., & Borychowski, M. (2020). Food insecurity among small-scale farmers in Poland. *Agriculture, 10*(7), 295. https://doi.org/10.3390/agriculture10070295

Popescu, A., Tindeche, C., Marcuță, A., Marcuță, L., Honțuș, A., & Angelescu, C. (2021). Labor force in the European Union agriculture-Traits and tendencies. *Economic Analysis, 20*, 27.

Poulton, C., Dorward, A., & Kydd, J. (2005, June 26–29). The future of small farms: New directions for services, institutions and intermediation (w:) The future of small farms. In *Proceedings of a Research Workshop*, Wye, UK, pp. 223–251.

Povellato, A., Bosello, F., & Giupponi, C. (2007). Cost-effectiveness of greenhouse gases mitigation measures in the European agro-forestry sector: A literature survey. *Environmental Science & Policy, 10*, 474e490.

Power, E. F., Kelly, D. L., & Stout, J. C. (2013). Impacts of organic and conventional dairy farmer attitude, behaviour and knowledge on farm biodiversity in Ireland. *Journal for Nature Conservation, 21*(5), 272–278.

Primdahl, B., Peco, J., Schramek, E., & Andersen, J. J. (2003). Oñate, Environmental effects of agri-environmental schemes in Western Europe. *Journal of Environmental Management, 67*(2), 129–138. https://doi.org/10.1016/S0301-4797(02)00192-5

Raggi, M., Viaggi, D., Bartolini, F., & Furlan, A. (2015). The role of policy priorities and targeting in the spatial location of participation in Agri-Environmental Schemes in Emilia-Romagna (Italy). *Land Use Policy, 47*, 78–89.

Rajan, R., & Ramchara, R. (2012). *The anatomy of a credit crisis: The boom and bust in farm land prices in the United States in the 1920s*. National Bureau of Economic Research, Working Paper 18027.

Randall, A. (2002). Valuing the outputs of multifunctional agriculture. *European Review of Agricultural Economics, 29*(3), 289–307.

Rao, D. S., O'donnell, C. J., & Battese, G. E. (2003). *Metafrontier functions for the study of inter-regional productivity differences*.

Ray, S. C. (2008). The directional distance function and measurement of super-efficiency: An application to airlines data. *Journal of the Operational Research Society, 59*(6), 788–797.

Reid, C. T. (2011). The privatisation of biodiversity? Possible new approaches to nature conservation law in the UK. *Journal of Environmental Law, 23*, 2.

Repar, N., Jan, P., Dux, D., Nemecek, T., & Doluschitz, R. (2017). Implementing farm-level environmental sustainability in environmental performance indicators: A combined global-local approach. *Journal of Cleaner Production, 140*, 692–704.

Reytar, K., Hanson, C., & Henninger, N. (2014). *Indicators of sustainable agriculture: A scoping analysis. Working paper of installment 6 of creating sustainnable food future*. Word Resources Institute.

Rezaei-Moghaddam, K., & Karami, E. (2008). A multiple criteria evaluation of sustainable agricultural development models using AHP. *Environment, Development and Sustainability, 10*, 407–426.

Ring, I., Hansjürgens, B., Elmqvist, T., Widmer, H., & Sukhdev, P. (2010). Challenges in framing the economics of ecosystems and biodiversity: The TEEB initiative. *Current Opinion in Environmental Sustainability, 2*, 15–26.

Ring, I., & Schröter-Schlaack, C. (eds). (2011) *Instrument mixes for biodiversity policies, POLICYMIX report, issue 2/2011*. Helmholtz Centre for Environmental Research - UFZ. http://policymix.nina.no/Portals/policymix/POLICYMIX%20Report_No%202_2011.pdf.

Rizov, M., Pokrivcak, J., & Ciaian, P. (2013). CAP subsidies and productivity of the EU farms. *Journal of Agricultural Economics, 64*(3), 537–557.

Robinson, J. (1948). *Economics of imperfect competition*. Macmillan.

Robinson, G. M. (2009). Towards sustainable agriculture: Current debates. *Geography Compass, 3*, 1757–1773.

Rosen, S. (1974). Hedonic prices and implicit markets: Product differentiation in pure competition. *Journal of Political Economy, 82*, 34–55.

Runciman, W. (1966). *Relative deprivation and social injustice*. University of California Press.

Rundlöf, M., Smith, H. G., & Birkhofer, K. (2016). Effects of organic farming on biodiversity. In *eLS* (pp. 1–7). John Wiley & Sons, Ltd. https://doi.org/10.1002/9780470015902.a0026342

Rutherford, T. F., Whalley, J., & Wigle, R. M. (1990). Capitalization, conditionality, and dilution: Land prices and the US wheat program. *Journal of Policy Modeling, 12*(3), 605–622.

Ruto, E., & Garrod, G. (2009). Investigating farmers' preferences for the design of agri-environment schemes: A choice experiment approach. *Journal of Environmental Planning and Management, 52*(5), 631–647.

Salois, M. J., Livanis, G. T., & Moss, C. B. (2006, February 5–8). *Estimation of production functions using average data, selected paper prepared for presentation at the southern agricultural economics association annual meetings*, Orlando, Florida.

Samuelson, P. A. (1967). Pitfalls in the analysis of public goods. *The Journal of Law and Economics, 10*, 199–204.

Santos, R., Clemente, P., Antunes, P., Schröter-Schlaack, C., & Ring, I. (2011). Offsets, habitat banking and tradable permits for biodiversity conservation. In I. Ring & C. Schröter-Schlaack (Eds.), *Instrument Mixes for Biodiversity Policies, POLICYMIX Report, Issue 2/2011* (pp. 59–88). Helmholtz Centre for Environmental Research -UFZ. http://policymix.nina.no/Portals/policymix/POLICYMIX%20Report_No%202_2011.pdf.

Santos, J. L., Madureira, L., Ferreira, A. C., Espinosa, M., & Gomez, P. S. (2016). Building an empirically-based framework to value multiple publicgoods of agriculture at broad supranational scales. *Land Use Policy, 53*, 56–70. https://doi.org/10.1016/j.landusepol.2015.12.001

Santos, R., Schröter-Schlaack, C., Antunes, P., Ring, I., & Clemente, P. (2015). Reviewing the role of habitat banking and tradable development rights in the conservation policy mix. *Environmental Conservation, 42*(4), 294–305. https://doi.org/10.1017/S0376892915000089

Scarpa, R., Campbell, D., & Hutchinson, G. (2007). Benefit estimates for landscape improvements: Sequential bayesian design and respondents' rationality in a choice experiment. *Land Economics, 83*, 617–634.

Schaltegger, S., & Sturm, A. (1990). Ökologische rationalität: Ansatzpunkte zur ausgestaltung von ökologieorientierten managementinstrumenten. *die Unternehmung, 44*(4), 273–290.

Schläpfer, F., et al. (2015). Valuation of landscape amenities: A hedonic pricing analysis of housing rents in urban, suburban and periurban Switzerland. *Landscape and Urban Planning, 141*, 24–40.

Schultz, T. W. (1975). The value of the ability to deal with disequilibria. *Journal of Economic Literature, 13*(3), 827–846.

Schulz, N., Breustedt, G., & Latacz-Lohmann, U. (2014). Assessing farmers' willingness to accept "greening": Insights from a discrete choice experiment in Germany. *Journal of Agricultural Economics, 65*(1), 26–48.

Sebaldt, M. (2002). *Sustainable development—Utopie oder realistische vision?* Verlag Kovac.

Sen, A. K. (1992). *Inequality reexamined*. Oxford Press.

Setterfield, M. (1997). A model of institutional hysteresis. *Journal of Economic Issues, 27*, 755–775.

Severini, S., & Tantari, A. (2013). The impact of agricultural policy on farm income concentration: The case of regional implementation of the CAP direct payments in Italy. *Agricultural Economics, 44*(3), 275–286.

Shen, Z. Y., Balezentis, T., Chen, X. L., & Valdmanis, V. (2018). Green growth and structural change in Chinese agricultural sector during 1997–2014. *China Economic Review, 51*, 83–96.

Sherrick, B., & Kuethe, T., (2014). *Impact of recent changes in the illinois farmland assessment act, farmdoc daily, dostępne w Internecie*. http://farmdocdaily.illinois.edu/2014/01/impactsrecent-changes-illinois-farmland-assessment-act.html (28.02.2014).

Shestalova, V. (2003). Sequential Malmquist indices of productivity growth: An application to OECD industrial activities. *Journal of Productivity Analysis, 19*(2), 211–226.

Shi, Y. J., Phipps, T. T., & Colyer, D. (1997). Agricultural land values under urbanizing influences. *Land Economics, 73*(1), 90–100.

Shields, D. A. (2015). *Federal crop insurance: Background* (CRS Report for Congress 7-5700, R40532). Congressional Research Service.

Simar, L., & Wilson, P. W. (1998). Sensitivity analysis of efficiency scores: How to bootstrap in nonparametric frontier models. *Management Science, 44*, 49–61.

Simar, L., & Wilson, P. W. (1999). Estimating and bootstrapping Malmquist indices. *European Journal of Operational Research, 115*, 459–471.

Simar, L., & Wilson, P. W. (2000). A general methodology for bootstrapping in non-parametric frontier models. *Journal of Applied Statistics, 27*, 779–802.

Simar, L., & Wilson, P. (2007). Estimation and inference in two-stage, semi-parametric models of production processes. *Journal of Econometrics, 136*(1), 31–64.

Šimpachová Pechrová, M., Šimpach, O., Medonos, T., Spěšná, D., & Delín, M. (2018). What are the motivation and barriers of young farmers to enter the sector? *AGRIS on-line Papers in Economics and Informatics, 10*(4), 79–87.

Siudek, T., & Zawojska, A. (2016). Foreign labour in agricultural sectors of some EU countries. In *The 160th EAAE Seminar 'Rural Jobs and the CAP'*, Warsaw, Poland, December 1–2, dostepne

Smith, P., Bustamante, M., Ahammad, H., et al. (2014). Agriculture, Forestry and Other Land Use (AFOLU). In O. Edenhofer, R. Pichs-Madruga, Y. Sokona, et al. (Eds.), *Climate change 2014: Mitigation of climate change. Contribution of working group III to the fifth assessment report of the intergovernmental panel on climate change* (pp. 811–922). Cambridge University Press.

Smith, P., Haberl, H., Popp, A., Erb, K. H., Lauk, C., Harper, R., et al. (2013). How much land-based greenhouse gas mitigation can be achieved without compromising food security and environmental goals? *Global Change Biology, 19*(8), 2285–2302.

Smith, P., Martino, D., Cai, Z., Gwary, D., Janzen, H., Kumar, P., McCarl, B., Ogle, S., O'Mara, F., Rice, C., Scholes, B., Sirotenko, O., Howden, M., McAllister, T., Pan, G., Romanenkov, V., Schneider, U., Towprayoon, S., Wattenbach, M., & Smith, J. (2008). Greenhouse gas mitigation in agriculture. *Philosophical Transactions of the Royal Society B: Biological Sciences, 363*(1492), 789–813. Report to the Committee on Climate Change. SAC.

Soheili-Fard, F., Kouchaki-Penchah, H., Raini, M. G. N., & Chen, G. (2018). Cradle to grave environmental-economic analysis of tea life cycle in Iran. *Journal of Cleaner Production, 196*, 953–960. https://doi.org/10.1016/j.jclepro.2018.06.083

Song, J., & Chen, X. (2019). Eco-efficiency of grain production in China based on water footprints: A stochastic frontier approach. *Journal of Cleaner Production, 236*, 117685.

Soteriades, A. D., Faverdin, P., March, M., & Stott, A. W. (2015). Improving efficiency assessments using additive data envelopment analysis models: An application to contrasting dairy farming systems. *Agricultural and Food Science, 24*(3), 235–248.

Staniszewski, J. (2018). Attempting to measure sustainable intensification of agriculture in countries of the European Union. *Journal of Environmental Protection and Ecology, 19*(2), 949–957.

Stanton, K. R. (2002). Trends in relationship lending and factors affecting relationship lending efficiency. *Journal of Banking & Finance, 26*, 127e152.

Stark, O., & Fan, S. (2011). Migration for degrading work as an escape from humilitaion. *Journal of Economic Behavior and Organization, 3*(77), 241–247.

Stavins, R. N. (1995). Transaction costs and tradeable permits. *Journal of Environmental Economics and Management, 29*, 133e148.

Stępień, S., Czyżewski, B., Sapa, A., Borychowski, M., Poczta, W., & Poczta-Wajda, A. (2021). Eco-efficiency of small-scale farming in Poland and its institutional drivers. *Journal of Cleaner Production, 279*, 123721.

Stępień, S., & Polcyn, J. (2019, May 9–10). Risk management in small family farms in Poland. In *Proceedings of the International Scientific Conference Economic Science for Rural Development*, Jelgava, Latvia, No 50, pp. 382–388.

Stoate, C., Baldi, A., Beja, P., Boatman, N. D., Herzon, I., van Doorn, A., de Snoo, G. R., Rakosy, L., & Ramwell, C. (2009). Ecological impacts of early 21st century agricultural change in Europe-a review. *Journal of Environmental Management, 91*(1), 22–46.

Strijker, D. (2005). Marginal lands in Europe - causes of decline. *Basic and Applied Ecology, 6*(2), 99–106.

Stubbs, M. (2013). *Conservation reserve program (CRP): Status and issues* (Report for Congress R42783). Congressional Research Service.

Subić, J., Jeloćnik, M., & Jovanović, M. (2013). Evaluation of social sustainability of agriculture within the Carpathians in the Republic of Serbia. *Scientific Papers Series. Management, Economic Engineering in Agriculture and Rural Development, 13*, 411–416.

Sutherland, L. A., Toma, L., Barnes, A. P., Matthews, K. B., & Hopkins, J. (2016). Agri-environmental diversification: Linking environmental, forestry and renewable energy engagement on Scottish farms. *Journal of Rural Studies, 47*, 10–20.

Swinnen, J. (2018). *The political economy of agricultural and food policies* (p. 254). Palgrave Macmillan US.

Tait, J., & Morris, D. (2000). Sustainable development of agricultural systems: Competing objectives and critical limits. *Futures, 32*, 247–260.

Tappin, B. M., & Capraro, V. (2018). Doing good vs. avoiding bad in prosocial choice: A refined test and extension of the morality preference hypothesis. *Journal of Experimental Social Psychology, 79*, 64–70.

Tassone, V. C., Wesseler, J., & Nesci, F. S. (2004). Diverging incentives for afforestation from carbon sequestration: An economic analysis of the EU afforestation program in the south of Italy. *Forest Policy and Economics, 6*, 567–578.

Tavana, M., Mirzagoltabar, H., Mirhedayatian, S. M., Farzipoor Saen, R., & Azadi, M. (2013). A new network epsilon-based DEA model for supply chain performance evaluation. *Computers & Industrial Engineering, 66*(2), 501–513. https://doi.org/10.1016/j.cie.2013.07.016

Tchayanov, A. (1966). *The theory of peasant economy*. The University of Wisconsin Press.

Thompson, J., & Scoones, I. (2009). Addressing the dynamics of agri-food systems: An emerging agenda for social science research. *Environmental Science and Policy, 12*, 386–397.

Tietenberg, T. (2004). *Environmental economics and policy*. Pearson Addison Wesley.

Timo Kuosmanen, T., & Kortelainen, M. (2006). *Valuing environmental factors in cost-benefit analysis using data envelopment analysis*. SIEV - Sustainability indicators and environmental, valuationsocial science research network electronic paper collection. http://ssrn.com/abstract=912464

Tone, K. (2001). A slacks-based measure of efficiency in data envelopment analysis. *European Journal of Operational Research, 130*(3), 498–509.

Tone, K. (2004). *Dealing with undesirable outputs in DEA: A slacks-based measure (SBM) approach*. GRIPS research report series. National Graduate Institute for Policy Studies.

Tone, K., & Tsutsui, M. (2009). Network DEA: A slacks-based measure approach. *European Journal of Operational Research, 197*(1), 243–252. https://doi.org/10.1016/j.ejor.2008.05.027

Tscharntke, T., Klein, A. M., Kruess, A., Steffan-Dewenter, I., & Thies, C. (2005). Landscape perspectives on agricultural intensification and biodiversity - ecosystem service management. *Ecology Letters, 8*(Aug.), 857–874.

Tuck, S. L., Winqvist, C., Mota, F., Ahnström, J., Turnbull, L. A., & Bengtsson, J. (2014). Land-use intensity and the effects of organic farming on biodiversity: A hierarchical meta-analysis. *Journal of Applied Ecology, 51*(3), 746–755.

Tweeten, L., & Zulauf, C. (2008). Farm price and income policy: Lessons from history. *Agribusiness, 24*(2), 145–160.

Twisk, J. W. R. (2006). *Applied multilevel analysis. Practical guides to biostatistics and epidemiology* (pp. 30–33). Cambridge University Press.

U.S. Congress. (1990, 28 November). *Food, Agriculture, Conservation, and Trade Act of 1990*. Public Law 101–624: U.S. Farm Bill.

UCS (Union of Concerned Scientists). (2005). http://www.ucsusa.org/food_and_environment/antibiotics_ and_food/terms-frequently-used-in-discussions-of-antibiotic-resistance.html.

Ullah, A., Perret, S. R., Gheewala, S. H., & Soni, P. (2016). Eco-efficiency of cotton-cropping systems in Pakistan: An integrated approach of life cycle assessment and data envelopment analysis. *Journal of Cleaner Production, 134,* 623–632.

Ullah, A., Silalertruksa, T., Pongpat, P., & Gheewala, S. H. (2019). Efficiency analysis of sugarcane production systems in Thailand using data envelopment analysis. *Journal of Cleaner Production, 238,* 117877.

UNFCCC United Nations Framework Convention on Climate Change 2021. https://unfccc.int

Urdiales, M. P., Lansink, A. O., & Wall, A. (2016). Eco-efficiency among dairy farmers: The importance of socio-economic characteristics and farmer attitudes. *Environmental and Resource Economics, 64*(4), 559–574. https://doi.org/10.1007/s10640-015-9885-1

USDA. (2017). United States Department of Agriculture Economic Research Service. In *Methodology for calculation of total factor productivity (TFP).* Available at https://www.ers.usda.gov/data-products/international-agriculturalproductivity/documentation-and-methods/ (Accessed on 31 August 2022).

Uthes, S., & Matzdorf, B. (2013). Studies on agri-environmental measures: A survey of the literature. *Environmental Management, 51,* 251–266.

Valenti, D., Bertoni, D., Cavicchioli, D., & Olper, A. (2021). The capitalization of CAP payments into land rental prices: A grouped fixed-effects estimator. *Applied Economics Letters, 28*(3), 231–236.

Van Herck, K., Swinnen, J., & Vranken, L. (2013). Capitalization of direct payments in land rents: Evidence from New EU Member States. *Eurasian Geography and Economics, 54*(4), 423–443.

Van Passel, S., Nevens, F., Mathijs, E., & Van Huylenbroeck, G. (2007). Measuring farm sustainability and explaining differences in sustainable efficiency. *Ecological Economics, 62*(1), 149–161.

van Passel, S., Van Huylenbroeck, G., Lauwers, L., & Mathijs, E. (2009). Sustainable value assessment of farms using frontier efficiency benchmarks. *Journal of Environmental Management, 90*(10), 3057–3069.

van Zeijts, H., Overmars, K., van der Bilt, W., Schulp, N., Notenboom, J., Westhoek, H., Helming, J., Terluin, I., & Janssen, S. (2011). *Greening the common agricultural policy: Impacts on farmland biodiversity on an EU scale, policy studies.* PBL Netherlands Environmental Assessment Agency.

Vanslembrouck, I., et al. (2005, March). Impact of agriculture on rural tourism: A hedonic pricing approach. *Journal of Agricultural Economics, 56*(1), 17–30.

Velten, S., Leventon, J., Jager, N., & Newig, J. (2015). What is sustainable agriculture? A systematic review. *Sustainability, 2015*(7), 7833–7865. https://doi.org/10.3390/su7067833

Vergamini, D., Viaggi, D., & Raggi, M. (2020). Evaluating the potential contribution of multi-attribute auctions to achieve agri-environmental targets and efficient payment design. *Ecological Economics, 176*(C), 106756. Elsevier.

Vergopoulos, K. (1978). Capitalism and peasant productivity. *The Journal of Peasant Studies, 5*(4), 446–465.

Volk, M., Liersch, S., & Schmidt, G. (2009). Towards the implementation of the European water framework directive? *Land Use Policy, 26*(3), 580–588.

Wallace, A. (1994). Sense with sustainable agriculture. *Communications in Soil Science and Plant Analysis, 25*(1-2), 5–13. https://doi.org/10.1080/00103629409368998

Wang, X., Ding, H., & Liu, L. (2019). Eco-efficiency measurement of industrial sectors in China: A hybrid super-efficiency DEA analysis. *Journal of Cleaner Production, 229,* 53–64.

Wang, B. Y., & Zhang, W. G. (2018). Cross-provincial differences in determinants of agricultural ecoefficiency in China: An analysis based on panel data from 31 provinces in 1996–2015. *Chinese Rural Economy, 1,* 46–62.

Wanyama, F. O. (2014). *Cooperatives and the sustainable development goals a contribution to the post-2015 development debate.* International Labour Organization.

Ward, N. (1993). The agricultural treadmill and the rural environment in the post-productivist era. *Sociologia Ruralis, 33*(3–4), 348–364.

Wasson, J. R., McLeod, D. M., Bastian, C. T., & Rashford, B. S. (2013). The effects of environmental amenities on agricultural land values. *Land Economics, 89*(3), 466–478.

Wauters, E., Bielders, C., Poesen, J., Govers, G., & Mathijs, E. (2010). Adoption of soil conservation practices in Belgium: An examination of the theory of planned behaviour in the agri-environmental domain. *Land Use Policy, 27*(1), 86–94. https://doi.org/10.1016/j.landusepol.2009.02.009

Weber, J. G., & Key, N. (2014). Dowealth gains from land appreciation cause farmers to expand acreage or buy land? *American Journal of Agricultural Economics, 96*(5), 1334–1348.

Weingaertner, C., & Moberg, Å. (2011). Exploring social sustainability: Learning from perspectives on urban development and companies and products. *Sustainable Development, 22*, 122–133. Available online: https://www.diva-portal.org/smash/get/diva2:378611/FULLTEXT02.pdf

Weis, T. (2011). *Światowa gospodarka żywnościowa. Batalia o przyszłość rolnictwa, Polska Akcja Humanitarna.*

Weitzman, M. L. (1974). Prices vs. quantities. *The Review of Economic Studies, 41*(4), 477–491.

Wettemann, P. J. C., & Latacz-Lohmann, U. (2017). An efficiency-based concept to assess potential cost and greenhouse gas savings on German dairy farms. *Agricultural Systems, 152*, 27–37.

Whitehead, J. C., & Blomquist, G. C. (2006). The use of contingent valuation in benefit-cost analysis. In A. Alberini & J. R. Kahn (Eds.), *Handbook of Contingent Valuation.* Edward Elgar Publishing.

Wilkes, A., & Zhang, L. (2016). *Stepping stones towards sustainable agriculture in China: An overview of challenges, policies and responses.* International Institute for Environment and Development.

William, W. C., Lawrence, M. S., & Kaoru, T. (2007). *Data envelopment analysis: A comprehensive text with models, applications, references and DEA-solver software.* Springer.

Williamson, O. E. (1985). *The economic institutions of capitalism: Firms, markets, relational contracting.* University of Illinois at Urbana-Champaign's Academy for Entrepreneurial Leadership Historical Research Reference in Entrepreneurship, Available at SSRN https://ssrn.com/abstract=1496720

Wissel, S., & Wätzold, F. (2010). a conceptual analysis of the application of tradable permits to biodiversity conservation. *Conservation Biology, 24*(2), 404–411.

Wollenberg, E., Richards, M., Smith, P., Havlík, P., Obersteiner, M., Tubiello, F. N., et al. (2016). Reducing emissions from agriculture to meet the 2 C target. *Global Change Biology, 22*(12), 3859–3864.

Wooldridge, J. (2002). *Econometric analysis of cross section and panel data.* The MIT Press.

World Business Council for Sustainable Development. (2000). *Ecoefficiency—Creating more value with less impact.* WBCSD.

World Value Survey. (2021). https://www.worldvaluessurvey.org/wvs.jsp

Woś, A. (2003). *Konkurencyjność polskiego sektora żywnościowego* (p. 57). Synteza, IERiGŻ.

Wozniak, G. D. (1993). Joint information acquisition and new technology adoption: Late versus early adoption. *The Review of Economics and Statistics*, 438–445. https://doi.org/10.2307/2109457

Xie, H. L., Chen, Q. R., Wang, W., & He, Y. F. (2018). Analyzing the green efficiency of arable land use in China. *Technological Forecasting and Social Change, 133*(8), 15–28.

Xing, Z., Wang, J., & Zhang, J. (2018). Expansion of environmental impact assessment for eco-efficiency evaluation of China's economic sectors: An economic input-output based frontier approach. *Science of the Total Environment, 635*, 284–293.

Yang, H., & Pollitt, M. (2009). Incorporating both undesirable outputs and uncontrollable variables into DEA: The performance of Chinese coal-fired power plants. *European Journal of Operational Research, 197*(3), 1095–1105.

Yang, L., Ouyang, H., Fang, K., Ye, L., & Zhang, J. (2015). Evaluation of regional environmental efficiencies in China based on super-efficiency-DEA. *Ecological Indicators, 51*, 13–19.

Yang, Q., Wang, J., Li, C., & Liu, X. P. (2019). The spatial differentiation of agricultural green total factor productivity and its driving factor recognition in China. *The Journal of Quantitative & Technical Economics., 10*, 21–37.

Zawalińska, K., Majewski, E., & Wąs, A. (2016). Long-term changes in the incomes of the Polish agriculture compared to the European Union countries. *Annals of Polish Association of Agricultural Economists, XVII*(6), 346–354.

Zegar, J. S. (2009). Kwestia koncentracji ziemi w polskim rolnictwie indywidualnym. *Roczniki Nauk Rolniczych, 96*(4), 256–266.

Zegar, J. S. (2012). *Współczesne wyzwania rolnictwa* (pp. 49–53). Wydawnictwo Naukowe PWN.

Zhang, B., Fu, Z., Wang, J., Tang, X., Zhao, Y., & Zhang, L. (2017). Effect of householder characteristics, production, sales and safety awareness on farmers' choice of vegeTable marketing channels in Beijing, China. *British Food Journal, 119*(6), 1216–1231. https://doi.org/10.1108/BFJ-08-2016-0378

Zhang, Y., Xing, J., Zhang, T., Zhao, H., & Shi, Y. (2020). *Organic agriculture in China*. Organic and Beyond Corporation (OABC).

Zhu, J. (2008). *'Multiplier and slack-based models', quantitative models for performance evaluation and benchmarking: International series in operations research & management science* (Vol. 126, pp. 43–62). Springer.